TEST BANK
to accompany
GENERAL CHEMISTRY

SIXTH EDITION

Whitten • Davis • Peck

Cheri E. Bishop
Houston, Texas

Frank L. Kolar
Texas A&M University

Saunders College Publishing
A Division of Harcourt College Publishers

Fort Worth   Philadelphia   San Diego   New York   Orlando   Austin
San Antonio   Toronto   Montreal   London   Sydney   Tokyo

Copyright © 2000, 1996, 1992 by Harcourt, Inc.

All rights reserved. No parts of this publication may be reproduced or transmitted in any form or by any other means, electronic or mechanical, including photocopy, recording, or any other information storage and retrieval system, without permission in writing from the publisher, except that, until further notice, the contents or parts thereof may be reproduced for instructional purposes by users of GENERAL CHEMISTRY AND GENERAL CHEMISTRY WITH QUALITATIVE ANALYSIS, Sixth Edition, by Kenneth W. Whitten, Raymond E. Davis, and M. Larry Peck.

Printed in the United States of America

ISBN 0-03-021234-0

0 1 2   128   7 6 5 4 3 2

Preface

This test bank was patterned after questions in the test banks for the second and fifth editions of the textbook. There are multiple-choice questions for most sections of each chapter. The answers are given before the number of each question.

Many questions are marked with an asterisk (*) between the answer and the number of the question. These are to attract the attention of the instructor. They serve several purposes. They may indicate questions that are more challenging than most questions. They may also indicate questions, that while relatively easy, might be considered "tricky" by some students. It is assumed that the students will be supplied with a periodic table or table of constants for each test. An asterisk may indicate that the question requires a constant (that the instructor may or may not expect the student to know) in order to solve the problem.

Finally the asterisk may indicate that the question is based on material from the "Enrichment" or "Chemistry in Use" sections or end of chapter questions.

I wish to thank the textbook authors for making possible this opportunity.

I thank Dr. Peck for his constant encouragement and advice.

I thank Marc Sherman for his patience.

I especially thank my co-author without whose abilities and hard work, the previous version of this work would never have been completed and thus this version would not have been possible.

Frank L. Kolar

Harcourt, Inc

Table of Contents

1	The Foundations of Chemistry	1
2	Chemical Formulas and Composition Stoichiometry	15
3	Chemical Equations and Reaction Stoichiometry	25
4	Some Types of Chemical Reactions	39
5	The Structure of Atoms	65
6	Chemical Periodicity	85
7	Chemical Bonding	101
8	Molecular Structure and Covalent Bonding Theories	117
9	Molecular Orbitals in Chemical Bonding	133
10	Reactions in Aqueous Solutions I: Acids, Bases, and Salts	143
11	Reactions in Aqueous Solutions II: Calculations	157
12	Gases and the Kinetic-Molecular Theory	171
13	Liquids and Solids	195
14	Solutions	209
15	Chemical Thermodynamics	219
16	Chemical Kinetics	237
17	Chemical Equilibrium	257
18	Ionic Equilibria I: Acids and Bases	277
19	Ionic Equilibria II: Buffers and Titration Curves	297
20	Ionic Equilibria III: The Solubility Product Principle	309
21	Electrochemistry	321
22	Metals I: Metallurgy	337
23	Metals II: Properties and Reactions	345
24	Some Nonmetals and Metalloids	353
25	Coordination Compounds	363
26	Nuclear Chemistry	375
27	Organic Chemistry I: Formulas, Names, and Properties	385
28	Organic Chemistry II: Shapes, Selected Reactions, and Biopolymers	401

1 The Foundations of Chemistry

Matter and Energy

d 1. Which of the following statements is **incorrect**?

    (a) A body in motion possesses kinetic energy because of its motion.
    (b) An object possesses potential energy because of its position or composition.
    (c) Nuclear energy is an important kind of potential energy.
    (d) Chemical reactions and physical changes that release energy to their surroundings are endothermic.
    (e) The Law of Conservation of Matter and Energy states that the combined amount of matter and energy in the universe is fixed.

States of Matter

a 2. Matter is classified into three states. Which answer contains **all** of the following statements that are **true** and no false statements?

    I. The gaseous state of a substance is less dense than the liquid state.
    II. Substances in the solid state are rigid and have definite shapes.
    III. Substances in the liquid or gaseous state are easily compressed.
    IV. Substances in the gaseous state fill any container completely.
    V. Substances in the liquid state assume the volume of the container.

    (a) I, II, and IV   (b) I, II, and V   (c) I, II, and III
    (d) II, IV, and V   (e) I and II

Chemical and Physical Properties

c 3. All of the following are properties of antimony. Which one is **not** a physical property?

    (a) It is a solid at room temperature.
    (b) It has both yellow and gray forms (allotropes) in the solid state.
    (c) It burns in an atmosphere of chlorine.
    (d) It is one of the few substances that expands upon freezing.
    (e) The gray form melts at 631°C.

a 4. Which response lists all of the following properties of sulfur that are **physical** properties and no other properties?

    I. It reacts with hydrogen when heated.
    II. It is a yellow solid at room temperature.
    III. It is soluble in carbon disulfide.
    IV. Its density is 2.97 g/cm^3.
    V. It melts at 112°C.

    (a) II, III, IV, and V   (b) II, IV, and V   (c) I
    (d) II, III, and IV   (e) III, IV, and V

e 5.   The following properties describe zinc. Which one(s) is (are) **chemical** properties?

    I.   It is bluish-white metal.
    II.   It corrodes upon prolonged contact with moist air.
    III. Its density is 7.14 g/cm^3.
    IV. It melts at 419°C.
    V.   It conducts electricity.

(a) IV and V   (b) IV   (c) V
(d) II, IV, and V   (e) II

d 6.   Which response includes all the following that are **chemical** properties of carbon and no physical properties?

    I.   It is a solid at room temperature and atmospheric pressure.
    II.   It undergoes combustion to produce carbon dioxide and water.
    III. It boils at 4200°C.
    IV. It is not attracted strongly by a magnet.
    V.   Its density is 2.25 g/cm^3.

(a) I, III, and IV   (b) II and IV   (c) I and V
(d) II   (e) III, IV, and V

b 7.   The following statements describe some physical and chemical properties of sucrose (table sugar). Which response includes all that describe **chemical** properties, and none that describe physical properties?

    I.   It is a colorless solid.
    II.   It chars or blackens when heated gently.
    III. Its density is 1.6 g/mL.
    IV. It ignites and burns with a yellow flame when heated strongly.
    V.   It is usually in the form of small crystals although it can occur as a powder.

(a) I, III, and IV   (b) II and IV   (c) II, IV, and V
(d) I and V   (e) another combination is the answer

b 8.   All of the following are properties of oxygen. Which one is a **physical** property?

(a)   It supports combustion.
(b)   It condenses to a liquid at –219°C and atmospheric pressure.
(c)   It causes iron to rust.
(d)   It reacts with calcium to produce calcium oxide, CaO.
(e)   All of these are physical properties.

b 9.   Which of the following is an **intensive** property of matter?

(a) mass   (b) density   (c) volume   (d) weight   (e) heat capacity

a 10. Which of the following is an **extensive** property of matter?

    (a) weight   (b) melting point   (c) specific gravity
    (d) color   (e) density

Chemical and Physical Changes

b 11. Which response includes all of the following that involve **chemical** changes, and none that involve physical changes?

    I.   souring of milk   II.   melting of silver
    III.   digestion of food   IV.   burning of coal
    V.   condensation of steam

    (a) II and V   (b) I, III, and IV   (c) I, II, and III
    (d) IV and V   (e) III, IV, and V

b 12. Which answer includes all of the following that are **physical** changes and no chemical changes?

    I.   the electrolysis of molten sodium chloride to produce metallic sodium and gaseous chlorine
    II.   the solidification of mercury by cooling
    III. the reaction of hydrochloric acid with calcium carbonate
    IV. the mixing of water with ethyl alcohol
    V.   cutting an iron bar into small pieces

    (a) I and III   (b) II, IV, and V   (c) I and IV
    (d) II, III, and V   (e) IV

e 13. Which answer includes all the following that are **chemical** changes and no physical changes?

    I.   freezing of water
    II.   rusting of iron
    III. dropping a piece of iron into hydrochloric acid (H_2 is produced)
    IV. burning a piece of wood
    V.   emission of light by a kerosene oil lamp

    (a) III and IV   (b) II and V   (c) I, II, III, IV, and V
    (d) II, III, and V   (e) II, III, IV, and V

b 14. Which response lists all the **physical** changes but no chemical changes?

    I.   evaporation of water   II.   fermentation of sugar
    III. dissolving sugar in water   IV. melting of gold
    V.   changing of color of leaves

    (a) I, II, and III   (b) I, III, and IV   (c) III, IV, and V
    (d) I and IV   (e) II and V

Mixtures, Substances, Compounds, and Elements

e 15. Which answer lists all the substances below that are **compounds** and not any elements or mixtures?

　　I.　ethyl alcohol　　　II.　neon　　　　　III.　sulfur
　　IV.　water　　　　　　V.　crude oil

　　(a) I, II, and III　　　　(b) I, IV, and V　　　　(c) IV and V
　　(d) II, III, and V　　　　(e) I and IV

d 16. Which statement is **false**?

　　(a)　A compound is a substance that can be decomposed by chemical means into simpler substances.
　　(b)　All samples of a particular pure substance have the same composition and properties.
　　(c)　An example of a homogeneous mixture is one prepared by mixing two liquids, ethyl alcohol (grain alcohol) and water.
　　(d)　An example of a heterogeneous mixture is one prepared by dissolving the solid, sodium chloride (table salt), in the liquid, water.
　　(e)　Different mixtures of the same two substances can have different compositions.

e 17. Which response includes all the compounds listed below and only the compounds?

　　I.　ethyl alcohol　　　II.　air　　　　　　III.　mercury
　　IV.　steam　　　　　　V.　calcium fluoride

　　(a) I, II, and IV　　　　(b) III and V　　　　　(c) II, IV, and V
　　(d) I, III, and IV　　　　(e) another one or another combination

d 18. Choose the response that includes all things listed below that are pure substances and no others.

　　I.　steel　　　　　II.　sugar water　　　III.　oxygen
　　IV.　steam　　　　V.　sodium nitrite　　VI.　gunpowder

　　(a) II, III, and VI　　　(b) I and III　　　　　(c) III and V
　　(d) III, IV, and V　　　(e) II and VI

c 19. Which mixture is incorrectly labelled?

　　(a) homogeneous – salt dissolved in water
　　(b) homogeneous – gasoline
　　(c) heterogeneous – the oxygen and nitrogen in a scuba tank
　　(d) homogeneous – aqueous solution of alcohol
　　(e) heterogeneous – bauxite (an aluminum ore)

e 20. A sample of matter that can be decomposed into three different elements

(a) must be a solution.
(b) must be a compound.
(c) must be a heterogeneous mixture.
(d) must be a homogeneous mixture.
(e) Could be any of the preceding four answers.

c 21. Which of the following are heterogeneous mixtures?

I. sugar
II. sugar dissolved in water
III. an aqueous suspension of magnesium hydroxide (milk of magnesia)
IV. samples of nitrogen and oxygen in the same container
V. samples of argon and iron in the same container

(a) III only
(b) I, II, and IV
(c) III and V
(d) III and IV
(e) II, IV, and V

e *22. Trying to identify a sample of a pure substance, a student makes the following observations:

I. It has a mass of 5400 g.
II. It is 10. cm long, 10. cm wide, and its height is 20. cm.
III. It is a shiny solid at room temperature.
IV. It dissolves in hydrochloric acid.
V. It melts at 660°C.
VI. It is a good conductor of electricity.

Which response includes all of these observations that, **individually or in combination**, would be helpful in identifying the substance of which the sample is composed?

(a) I, III, IV, and V
(b) II, IV, V, and VI
(c) III and IV
(d) III, IV, and VI
(e) I, II, III, IV, V, and VI

a 23. What is the symbol for the element potassium?

(a) K   (b) P   (c) Pt   (d) W   (e) Po

b 24. The element whose symbol is Cl is

(a) carbon.   (b) chlorine.   (c) cobalt.   (d) calcium.   (e) californium.

c 25. What is the symbol for the element manganese?

(a) Mg   (b) Fe   (c) Mn   (d) Mo   (e) Hg

d 26. What is the symbol for the element copper?

(a) C   (b) Co   (c) Cm   (d) Cu   (e) Cr

c 27. What is the symbol for the element gold?

(a) Hg   (b) At   (c) Au   (d) Gd   (e) Ag

d 28. The element whose symbol is Ca is

(a) carbon. (b) cadmium. (c) cobalt. (d) calcium. (e) californium.

c 29. The element whose symbol is Br is

(a) boron. (b) barium. (c) bromine. (d) berkelium. (e) beryllium.

e 30. The element whose symbol is Sr is

(a) silver. (b) sulfur. (c) tin. (d) samarium. (e) strontium.

e 31. Which symbol – name combination is **incorrect**?

(a) Mn – manganese (b) Zn – zinc (c) Sn – tin
(d) Sr – strontium (e) Pd – lead

a 32. Which name – symbol combination is **wrong**?

(a) silver – Au (b) krypton – Kr (c) zinc – Zn
(d) platinum – Pt (e) tungsten – W

b 33. Which name – symbol combination is **incorrect**?

(a) aluminum – Al (b) cadmium – Ca (c) cobalt – Co
(d) magnesium – Mg (e) chromium – Cr

Measurements in Chemistry

d 34. Which one of the following lists units of volume in **increasing** order, i.e., from smallest to largest?

(a) microliter < kiloliter < centiliter < liter < milliliter
(b) milliliter < centiliter < microliter < kiloliter < liter
(c) centiliter < milliliter < liter < kiloliter < microliter
(d) microliter < milliliter < centiliter < liter < kiloliter
(e) milliliter < centiliter < liter < kiloliter < microliter

c 35. Below is a list of physical properties and the fundamental unit used in the SI system to measure each. Which unit is **not** the correct **fundamental** unit for the property?

	physical property	unit
(a)	length	meter
(b)	time	second
(c)	mass	gram
(d)	temperature	kelvin
(e)	electric current	ampere

d 36. Below is a list of common prefixes used in the SI and metric systems. Included with each is an abbreviation and meaning. Which set contains an **error**?

(a) mega- M 10^6
(b) deci- d 10^{-1}
(c) centi- c 10^{-2}
(d) micro- m 10^{-6}
(e) kilo- k 10^3

a 37. Below is a list of common prefixes used in the SI and metric systems. Included with each is an abbreviation and meaning. Which set contains an **error**?

(a) mega- M 10^{-6}
(b) deci- d 10^{-1}
(c) nano- n 10^{-9}
(d) milli- m 10^{-3}
(e) kilo- k 10^3

Units of Measurement

c 38. One of the following is a reasonable approximation of the **diameter** of a quarter (U.S. coin). Which one?

(a) 24 Mm (b) 0.24 m (c) 24 mm (d) 24 cm (e) 0.0024 km

a 39. Which of the following volumes is closest to one **pint**?

(a) 500. mL (b) 50. dL (c) 50. cm³ (d) 50. mL (e) 5000. cm³

c 40. Which one of the following is **closest** to one centimeter?

(a) the diameter of an apple
(b) the width of a standard door
(c) the diameter of a dime
(d) the length of a football field
(e) the diameter of a typical apple pie

Use of Numbers

a 41. Which of the following numbers has 4 significant figures?

(a) 0.04309 (b) 0.0430 (c) 0.0431 (d) 0.43980 (e) 0.043090

b 42. The sum 2.834 + 5.71520 + 2.12 + 178.1 + 250.2619 expressed to the proper number of significant figures is:

(a) 439 (b) 439.0 (c) 439.03 (d) 439.031 (e) 439.0311

b 43. Perform the indicated mathematical operations and round off the answer to the proper number of significant figures:

$(12.67 \times 4.23) \div 23.42 =$

(a) 2.3 (b) 2.29 (c) 2.288 (d) 2.88 (e) 2.2884

d 44. Perform the indicated mathematical operations and express the answer in scientific notation rounded off to the proper number of significant figures:

$$(8.001 \times 10^2) \times (2.88 \times 10^3 \div 2.4 \times 10^{-3}) =$$

(a) 9.6×10^2
(b) 9.60×10^8
(c) 9.601×10^8
(d) 9.6×10^8
(e) 9.6×10^{-2}

The Unit Factor Method (Dimensional Analysis)

d 45. How many millimeters are there in 25 feet?

(a) 7.6×10^2 mm
(b) 2.6×10^3 mm
(c) 1.0×10^2 mm
(d) 7.6×10^3 mm
(e) 1.2×10^3 mm

d 46. What is the height in centimeters of a person who is 6 feet 10 inches tall?

(a) 2.0×10^4 cm
(b) 2.1×10^{-1} cm
(c) 5.3×10^3 cm
(d) 2.1×10^2 cm
(e) 3.3×10^2 cm

b 47. If 5.76×10^{13} neon atoms (spherical) were laid in a line, each touching the next, the line would measure 2.54 miles. What is the diameter of a neon atom in Å?

(a) 0.92 Å   (b) 0.71 Å   (c) 1.86 Å   (d) 1.44 Å   (e) 1.74 Å

c 48. Convert 25.2 kilometers to inches.

(a) 99.2 in
(b) 6.40×10^6 in
(c) 9.92×10^5 in
(d) 9.92×10^{-1} in
(e) 6.40×10^3 in

c 49. How many milliliters are there in 1.0 microliter?

(a) 1.0×10^{-1} mL
(b) 1.0×10^{-2} mL
(c) 1.0×10^{-3} mL
(d) 1.0×10^{-4} mL
(e) 1.0×10^{-5} mL

a 50. Convert 175 milliliters to gallons.

(a) 0.0462 gal
(b) 0.0414 gal
(c) 0.740 gal
(d) 0.164 gal
(e) 0.660 gal

d 51. Express 39.34 square yards in square centimeters.

(a) 0.6417 cm^2
(b) 17280 cm^2
(c) 1.835×10^6 cm^2
(d) 3.289×10^5 cm^2
(e) 7.632×10^6 cm^2

a 52. What is the area in square millimeters of a rectangle that is 8.632 cm long and 26.41 mm wide?

(a) 2280 mm^2
(b) 3.060 mm^2
(c) 22.80 mm^2
(d) 0.3060 mm^2
(e) 30.60 mm^2

a 53. What is the area (in mm²) of a rectangular surface that is 0.640 inch wide and 1.14 inches long?

(a) 471 mm² (b) 328 mm² (c) 84.2 mm²
(d) 242 mm² (e) 680 mm²

b 54. What is the area (in mm²) of a rectangle that is 6.04 cm wide and 8.16 inches long?

(a) 19.4 mm² (b) 1.25 x 10⁴ mm² (c) 1.94 x 10³ mm²
(d) 1.25 x 10³ mm² (e) 125 mm²

e 55. What is 16.7 cubic feet expressed in cubic centimeters?

(a) 1.55 x 10⁴ cm³ (b) 510 cm³ (c) 0.159 cm³
(d) 1.76 x 10³ cm³ (e) 4.73 x 10⁵ cm³

d 56. Express 21.5 cubic centimeters in quarts.

(a) 2.03 x 10⁻² qt (b) 2.26 x 10⁴ qt (c) 9.24 x 10² qt
(d) 2.27 x 10⁻² qt (e) 2.02 x 10⁴ qt

b 57. Express 10.5 dm³ in quarts.

(a) 5.38 qt (b) 11.1 qt (c) 16.2 qt (d) 19.6 qt (e) 9.93 qt

d 58. Assuming a magnesium atom is spherical, calculate its volume in nm³. The **diameter** of a magnesium atom is 3.20 Å. The volume of a sphere is V = (4/3) π r³. 1 Å = 1 x 10⁻¹⁰ m and 1 nm = 1 x 10⁻⁹ m (Both of these relationships are exact.) π = 3.14

(a) 5.57 x 10³ nm³ (b) 2.34 x 10⁻²² nm³ (c) 5.57 x 10⁻²⁴ nm³
(d) 1.71 x 10⁻² nm³ (e) 5.57 x 10⁻³ nm³

b 59. Assuming that a lithium atom is spherical, calculate its volume in cm³. The volume of a sphere is given by V = (4/3) π r³. The radius of a lithium atom is 1.52 Å. 1 Å = 10⁻⁸ cm and π = 3.14.

(a) 4.78 x 10⁻⁸ cm³ (b) 1.47 x 10⁻²³ cm³ (c) 6.14 x 10⁻²² cm³
(d) 3.06 x 10⁻¹⁷ cm³ (e) 4.68 x 10⁻²⁴ cm³

b 60. Assume that a calcium atom is spherical, and its radius is 1.97 x 10⁻⁸ cm. What is the volume of a calcium atom expressed in cubic inches? Volume of a sphere: V = (4/3) π r³ and π = 3.14.

(a) 5.24 x 10⁻²² in³ (b) 1.95 x 10⁻²⁴ in³ (c) 3.25 x 10⁻⁸ in³
(d) 68.6 in³ (e) 8.16 x 10⁻¹² in³

b 61. How many kilograms of lead are there in 749 pounds of lead?

(a) 95.1 kg (b) 340 kg (c) 760 kg (d) 28.0 kg (e) 1650 kg

d 62. Express 8.37 x 10⁵ milligrams in ounces.

(a) 2.27 oz   (b) 0.100 oz   (c) 8.51 oz   (d) 29.5 oz   (e) 5.81 oz

c 63. If an airplane's speed is 850 mi/h, what is its speed in m/s?

(a) 40 m/s   (b) 1360 m/s   (c) 380 m/s   (d) 528 m/s   (e) 4.0 x 10⁵ m/s

a 64. The mileage rating of an automobile that is 12.0 kilometers per liter could also be expressed as _____ miles per gallon.

(a) 28.2   (b) 31.6   (c) 32.0   (d) 32.6   (e) 73.1

c 65. The 1970 standard established by the U.S. government for carbon monoxide emission for automobiles limited exhaust to 23.0 grams of CO per vehicle-mile. Assume that in a given metropolitan area there are 82,700 automobiles, driven an average of 13.5 miles per 24-hour period. How much CO (in tons) could legally be discharged into the area's atmosphere per day?

(a) 270 tons/day   (b) 0.155 tons/day   (c) 28.3 tons/day
(d) 0.0535 tons/day   (e) 39.0 tons/day

Percentage

d 66. Dark dental 22 carat gold is an alloy consisting of 92% Au, 4.9% Ag, and 3.1% Cu. If a patient leaves the dentist's office with 3.25 g of dark dental gold in her mouth, what mass of each element does she have in her mouth? (Note: round off may result in total mass not quite equaling 3.25 g.)

(a)   2.8 g Au   0.25 g Ag   0.15 g Cu
(b)   2.9 g Au   0.10 g Ag   0.16 g Cu
(c)   2.3 g Au   0.75 g Ag   0.20 g Cu
(d)   3.0 g Au   0.16 g Ag   0.10 g Cu
(e)   2.9 g Au   0.25 g Ag   0.10 g Cu

d 67. Stainless 316 is a steel alloy containing 17% Cr, 12% Ni, 3.0% Mo, and 0.10% C with the rest being Fe. What mass of iron (in kg) would be contained in 1.00 metric tonne of this alloy? 1 metric tonne = 10^3 kg.

(a) 68 kg   (b) 760 kg   (c) 580 kg   (d) 680 kg   (e) 780 kg

Density and Specific Gravity

c 68. A cube of metal is 1.42 centimeters on an edge. Its mass is 16.3 grams. What is its density?

(a) 4.68 g/cm³   (b) 4.30 g/cm³   (c) 5.69 g/cm³
(d) 6.14 g/cm³   (e) 4.86 g/cm³

b 69. A metal cube having a mass of 112 grams is dropped into a graduated cylinder containing 30.00 mL of water. This causes the water level to rise to 39.50 mL. What is the density of the cube?

(a) 2.86 g/mL (b) 11.8 g/mL (c) 10.8 g/mL
(d) 3.74 g/mL (e) 10.6 g/mL

c 70. What is the specific gravity of nickel if 2.35 cm³ of nickel has the same mass as 20.9 mL of water at room temperature?

(a) 0.112 (b) 2.14 (c) 8.89 (d) 19.7 (e) 49.2

b 71. The density of mercury is 13.6 g/cm³. What is the mass of 6.50 cm³ of mercury?

(a) 0.478 g (b) 88.4 g (c) 18.9 g (d) 2.38 g (e) 1.10 x 10² g

d 72. The density of octane is 0.702 g/cm³. What is the mass of 65.0 mL of octane?

(a) 110 g (b) 92.6 g (c) 22.5 g (d) 45.6 g (e) 1.08 x 10⁻² g

a 73. The specific gravity of ethyl chloride, an external painkiller, is 1.37 at 10°C. What is the mass of 47.4 mL of the liquid?

(a) 64.9 g (b) 0.346 g (c) 34.6 g (d) 52.5 g (e) 56.6 g

d 74. What is the mass of 35.0 mL of a liquid with a specific gravity of 2.64?

(a) 35.0 g (b) 13.3 g (c) 26.2 g (d) 92.4 g (e) 0.0754 g

d 75. What is the volume of a 2.50-gram block of metal whose density is 6.72 grams per cubic centimeter?

(a) 16.8 cm³ (b) 2.69 cm³ (c) 0.0595 cm³
(d) 0.372 cm³ (e) 1.60 cm³

d 76. What volume is occupied by 14.3 g of mercury? Density = 13.6 g/mL.

(a) 37.2 mL (b) 0.236 mL (c) 193 mL (d) 1.05 mL (e) 4.82 mL

a 77. What is the volume of a 58.5 gram sample of a liquid with a specific gravity of 1.24?

(a) 47.2 mL (b) 63.2 mL (c) 72.8 mL (d) 20.2 mL (e) 35.2 mL

d 78. Bromine, a brick red liquid, has a specific gravity of 3.12. What is the volume occupied by 25.0 grams of bromine?

(a) 11.7 mL (b) 78.0 mL (c) 32.6 mL (d) 8.01 mL (e) 2.48 mL

d 79. Which of the following statements about density is incorrect?

    (a)   The densities of gases are usually expressed in units of g/L.
    (b)   The intensive property density can be calculated from the two extensive properties: mass and volume.
    (c)   The densities of liquids are usually expressed in units of g/mL (or g/cm^3).
    (d)   If oil and water are placed in a container, they form two layers with oil as the top layer because it has the greater density.
    (e)   Densities of gases change greatly with changes in temperature and pressure.

Heat and Temperature

a 80. The freezing point of argon is –189°C. What is its freezing point on the absolute temperature scale?

(a) 84 K   (b) –136 K   (c) 461 K   (d) 112 K   (e) 72 K

c 81. Liquid propane boils at 231K. What is its boiling point in °C ?

(a) 42°C   (b) 315 °C   (c) –42°C   (d) 504°C   (e) –231°C

c 82. Methane, CH$_4$, boils at –162°C. What is the boiling point of methane in °F?

(a) –58°F   (b) –108°F   (c) –260°F   (d) –323°F   (e) –464°F

e 83. Copper melts at 1083°C. What is its melting temperature in °F ?

(a) 1324°F   (b) 583°F   (c) 619°F   (d) 797°F   (e) 1981°F

e 84. The normal boiling point of zinc is 788°F. What is this in °C ?

(a) 251°C   (b) 1450°C   (c) 724°C   (d) 962°C   (e) 420°C

b 85. If normal body temperature is 98.6°F, what is normal body temperature in °C ?

(a) 45.0°C   (b) 37.0°C   (c) 20.0°C   (d) 52.6°C   (e) 25.6°C

d 86. The freezing point of the noble gas, argon, is –308°F. What is this temperature in kelvins?

(a) 120 K   (b) 462 K   (c) –249 K   (d) 84 K   (e) 462 K

b 87. The normal boiling point of radon is –95.8°F. What is this temperature in kelvins?

(a) 190 K   (b) 202 K   (c) 224 K   (d) 249 K   (e) 294 K

a 88. What is the boiling point of liquid nitrogen, 77 K, on the Fahrenheit scale?

(a) –321°F   (b) 171°F   (c) –168°F   (d) 14.0°F   (e) –144°F

b 89. Which of the following are sets of equivalent temperatures?

I.	77°F	25°C	278 K
II.	86°F	30°C	303 K
III.	194°F	90°C	363 K
IV.	68°F	22°C	251 K

(a) I and II (b) II and III (c) I and III (d) II and IV (e) III and IV

Heat Transfer and the Measurement of Heat

d 90. How many kilojoules are equivalent to 565 calories?

(a) 130 kJ (b) 930 kJ (c) 0.135 kJ (d) 2.36 kJ (e) 2360 kJ

a 91. How much heat is released as the temperature of 9.52 grams of iron is decreased from 50.0°C to 10.0°C? The specific heat of iron is 0.444 J/g·°C

(a) 169 J (b) 211 J (c) 42.3 J (d) 376 J (e) 1.50 x 10^5 J

a 92. The specific heat of aluminum is 0.900 J/g·°C. How many joules of heat are absorbed by 15.0 g of Al if it is heated from 20.0°C to 60.0°C?

(a) 540 J (b) 270 J (c) 812 J (d) 2.40 J (e) 1.17 x 10^4 J

c 93. If 5.0 g of copper cools from 35.0°C to 22.6°C and loses 23.6 joules of heat, what is the specific heat of copper?

(a) 0.076 J/g·°C (b) 3.8 x 10^2 J/g·°C (c) 0.38 J/g·°C
(d) 0.62 J/g·°C (e) 76 J/g·°C

b *94. A 10.0 kg piece of metal at 50.0°C is placed in 1000. g of water at 10.0°C in an insulated container. The metal and water come to the same temperature at 30.6°C. What is the specific heat of metal? The specific heat of water is 4.18 J/g·°C.

(a) 0.0686 J/g·°C (b) 0.444 J/g·°C (c) 0.721 J/g·°C
(d) 0.124 J/g·°C (e) 0.0216 J/g·°C

d 95. The same amount of heat is added to a 25-g sample of each of the following metals. If each of the metals was initially at 20.0°C, which metal will reach the highest temperature?

	Metal	Specific Heat
(a)	beryllium	1.82 J/g·°C
(b)	calcium	0.653 J/g·°C
(c)	copper	0.385 J/g·°C
(d)	gold	0.129 J/g·°C
(e)	nickel	0.444 J/g·°C

c *96. If 100. grams of liquid water at 100.0°C and 200. grams of water at 20.0°C are mixed in an insulated container, what will be the final temperature of the water? The specific heat of water is 4.18 J/g J/g·°C.

(a) 41.9°C (b) 44.2°C (c) 46.7°C (d) 48.3°C (e) 45.1°C

d 97. How many grams of water at 25°C must be mixed in an insulated container with 250. grams of water at 93°C if the final temperature of the combined water is to be 75°C?

(a) 45 g (b) 376 g (c) 120 g (d) 90 g (e) 140 g

a *98. If a 10.0 g ball of iron at 160.0°C is dropped into 50.0 g of water at 20.0°C in an insulated container, what will be the final temperature of the water? The specific heat of iron is 0.444 J/g·°C and that of water is 4.18 J/g·°C.

(a) 23°C (b) 38°C (c) 48°C (d) 82°C (e) 110°C

e *99. What is the final temperature of the chromium and water combination when 50.0 g of chromium at 15°C (specific heat = 0.448 J/g·°C) is added to 25 mL of water (specific heat = 4.18 J/g·°C) at 45°C?

(a) 20°C (b) 25°C (c) 30°C (d) 35°C (e) 40°C

2 Chemical Formulas and Composition Stoichiometry

Atoms and Molecules

c 1. Which of the following statements is **not** an idea from Dalton's Atomic Theory?

(a) An element is composed of extremely small indivisible particles called atoms.
(b) All atoms of a given element have identical properties which differ from those of all other elements.
(c) Atoms can only be transformed into atoms of another element by nuclear reactions.
(d) Compounds are formed when atoms of different elements combine with each other in small whole-number ratios.
(e) The relative numbers and kind of atoms are constant in a given compound.

e 2. Atoms consist principally of what three fundamental particles?

(a) electrons, positrons, and neutrons
(b) elements, positrons, and neutrons
(c) elements, protons, and neutrons
(d) electrons, protons, and molecules
(e) electrons, protons, and neutrons

c 3. Which of the following statements is **incorrect**?

(a) A molecule is the smallest particle of a compound that can have a stable independent existence.
(b) Molecules that consist of more than one atom are called polyatomic molecules.
(c) The atomic number of an element is defined as the number of neutrons in the nucleus.
(d) Molecules of compounds are composed of more than one kind of atom.
(e) The charge on an electron is negative, and the charge on a proton is positive.

Chemical Formulas

d 4. There are two different common crystalline forms of carbon — diamond and graphite. A less common form called fullerene, C_{60}, also exists. Different forms of the same element in the same physical state are called

(a) isotopes. (b) isomers. (c) alloforms.
(d) allotropes. (e) structural formulas.

c 5. The formula for oxalic acid is $(COOH)_2$. How many atoms does each molecule contain?

(a) 6 (b) 4 (c) 8 (d) 10 (e) 5

e 6. If a sample of propane, C_3H_8, contains a total of 6.0×10^3 atoms of carbon, how many molecules of propane are in the sample?

(a) 6.0×10^3 (b) 3.0×10^3 (c) 8.0×10^3 (d) 1.1×10^4 (e) 2.0×10^3

b 7. Name the molecular compound, HNO₃.

(a) ammonia (b) nitric acid (c) nitrous acid
(d) nitric oxide (e) nethane

d 8. Name the molecular compound, H₂SO₄.

(a) hydrogen persulfide (b) sulfurous acid (c) sulfur tetroxide
(d) sulfuric acid (e) hydrogen persulfate

a 9. Name the molecular compound, CH₃COCH₃.

(a) acetone (b) ethanol (c) diethyl ether
(d) propane (e) ethyl alcohol

a 10. What is the molecular formula for acetic acid?

(a) CH₃COOH (b) CH₃COCH₃ (c) CH₃CH₂OH
(d) CH₃CH₂CO₂H (e) CH₃CH₂OCH₂CH₃

e 11. What is the molecular formula for ethanol?

(a) CH₃COOH (b) CH₃COCH₃ (c) CH₃CH₂OCH₂CH₃
(d) CH₃CH₂CO₂H (e) CH₃CH₂OH

d 12. What is the molecular formula for hydrogen peroxide?

(a) H₂O (b) H₂SO₄ (c) H₃O₂ (d) H₂O₂ (e) H₂O₃

c 13. A compound contains only calcium and fluorine. A sample of the compound is determined to contain 2.00 g of calcium and 1.90 g of fluorine. According to the Law of Definite Proportions, how much calcium should another sample of this compound contain if it contains 2.85 g of fluorine?

(a) 2.71 g (b) 4.00 g (c) 3.00 g (d) 4.50 g (e) 6.00 g

Ions and Ionic Compounds

a 14. Which of the following statements is **incorrect**?

(a) A molecule of potassium chloride, KCl, consists of one K⁺ ion and one Cl⁻ ion.
(b) Ions that possess a positive charge are called cations.
(c) Polyatomic ions are groups of atoms that have an electric charge.
(d) It is acceptable to use formula unit to refer to either an ionic compound or a molecular compound.
(e) Ions that possess a negative charge are called anions.

d 15. What is the correct classification for SCN⁻?

(a) polyatomic molecule (b) monatomic cation (c) polyatomic cation
(d) polyatomic anion (e) monatomic anion

a 16. Each response below lists an ion by name and by chemical symbol or formula. Also each ion is classified as monatomic or polyatomic and as a cation or anion. Which response contains an **error**?

(a) hydroxide OH^- monatomic anion
(b) carbonate CO_3^{2-} polyatomic anion
(c) ammonium NH_4^+ polyatomic cation
(d) magnesium Mg^{2+} monatomic cation
(e) sulfite SO_3^{2-} polyatomic anion

c 17. Each response below lists an ion by name and by chemical symbol or formula. Also each ion is classified as monatomic or polyatomic and as a cation or anion. Which response contains an **error**?

(a) phosphate PO_4^{3-} polyatomic anion
(b) sulfite SO_3^{2-} polyatomic anion
(c) nitrite NO_3^- polyatomic anion
(d) iron(II) Fe^{2+} monatomic cation
(e) bromide Br^- monatomic anion

Names and Formulas of Some Ionic Compounds

c 18. What is the formula for zinc sulfide?

(a) ZnS_2 (b) $ZnSO_4$ (c) ZnS (d) $ZnSO_3$ (e) Zn_2S

d 19. What is the formula for iron(III) oxide?

(a) FeO_3 (b) FeO_2 (c) Fe_3O_2 (d) Fe_2O_3 (e) Fe_3O

e 20. What is the formula for copper(II) nitrate?

(a) $CuNO_3$ (b) Cu_2NO_3 (c) $CuNO_2$ (d) Cu_2NO_2 (e) $Cu(NO_3)_2$

c 21. Choose the name-formula pair that does not correctly match.

(a) magnesium phosphate $Mg_3(PO_4)_2$
(b) ferrous sulfite $FeSO_3$
(c) silver carbonate $AgCO_3$
(d) potassium fluoride KF
(e) cupric bromide $CuBr_2$

b 22. Choose the name-formula pair that does not correctly match.

(a) aluminum phosphate $AlPO_4$
(b) calcium acetate $CaCH_3COO$
(c) ammonium sulfide $(NH_4)_2S$
(d) magnesium hydroxide $Mg(OH)_2$
(e) zinc carbonate $ZnCO_3$

d 23. From the following ionic compounds, choose the name-formula pair that is not correctly matched.

（a) sodium sulfide Na_2S
(b) ammonium nitrate NH_4NO_3
(c) zinc hydroxide $Zn(OH)_2$
(d) sodium sulfate Na_2SO_3
(e) calcium oxide CaO

e 24. From the following compounds choose the name-formula pair that is incorrectly matched.

(a) sodium sulfite Na_2SO_3
(b) ammonium fluoride NH_4F
(c) copper(II) carbonate $CuCO_3$
(d) ferric chloride $FeCl_3$
(e) cuprous sulfide Co_2S

Atomic Weights

b 25. Which element has a mass that is 7.30 times that of carbon–12?

(a) Mg (b) Sr (c) Ca (d) Br (e) Rb

b *26. If 6.653 g of calcium combines exactly with 26.53 g of bromine and 12.29 g of zinc combines exactly with 30.04 g of bromine, what is the mass ratio of one atom of calcium to one atom of zinc?

(a) 1.68 (b) 0.613 (c) 0.541 (d) 1.85 (e) 0.133

d *27. The molecular formula for a compound is CX_4. If 2.819 g of this compound contains 0.102 g of carbon, what is the atomic weight of X?

(a) 320 (b) 160 (c) 35.5 (d) 79.9 (e) 39.9

The Mole

a 28. Calculate the number of moles of oxygen atoms in 35.2 grams of oxygen.

(a) 2.20 moles (b) 4.42 moles (c) 0.54 moles
(d) 2.57 moles (e) 1.13 moles

e 29. Calculate the mass of one bromine atom.

(a) 2.654×10^{-22} g (b) 6.022×10^{23} g (c) 1.661×10^{-24} g
(d) 4.812×10^{25} g (e) 1.327×10^{-22} g

c 30. Calculate the number of atoms in 40.5 g of aluminum.

(a) 900 (b) 2.5×10^{-24} (c) 9.0×10^{23}
(d) 6.6×10^{26} (e) 1.8×10^{-21}

Formula Weights, Molecular Weights, and Moles

c 31. Calculate the formula weight of Na_2SO_4.

    (a) 193.2 amu   (b) 119.1 amu   (c) 142.1 amu
    (d) 215.3 amu   (e) 185.1 amu

a 32. What is the formula weight of $Fe(NH_4)_2(SO_4)_2 \cdot 6H_2O$?

    (a) 392 amu   (b) 384 amu   (c) 412 amu
    (d) 376 amu   (e) 436 amu

e 33. What is the mass of 2.2×10^9 CO_2 molecules?

    (a) 9.7×10^{10} g   (b) 1.0×10^{-12} g   (c) 1.2×10^6 g
    (d) 4.4×10^{-14} g   (e) 1.6×10^{-13} g

e 34. What is the mass of 0.432 moles of $C_8H_9O_4$?

    (a) 86.9 g   (b) 391 g   (c) 169 g   (d) 113.8 g   (e) 73.0 g

d 35. How many grams of $CaCl_2$ equal 3.40 moles of $CaCl_2$?

    (a) 130 g   (b) 159 g   (c) 306 g   (d) 377 g   (e) 239 g

a 36. How many moles of $POCl_3$ are there in 10.0 grams of $POCl_3$?

    (a) 6.51×10^{-2} mol   (b) 3.68×10^{-1} mol   (c) 4.09×10^{-2} mol
    (d) 1.21×10^{-1} mol   (e) 1.17×10^{-3} mol

c 37. Calculate the number of moles of $KMnO_4$ in 25.5 grams of $KMnO_4$.

    (a) 0.0161 mol   (b) 0.879 mol   (c) 0.161 mol
    (d) 1.88 mol   (e) 0.632 mol

d 38. How many molecules are contained in 5.00 grams of NH_3?

    (a) 5.42×10^{22} (b) 3.00×10^{24} (c) 3.40×10^{22}
    (d) 1.77×10^{23} (e) 9.45×10^{22}

d 39. How many atoms of hydrogen are there in 88 g of C_6H_6?

    (a) 3.6×10^{23} (b) 6.8×10^{22} (c) 2.6×10^{23}
    (d) 4.1×10^{24} (e) 5.3×10^{23}

e 40. A 12.0-gram sample of $Cr_2(SO_4)_3$ contains how many sulfur atoms?

    (a) 1.84×10^{22} (b) 1.53×10^{21} (c) 4.82×10^{21}
    (d) 6.67×10^{22} (e) 5.52×10^{22}

b 41. How many hydrogen atoms are present in 7.50 g of $(NH_4)_2Fe(SO_4)_2 \cdot 6H_2O$?

(a) 5.16×10^{22} (b) 2.30×10^{23} (c) 1.15×10^{22}
(d) 4.65×10^{23} (e) 3.35×10^{24}

c 42. Which of the following is **not** a correct description of 16.0 grams of methane, CH_4?

(a) It is one mole of methane.
(b) It is the amount of methane that contains 12.0 g of carbon.
(c) It is $16.0 \times 6.02 \times 10^{23}$ molecules of methane.
(d) It is the amount of methane that contains 4.0 grams of hydrogen.
(e) It is the amount of methane that contains $4 \times 6.02 \times 10^{23}$ hydrogen atoms.

c *43. A sample of ethane, C_2H_6, contains a total of $16N$ atoms, where $N = 6.02 \times 10^{23}$. How much C_2H_6 is in the sample?

(a) 2.0 g (b) 30 g (c) 60 g (d) 16 mol (e) 4 mol

d 44. Suppose you have a 100-gram sample of each of the following compounds. Which sample contains the smallest number of moles of compound?

(a) NH_3 (b) $MgCl_2$ (c) H_3PO_4 (d) $CrCl_3$ (e) $NaCl$

Percent Composition and Formulas of Compounds

d 45. What is the percent by mass of sulfur in $Al_2(SO_4)_3$?

(a) 9.38% (b) 18.8% (c) 24.6% (d) 28.1% (e) 35.4%

c 46. Calculate the percent by mass of nitrogen in ammonium carbonate, $(NH_4)_2CO_3$.

(a) 14.5% (b) 27.8% (c) 29.2% (d) 33.3% (e) 17.1%

c 47. Calculate the percent composition of K_2CO_3.

(a) % K = 58.2%   % C = 17.9%   % O = 23.9%
(b) % K = 28.2%   % C = 8.8%   % O = 35.9%
(c) % K = 56.6%   % C = 8.7%   % O = 34.7%
(d) % K = 39.4%   % C = 12.0%   % O = 48.4%
(e) % K = 35.1%   % C = 21.6%   % O = 43.2%

d 48. Calculate the percent composition of $FeSO_4$.

(a) % Fe = 27.9%   % S = 24.1%   % O = 48.0%
(b) % Fe = 16.2%   % S = 27.9%   % O = 55.8%
(c) % Fe = 33.2%   % S = 9.5%   % O = 57.2%
(d) % Fe = 36.8%   % S = 21.1%   % O = 42.1%
(e) % Fe = 25.1%   % S = 21.6%   % O = 53.2%

Derivation of Formulas from Elemental Composition

b 49. Analysis of a sample of a covalent compound showed that it contained 14.4% hydrogen and 85.6% carbon by mass. What is the empirical formula for this compound?

    (a) CH   (b) CH_2 (c) CH_3 (d) C_2H_3 (e) C_2H_5

d 50. What would be the empirical formula for a compound containing 33.36% calcium, 26.69% sulfur, and 39.95% oxygen?

    (a) $CaSO_4$ (b) $CaSO_2$ (c) $CaSO$ (d) $CaSO_3$ (e) $Ca_2S_2O_5$

b 51. A compound contains sulfur, oxygen, and chlorine. Analysis shows that it contains by mass 26.95% sulfur and 59.61% chlorine. What is the simplest formula for this compound?

    (a) $SOCl$ (b) $SOCl_2$ (c) SO_2Cl_2 (d) SO_2Cl (e) S_2OCl_2

b 52. A compound contains carbon, oxygen, and hydrogen. Analysis of a sample showed that it contained by mass 53.4% carbon and 11.1% hydrogen. What is the simplest formula for this compound?

    (a) CHO (b) C_2H_5O (c) C_2H_4O (d) CH_4O_2 (e) $C_2H_5O_2$

c 53. A 1.520-gram sample of a compound containing only nitrogen and oxygen is found to contain 0.560 grams of nitrogen and 0.960 grams of oxygen. What is the simplest formula for this compound?

    (a) NO   (b) N_2O (c) N_2O_3 (d) NO_2 (e) N_3O_2

c 54. A 4.628-g sample of an oxide of iron was found to contain 3.348 g of iron and 1.280 g of oxygen. What is simplest formula for this compound?

    (a) FeO   (b) Fe_2O_3 (c) Fe_3O_4 (d) FeO_2 (e) Fe_3O_2

a 55. A 2.086-g sample of a compound contains 0.884 g of cobalt, 0.482 g of sulfur, and 0.720 g of oxygen. What is its simplest formula?

    (a) $CoSO_3$ (b) $CoSO_4$ (c) $Co(SO_3)_2$
    (d) $Co(SO_4)_2$ (e) $Co_3(SO_4)_4$

a 56. Determine the simplest formula for a hydrocarbon if the complete combustion of a sample produces 5.28 g of CO_2 and 1.62 g of H_2O.

    (a) C_2H_3 (b) CH_2 (c) CH_3 (d) CH   (e) C_2H_5

b 57. Determine the simplest formula for a hydrocarbon if the complete combustion of a sample produces 3.96 g of CO_2 and 2.16 g of H_2O.

    (a) C_2H_3 (b) C_3H_8 (c) CH_3 (d) CH   (e) C_2H_5

d 58. The complete combustion of a hydrocarbon produced 352 mg of CO_2 and 216 mg of H_2O. What is the simplest formula of this hydrocarbon?

(a) CH   (b) CH_2 (c) C_2H_3 (d) CH_3 (e) C_3H_2

d 59. A compound is known to contain only carbon, hydrogen, and oxygen. If the complete combustion of a 0.150-g sample of this compound produces 0.225 g of CO_2 and 0.0614 g of H_2O, what is the empirical formula of this compound?

(a) C_3H_4 (b) CH_4O (c) C_3HO_3 (d) $C_3H_4O_3$ (e) $C_5H_7O_5$

Determination of Molecular Formulas

b 60. Glucose has a molecular weight of 180.2 g and an empirical formula CH_2O. What is its molecular formula?

(a) $C_8H_4O_5$ (b) $C_6H_{12}O_6$ (c) $C_{12}H_{22}O_{11}$
(d) $C_{10}H_{12}O_3$ (e) CH_2O

b 61. A compound contains, by mass, 87.5% nitrogen and 12.5% hydrogen. Its molecular weight is found to be 32 g/mol. What is its molecular formula?

(a) N_2H_6 (b) N_2H_4 (c) N_2H_5 (d) NH_3 (e) NH_2

d 62. A compound contains only carbon, hydrogen, and oxygen. Analysis of a sample showed that it contained 54.53% C and 9.15% H. Its molecular weight was determined to be approximately 88 g/mol. What is its molecular formula?

(a) C_2H_4O (b) C_4H_8O (c) C_4H_8 (d) $C_4H_8O_2$ (e) $C_4H_{12}O_2$

e 63. A compound contains, by mass, 81.7% carbon and 18.3% hydrogen. A 0.15 mole sample of this compound weighs 6.6 g. What is the molecular formula of this compound?

(a) C_2H_6 (b) CH_3 (c) C_3H_{12} (d) C_4H_{10} (e) C_3H_8

c 64. The complete combustion of a hydrocarbon produced 26.4 g of CO_2 and 5.40 g of H_2O. Another experiment determined the molecular weight of this hydrocarbon to be approximately 52 g/mol. What is its molecular formula?

(a) C_2H_4 (b) CH   (c) C_4H_4 (d) C_2H_2 (e) CH_4

a 65. The complete combustion of a 0.2864-g sample of a compound yielded 0.420 g of CO_2 and 0.172 g of H_2O. The molecular weight was determined to be approximately 60 g/mol. What is the molecular formula of this compound if it contains only carbon, hydrogen, and oxygen?

(a) $C_2H_4O_2$ (b) CH_2O (c) CH_4O_2 (d) $C_3H_6O_3$ (e) $C_{19}H_{38}O_{19}$

e 66. Which of the following sets illustrates the Law of Multiple Proportions?

(a) Li_2O, Na_2O, K_2O (b) KCl, $CaCl_2$, $ScCl_3$ (c) $_1^1H$, $_1^2H$, $_1^3H$
(d) O, O_2, O_3 (e) BrF, BrF_3, BrF_5

e *67. What is the ratio of the masses of oxygen that combine with 1.00 gram of lead in the compounds PbO, PbO_2, and Pb_2O_3?

(a) 1:2:2   (b) 1:2:1   (c) 2:4:4   (d) 6:12:8   (e) 2:4:3

Some Other Interpretations of Chemical Formulas

a 68. What mass of iron is contained in 86.6 grams of pyrite, FeS_2?

(a) 40.3 g   (b) 29.2 g   (c) 47.3 g   (d) 80.6 g   (e) 64.1 g

a 69. What mass of antimony is contained in 106 grams of stibnite, Sb_2S_3?

(a) 76.0 g   (b) 0.72 g   (c) 59.2 g   (d) 114 g   (e) 42.4 g

e *70. What mass of lead is present in 5.05 kg of mendipite, $PbCl_2 \cdot 2PbO$?

(a) 2.89 kg   (b) 4.05 kg   (c) 5.05 kg   (d) 3.68 kg   (e) 4.33 kg

b 71. What mass of cerussite, $PbCO_3$, would contain 25.0 grams of lead?

(a) 19.4 g   (b) 32.2 g   (c) 29.3 g   (d) 25.4 g   (e) 36.9 g

a 72. What mass of hematite, Fe_2O_3, would contain 24.0 kg of iron?

(a) 34.3 kg   (b) 68.3 kg   (c) 44.7 kg   (d) 30.5 kg   (e) 41.4 kg

b *73. What mass of fluoristan, SnF_2, would contain the same mass of tin as 306 grams of cassiterite, SnO_2?

(a) 295 g   (b) 318 g   (c) 278 g   (d) 367 g   (e) 335 g

b *74. What mass of $FeCl_3$ would contain the same **total** number of ions as 16.8 g of $Al_2(SO_4)_3$?

(a) 7.96 g   (b) 9.95 g   (c) 10.8 g   (d) 13.3 g   (e) 8.01 g

e *75. Heating $MgSO_4 \cdot 7H_2O$ at 150°C produces $MgSO_4 \cdot xH_2O$. If heating 24.4 g of pure $MgSO_4 \cdot 7H_2O$ at 150°C were to give 13.7 g of pure $MgSO_4 \cdot xH_2O$, calculate the value for x.

(a) 5   (b) 4   (c) 3   (d) 2   (e) 1

Purity of Samples

c 76. An ore of lead is 45.0% pure lead sulfide, PbS, and 55.0% impurities in which no other lead compounds are present. What mass of lead is contained in 150.0 grams of this ore?

(a) 71.4 g   (b) 67.5 g   (c) 58.5 g   (d) 9.05 g   (e) 18.0 g

a 77. What mass of fluorine is contained in 2.00 tons of cryolite that is 42.0% pure Na_3AlF_6? (No other compounds containing fluorine are present.)

(a) 912 lb   (b) 832 lb   (c) 456 lb   (d) 304 lb   (e) 152 lb

b 78. What mass of calcium metal could be obtained from one kg of limestone that is 50.0% pure $CaCO_3$? (No other calcium-containing compounds are present.)

(a) 0.05 kg   (b) 0.2 kg   (c) 0.4 kg   (d) 0.5 kg   (e) 0.1 kg

d 79. A dolomite ore contains 40.0% pure $MgCO_3 \cdot CaCO_3$. No other compounds of magnesium or calcium are present in the ore. What mass of magnesium and what mass calcium are contained in 100.0 grams of this ore?

(a) 18.3 g Mg – 21.7 g Ca   (b) 7.91 g Mg – 13.0 g Ca
(c) 8.70 g Mg – 31.3 g Ca   (d) 5.27 g Mg – 8.69 g Ca
(e) 34.5 g Mg – 5.30 g Ca

3 Chemical Equations and Reaction Stoichiometry

Chemical Equations

d 1. What scientific (natural) law serves as the basis for balancing chemical equations by requiring that there be no observable change in the quantity of matter during a chemical reaction?

(a) Law of Conservation of Energy
(b) Law of Multiple Proportions
(c) Law of Conservation of Matter
(d) Law of Definite Proportions
(e) Law of Constant Composition

b 2. What scientific law requires that subscripts in formulas should never be changed while balancing a chemical equation?

(a) Law of Multiple Proportions
(b) Law of Definite Proportions
(c) Law of Conservation of Matter
(d) Law of Conservation of Matter and Energy
(e) Law of Conservation of Energy

e 3. Balance the following equation with the **smallest whole number coefficients**. What is the coefficient for O_2 in the balanced equation?

$$C_4H_{10} + O_2 \rightarrow CO_2 + H_2O$$

(a) 9 (b) 5 (c) 15 (d) 6 (e) 13

e *4. Balance the following equation with the **smallest whole number coefficients**. What is the coefficient for O_2 in the balanced equation?

$$C_4H_9SO + O_2 \rightarrow CO_2 + SO_2 + H_2O$$

(a) 54 (b) 29 (c) 23 (d) 32 (e) 27

c 5. What is the coefficient for CO when the following equation is balanced with the **smallest whole number coefficients**?

$$Fe_2O_3 + CO \rightarrow CO_2 + Fe$$

(a) 5 (b) 6 (c) 3 (d) 4 (e) 1

b 6. Balance the following equation with the **smallest whole number coefficients**. What is the coefficient for HCl in the balanced equation?

$$SnS_2 + HCl \rightarrow H_2SnCl_6 + H_2S$$

(a) 4 (b) 6 (c) 3 (d) 12 (e) 2

a 7. Balance the following equation with the **smallest whole number coefficients**. What is the coefficient for H_2O in the balanced equation?

$$LiBF_4 + H_2O \rightarrow H_3BO_3 + HF + LiF$$

(a) 3   (b) 2   (c) 5   (d) 6   (e) 8

a 8. What is the coefficient for HBr when the following equation is balanced with the **smallest whole number coefficients**?

$$Br_2 + H_2O \rightarrow HBr + HBrO_3$$

(a) 5   (b) 7   (c) 8   (d) 3   (e) 6

b 9. Balance the following equation with the **smallest whole number coefficients**. What is the coefficient for NH_3 in the balanced equation?

$$Fe(NO_3)_3 + NH_3 + H_2O \rightarrow Fe(OH)_3 + NH_4NO_3$$

(a) 1   (b) 3   (c) 2   (d) 6   (e) 4

d 10. When heated lead nitrate decomposes according to the following equation. What is the coefficient for NO_2 when the this equation is balanced with the **smallest whole number coefficients**?

$$Pb(NO_3)_2 \rightarrow PbO + O_2 + NO_2$$

(a) 1   (b) 2   (c) 3   (d) 4   (e) 5

b 11. Balance the following equation with the **smallest whole number coefficients**. Choose the answer that is the sum of the coefficients in the balanced equation. Do not forget coefficients of "one".

$$C_7H_{16} + O_2 \rightarrow CO_2 + H_2O$$

(a) 23   (b) 27   (c) 29   (d) 30   (e) 32

d 12. Balance the following equation with the **smallest whole number coefficients**. Choose the answer that is the sum of the coefficients in the balanced equation. Do not forget coefficients of "one".

$$C_5H_6NS + O_2 \rightarrow CO_2 + H_2O + NO_2 + SO_2$$

(a) 33   (b) 35   (c) 37   (d) 39   (e) 41

b 13. Balance the following equation with the **smallest whole number coefficients**. Choose the answer that is the sum of the coefficients in the balanced equation. Do not forget coefficients of "one".

$$Cr + H_2SO_4 \rightarrow Cr_2(SO_4)_3 + H_2$$

(a) 7   (b) 9   (c) 11   (d) 13   (e) 15

e 14. Balance the following equation with the **smallest whole number coefficients**. Choose the answer that is the sum of the coefficients in the balanced equation. Do not forget coefficients of "one".

$$P_4 + Cl_2 \rightarrow PCl_5$$

(a) 7 (b) 9 (c) 11 (d) 13 (e) 15

a 15. What is the sum of **all** coefficients when the following equation is balanced, using the **smallest whole number coefficients**? Do not forget coefficients of "one".

$$CS_2 + Cl_2 \rightarrow CCl_4 + S_2Cl_2$$

(a) 6 (b) 8 (c) 10 (d) 12 (e) 14

a 16. Balance the following equation with the **smallest whole number coefficients**. Choose the answer that is the sum of the coefficients in the balanced equation. Do not forget coefficients of "one".

$$RbOH + H_3PO_4 \rightarrow Rb_3PO_4 + H_2O$$

(a) 8 (b) 10 (c) 12 (d) 4 (e) 6

b 17. Balance the following equation with the **smallest whole number coefficients**. Choose the answer that is the sum of the coefficients in the balanced equation. Do not forget coefficients of "one".

$$Cr_2(SO_4)_3 + RbOH \rightarrow Cr(OH)_3 + Rb_2SO_4$$

(a) 10 (b) 12 (c) 13 (d) 14 (e) 15

d 18. What is the sum of **all** coefficients when the following equation is balanced, using the **smallest whole number coefficients**? Do not forget coefficients of "one".

$$Cl_2O_7 + Ca(OH)_2 \rightarrow Ca(ClO_4)_2 + H_2O$$

(a) 6 (b) 8 (c) 10 (d) 4 (e) 5

b 19. Balance the following equation with the **smallest whole number coefficients**. Choose the answer that is the sum of the coefficients in the balanced equation. Do not forget coefficients of "one".

$$XeF_2 + H_2O \rightarrow Xe + HF + O_2$$

(a) 10 (b) 11 (c) 13 (d) 15 (e) 16

c 20. What is the sum of **all** coefficients when the following equation is balanced, using the **smallest whole number coefficients**? Do not forget coefficients of "one".

$$SiCl_4 + H_2O \rightarrow H_4SiO_4 + HCl$$

(a) 6 (b) 8 (c) 10 (d) 12 (e) 14

c 21. What is the sum of **all** coefficients when the following equation is balanced, using the **smallest whole number coefficients**? Do not forget coefficients of "one".

$$Ag + H_2S + O_2 \rightarrow Ag_2S + H_2O$$

(a) 9   (b) 10   (c) 11   (d) 12   (e) 14

a 22. Balance the following equation with the **smallest whole number coefficients**. Choose the answer that is the sum of the coefficients in the balanced equation. Do not forget coefficients of "one".

$$ZrO_2 + CCl_4 \rightarrow ZrCl_4 + COCl_2$$

(a) 6   (b) 8   (c) 10   (d) 12   (e) 14

b 23. Balance the following equation with the **smallest whole number coefficients**. Choose the answer that is the sum of the coefficients in the balanced equation. Do not forget coefficients of "one".

$$PBr_3 + H_2O \rightarrow H_3PO_3 + HBr$$

(a) 7   (b) 8   (c) 9   (d) 10   (e) 11

a 24. What is the sum of **all** coefficients when the following equation is balanced, using the **smallest whole number coefficients**? Do not forget coefficients of "one".

$$PtCl_4 + XeF_2 \rightarrow PtF_6 + ClF + Xe$$

(a) 16   (b) 22   (c) 24   (d) 26   (e) 32

c 25. Balance the following equation with the **smallest whole number coefficients**. Choose the answer that is the sum of the coefficients in the balanced equation. Do not forget coefficients of "one".

$$CuSO_4 + NH_3 + H_2O \rightarrow (NH_4)_2SO_4 + Cu(NH_3)_4(OH)_2$$

(a) 8   (b) 9   (c) 11   (d) 12   (e) 14

Calculations Based on Chemical Equations

d 26. Consider the following balanced equation.

$$2H_2 + O_2 \rightarrow 2H_2O$$

Which one of the following statements is **false**?

(a) One molecule of O_2 will react with 2 molecules of H_2.
(b) One mole of O_2 will react with 2 moles of H_2.
(c) The complete reaction of 32.0 g of O_2 will produce 2 moles of H_2O.
(d) The complete reaction of 2.0 g of H_2 will produce 36.0 g of H_2O.
(e) The amount of reaction that consumes 32.0 g of O_2 produces 36.0 g of H_2O.

a 27. How many molecules of O_2 would react with 56 C_2H_6 molecules according to the following balanced equation?

$$2C_2H_6 + 7O_2 \rightarrow 4CO_2 + 6H_2O$$

(a) 196 (b) 392 (c) 112 (d) 50 (e) 784

b 28. How many moles of CO_2 would be produced from 56 moles of O_2 according to the following balanced equation?

$$2C_2H_6 + 7O_2 \rightarrow 4CO_2 + 6H_2O$$

(a) 16 (b) 32 (c) 224 (d) 48 (e) 8

e 29. How many moles of O_2 are required to completely burn 1.10 moles of C_6H_6, according to the following equation?

$$2C_6H_6 + 15O_2 \rightarrow 12O_2 + 6H_2O$$

(a) 9.90 (b) 7.62 (c) 19.8 (d) 4.95 (e) 8.25

a 30. Propane (C_3H_8) burns in oxygen to form CO_2 and H_2O according to the following equation. How many grams of O_2 are required to burn 3.01×10^{23} propane molecules?

$$C_3H_8 + 5O_2 \rightarrow 3CO_2 + 4H_2O$$

(a) 80.0 g (b) 40.0 g (c) 160 g (d) 16.0 g (e) 64.0 g

b 31. How many moles of H_2O will be produced from the complete combustion of 2.4 grams of CH_4?

$$CH_4 + 2O_2 \rightarrow CO_2 + 2H_2O$$

(a) 0.15 (b) 0.30 (c) 1.5 (d) 3.0 (e) 6.0

e 32. How many moles of $CaCO_3$ would have to be decomposed to produce 129 grams of CaO?

$$CaCO_3 \rightarrow CaO + CO_2$$

(a) 1.75 (b) 1.86 (c) 2.00 (d) 2.25 (e) 2.30

c 33. How many grams of oxygen are required to burn 0.10 mole of C_3H_8?

$$C_3H_8 + 5O_2 \rightarrow 3CO_2 + 4H_2O$$

(a) 8.0 g (b) 12 g (c) 16 g (d) 32 g (e) 64 g

b 34. If sufficient acid is used to react completely with 72.9 g of magnesium, how much hydrogen will be produced?

$$2HCl + Mg \rightarrow MgCl_2 + H_2$$

(a) 4.5 g (b) 3.0 mol (c) 1.5 mol (d) 9.0 g (e) 6.0 mol

c 35. How many grams of O₂ are required to burn 18 grams of C₅H₁₂?

$$C_5H_{12} + 8O_2 \rightarrow 5CO_2 + 6H_2O$$

(a) 16 g (b) 32 g (c) 64 g (d) 80 g (e) 92 g

c 36. What mass of phosphoric acid, H₃PO₄, would actually react with 7.17 grams of LiOH?

$$3LiOH + H_3PO_4 \rightarrow Li_3PO_4 + 3H_2O$$

(a) 3.27 g (b) 6.53 g (c) 9.80 g (d) 19.6 g (e) 29.4 g

c 37. What mass of SiF₄ could be produced by the reaction of 15 g of SiO₂ with an excess of HF? The equation for the reaction is:

$$SiO_2 + 4HF \rightarrow SiF_4 + 2H_2O$$

(a) 1.04 g (b) 12 g (c) 26 g (d) 104 g (e) 52 g

e 38. What mass of SrF₂ can be prepared from the reaction of 10.0 g of Sr(OH)₂ with excess HF?

$$Sr(OH)_2 + 2HF \rightarrow SrF_2 + 2H_2O$$

(a) 9.67 g (b) 9.82 g (c) 10.0 g (d) 11.6 g (e) 10.3 g

e 39. What mass of Li₃PO₄ can be prepared from the complete reaction of 7.17 grams of LiOH with a stoichiometric amount of H₃PO₄?

$$3LiOH + H_3PO_4 \rightarrow Li_3PO_4 + 3H_2O$$

(a) 9.80 g (b) 9.34 g (c) 9.61 g (d) 10.4 g (e) 11.6 g

a *40. What is the **total mass of products** formed when 3.2 grams of CH₄ is burned in air?

$$CH_4 + 2O_2 \rightarrow CO_2 + 2H_2O$$

(a) 16 g (b) 36 g (c) 44 g (d) 80 g (e) 32 g

b 41. What mass of SnCl₄ must react with excess FeCl₂ to produce 70.0 grams of SnCl₂?

$$SnCl_4 + 2FeCl_2 \rightarrow SnCl_2 + 2FeCl_3$$

(a) 76.1 g (b) 96.2 g (c) 99.2 g (d) 106 g (e) 110 g

a 42 How many grams of KOH would have to be react with excess CO₂ to produce 69.1 grams of K₂CO₃?

$$2KOH + CO_2 \rightarrow K_2CO_3 + H_2O$$

(a) 56.1 g (b) 112 g (c) 28.1 g (d) 222 g (e) 84.2 g

b *43. A mixture of calcium oxide, CaO, and calcium carbonate, CaCO₃, that had a mass of 3.454 g was heated until all the calcium carbonate was decomposed according to the following equation. After heating, the sample had a mass of 3.102 g. Calculate the mass of CaCO₃ present in the original sample.

$$CaCO_3 \text{ (solid)} \rightarrow CaO \text{ (solid)} + CO_2 \text{ (gas)}$$

(a) 0.400 g (b) 0.800 g (c) 1.00 g (d) 1.60 g (e) 0.200 g

The Limiting Reactant Concept

d 44. The following statements apply to the interpretation of chemical equations. Not all of the statements are true. Which response includes all of the true statements, and no others?

I. The sum of the number of moles of the reactants must equal the sum of the number of moles of products in a balanced equation.
II. The sum of the number of grams of the reactants that react must equal the sum of the number of grams of the products produced by the reaction.
III. The following equation for the reaction involving hypothetical substances, A, B, C, and D, implies that the products C and D are **always** produced in a three to one mole ratio.

$$A + 2B \rightarrow 3C + D$$

IV. The equation shown in III implies that in any reaction involving A and B as reactants, A must be the limiting reactant.
V. The total number of atoms in the reactants that react must always equal the total number of atoms in the products produced by the reaction.

(a) I and V (b) I, II, and III (c) II, IV, and V
(d) II, III, and V (e) III, IV, and V

a 45. If 25.0 g of each reactant were used in performing the following reaction, which would be the limiting reactant?

$$3PbO_2 + Cr_2(SO_4)_3 + K_2SO_4 + H_2O \rightarrow 3PbSO_4 + K_2Cr_2O_7 + H_2SO_4$$

(a) PbO₂ (b) H₂O (c) K₂SO₄ (d) PbSO₄ (e) Cr₂(SO₄)₃

b 46. If 20.0 g of each reactant were used in performing the following reaction, which would be the limiting reactant?

$$2MnO_2 + 4KOH + O_2 + Cl_2 \rightarrow 2KMnO_4 + 2KCl + 2H_2O$$

(a) MnO₂ (b) KOH (c) O₂ (d) Cl₂ (e) KMnO₄

e 47. The thermite reaction is performed using 8.6 g Fe₂O₃ and 1.8 g powdered Al metal. Which reactant is in excess and by how much?

$$Fe_2O_3 + 2Al \rightarrow Al_2O_3 + 2Fe$$

(a) Al, 0.3 g (b) Fe₂O₃, 2.0 g (c) Al, 2.1 g
(d) Al, 1.1 g (e) Fe₂O₃, 3.3 g

d 48. What mass of ZnCl$_2$ can be prepared from the reaction of 1.69 grams of zinc with 1.10 grams of HCl?

$$Zn + 2HCl \rightarrow ZnCl_2 + H_2$$

(a) 2.30 g   (b) 2.27 g   (c) 2.45 g   (d) 2.06 g   (e) 4.11 g

d 49. What mass of SrF$_2$ can be prepared from the reaction of 8.05 g of Sr(OH)$_2$ with 3.88 g of HF?

$$Sr(OH)_2 + 2HF \rightarrow SrF_2 + 2H_2O$$

(a) 11.7 g   (b) 12.2 g   (c) 10.5 g   (d) 8.31 g   (e) 8.62 g

b 50. How many grams of nitric acid can be prepared from the reaction of 69.0 grams of nitrogen dioxide with 36.0 grams of water?

$$3NO_2 + H_2O \rightarrow 2HNO_3 + NO$$

(a) 252 g   (b) 63.0 g   (c) 116 g   (d) 84.0 g   (e) 76.0 g

d 51. What is the maximum amount of Ca$_3$(PO$_4$)$_2$ that can be prepared from 9.8 g of Ca(OH)$_2$ and 9.8 g of H$_3$PO$_4$?

$$3Ca(OH)_2 + 2H_3PO_4 \rightarrow Ca_3(PO_4)_2 + 6H_2O$$

(a) 6.8 g   (b) 8.6 g   (c) 10.3 g   (d) 13.7 g   (e) 15.5 g

a 52. How many grams of MnSO$_4$ can be prepared from the reaction of 15.0 g of KMnO$_4$ with 22.6 g of H$_2$SO$_3$?

$$2KMnO_4 + 5H_2SO_3 \rightarrow 2MnSO_4 + K_2SO_4 + 2H_2SO_4$$

(a) 14.3 g   (b) 15.4 g   (c) 15.9 g   (d) 16.6 g   (e) 17.4 g

d 53. A mixture of 13.1 g Zn and 22.0 g I$_2$ is reacted to completion in a closed, evacuated container. What are the contents of the container after this reaction?

$$Zn + I_2 \rightarrow ZnI_2$$

(a) 27.7 g of ZnI$_2$ and 5.7 g of Zn
(b) 63.9 g of ZnI$_2$ and 3.4 g of I$_2$
(c) 63.9 g of ZnI$_2$
(d) 27.7 g of ZnI$_2$ and 7.4 g of Zn
(e) 31.2 g of ZnI$_2$ and 3.9 g of I$_2$

Percent Yields from Chemical Reactions

c 54. What is the percent yield of CO$_2$ if the reaction of 10.0 grams of CO with excess O$_2$ produces 12.8 grams of CO$_2$?

$$2CO(g) + O_2(g) \rightarrow 2CO_2(g)$$

(a) 76.4%   (b) 78.1%   (c) 81.5%   (d) 84.4%   (e) 88.9%

b 55. The reaction of 5.0 g of fluorine with excess chlorine produced 5.6 g of ClF$_3$. What percent yield of ClF$_3$ was obtained?

$$Cl_2 + 3F_2 \rightarrow 2ClF_3$$

(a) 58%   (b) 69%   (c) 76%   (d) 86%   (e) 92%

c 56. What is the percent yield of elemental sulfur if 7.54 grams of sulfur are obtained from the reaction of 6.16 grams of SO$_2$ with an excess of H$_2$S?

$$2H_2S + SO_2 \rightarrow 2H_2O + 3S$$

(a) 72.6%   (b) 40.8%   (c) 81.5%   (d) 88.4%   (e) 91.4%

c 57. How many grams of PI$_3$ could be produced from 250. g of I$_2$ and excess phosphorus if the reaction gives a 98.5% yield?

$$P_4 + 6I_2 \rightarrow 4PI_3$$

(a) 246 g   (b) 254 g   (c) 266 g   (d) 270 g   (e) 286 g

b 58. Suppose 600. g of P$_4$ reacts with 1300. g of S$_8$. How many grams of P$_4$S$_{10}$ can be produced, assuming 80.0% yield based on the limiting reactant?

$$4P_4 + 5S_8 \rightarrow 4P_4S_{10}$$

(a) 4.62 x 10^2 g   (b) 1.44 x 10^3 g   (c) 2.16 x 10^3 g
(d) 4.92 x 10^3 g   (e) 6.50 x 10^2 g

e 59. If 6.6 g of fluorine reacts with 5.6 g chlorine to produce 8.5 g of chlorine trifluoride, what is the limiting reactant and the percent yield of chlorine trifluoride?

$$Cl_2 + 3F_2 \rightarrow 2ClF_3$$

(a) F$_2$, 45%   (b) Cl$_2$, 58%   (c) Cl$_2$, 53%
(d) F$_2$, 69%   (e) F$_2$, 79%

a 60. If a reaction of 27.5 g of Fe with 63.1 g Cl$_2$ produced 60.4 g of FeCl$_3$, what was the limiting reactant and the percent yield?

$$3Cl_2 + 2Fe \rightarrow 2FeCl_3$$

(a) Fe, 75.6%   (b) Fe, 62.8%   (c) Cl$_2$, 41.8%
(d) Cl$_2$, 63.0%   (e) Fe, 37.8%

b *61. If a reaction of 5.0 g of hydrogen with 5.0 g of carbon monoxide produced 4.5 g of methanol, what was the percent yield?

$$2H_2 + CO \rightarrow CH_3OH$$

(a) 11%   (b) 79%   (c) 96%   (d) 24%   (e) 63%

Sequential Reactions

b *62. Sulfuric acid is probably the most important industrial chemical because it is used in so many industrial processes to produce or purify other chemicals. It can be produced by a three step process. First, sulfur is burned in air to give sulfur dioxide. Second, the sulfur dioxide is converted to sulfur trioxide by passing the sulfur dioxide over a catalyst in the presence of oxygen at a high temperature. Third, the sulfur trioxide is reacted with water to form sulfuric acid. What mass of sulfuric acid would be formed from 1.00×10^3 kg of sulfur, if the three steps gave yields as listed below?

$$S + O_2 \rightarrow SO_2 \quad 92.5\%$$
$$2SO_2 + O_2 \rightarrow 2SO_3 \quad 72.6\%$$
$$SO_3 + H_2O \rightarrow H_2SO_4 \quad 98.2\%$$

(a) 4.03×10^3 kg (b) 2.02×10^3 kg (c) 3.06×10^3 kg
(d) 2.84×10^3 kg (e) 2.22×10^3 kg

e *63. Ethyl butyrate ($C_6H_{12}O_2$), "oil of pineapple", can be produced in the laboratory from butyric acid ($C_4H_8O_2$), a foul-smelling liquid, and ethyl alcohol (C_2H_6O). First, butyric acid is reacted with PCl_5 to produce butyryl chloride (C_4H_7ClO) which is then reacted with the ethyl alcohol to produce the ethyl butyrate. What mass of ethyl butyrate could be produced from 12.8 g of butyric acid if the first reaction gave a 65.8% yield of butyryl chloride and the second reaction gave a 88.4% yield of ethyl butyrate?

$$C_4H_8O_2 + PCl_5 \rightarrow C_4H_7ClO + HCl + POCl_3$$
$$C_4H_7ClO + C_2H_6O \rightarrow C_6H_{12}O_2 + HCl$$

(a) 8.80 g (b) 9.01 g (c) 8.47 g (d) 16.9 g (e) 9.81 g

Concentrations of Solutions

d 64. If 23.0 g of NaCl is dissolved in 300. g of H_2O, what is the percent by mass of NaCl in the resulting solution?

(a) 0.080% (b) 0.767% (c) 7.67% (d) 7.12% (e) 21.4%

e 65. What mass of silver nitrate, $AgNO_3$, is required to prepare 800. g of 3.50% solution of $AgNO_3$?

(a) 24.6 g (b) 26.7 g (c) 27.0 g (d) 25.5 g (e) 28.0 g

a 66. What mass of water is contained in 160. grams of 22.0% KCl solution?

(a) 125 g (b) 86.8 g (c) 35.2 g (d) 130 g (e) 112 g

b 67. What mass of 25.0% $Ba(NO_3)_2$ solution contains 40.0 grams of $Ba(NO_3)_2$?

(a) 117 g (b) 160 g (c) 321 g (d) 10.0 g (e) 62.5 g

Chapter 3 35

b 68. What mass of 30.0% Ca(NO$_3$)$_2$ solution contains 60.0 grams of water?

(a) 42.0 g (b) 85.7 g (c) 58.0 g (d) 14.3 g (e) 62.4 g

c 69. The density of a 7.50% solution of ammonium sulfate, (NH$_4$)$_2$SO$_4$, is 1.04 g/mL. What mass of (NH$_4$)$_2$SO$_4$ would be required to prepare 750. mL of this solution?

(a) 45.8 g (b) 54.0 g (c) 58.5 g (d) 62.4 g (e) 65.7 g

e 70. The density of a 34.0% solution of NaBr is 1.33 g/mL. What volume of the solution contains 25.7 g of NaBr?

(a) 101 mL (b) 6.57 mL (c) 32.8 mL (d) 105 mL (e) 56.8 mL

d 71. What volume of 40.0% NaNO$_3$ solution contains 0.15 mole of NaNO$_3$? Density = 1.32 g/mL.

(a) 42.0 mL (b) 3.86 mL (c) 9.60 mL (d) 24.1 mL (e) 38.2 mL

b 72. The molarity of a solution is defined as

(a) the number of moles of solute per kilogram of solvent.
(b) the number of moles of solute per liter of solution.
(c) the number of equivalent weights of solute per liter of solution.
(d) the number of moles of solute per kilogram of solution.
(e) the number of moles of solute per liter of solvent.

a 73. Calculate the molarity of a solution that contains 70.0 g of H$_2$SO$_4$ in 280. mL of solution.

(a) 2.55 M (b) 6.84 M (c) 8.62 M (d) 9.78 M (e) 11.84 M

d 74. What is the molarity of 850. mL of a solution containing 46.2 grams of NaBr?

(a) 0.495 M (b) 0.506 M (c) 0.516 M (d) 0.528 M (e) 0.545 M

e 75. What is the molarity of 175 mL of solution containing 2.18 grams of Na$_2$SO$_4$·10H$_2$O?

(a) 3.77 x 10^{-3} M (b) 6.44 x 10^{-1} M (c) 8.78 x 10^{-3} M
(d) 1.18 x 10^{-3} M (e) 3.87 x 10^{-2} M

a 76. What is the molarity of a barium chloride solution prepared by dissolving 2.50 g of BaCl$_2$·2H$_2$O in enough water to make 400. mL of solution?

(a) 2.56 x 10^{-2} M (b) 4.97 x 10^{-3} M (c) 4.09 x 10^{-2} M
(d) 7.31 x 10^{-3} M (e) 5.20 x 10^{-2} M

b 77. What mass of CaCl$_2$ must be dissolved in enough water to produce 2000. mL of 1.25 M CaCl$_2$?

(a) 174 g (b) 277 g (c) 90.7 g (d) 81.1 g (e) 310 g

d 78. What mass of Na₂SO₄ is required to prepare 400. mL of 1.50 M Na₂SO₄ solution?

(a) 213 g   (b) 56.8 g   (c) 71.4 g   (d) 85.2 g   (e) 8.52 x 10⁴ g

c 79. What volume of 0.250 M KOH solution contains 6.31 grams of KOH?

(a) 631 mL   (b) 28.1 mL   (c) 450 mL   (d) 2.22 mL   (e) 0.44 mL

b 80. The specific gravity of commercial nitric acid solution is 1.42 and it is 70.0% HNO₃ by mass. Calculate its molarity.

(a) 18.0 M   (b) 15.8 M   (c) 12.8 M   (d) 99.4 M   (e) 26.2 M

Dilution of Solutions

e 81. When a solution is diluted, what is the relationship of the number of moles of solute in the more concentrated initial volume of solution to the number of moles of solute in the less concentrated final volume of solution?

(a) The ratio of the numbers is directly proportional to the two volumes.
(b) The ratio of the numbers is inversely proportional to the two volumes.
(c) The number of moles of solute in the more concentrated initial volume is always greater.
(d) The number of moles of solute in the less concentrated final volume is always greater.
(e) The number of moles of solute in both solutions is the same.

c 82. How many mL of 18.4 M H₂SO₄ are needed to prepare 600. mL of 0.10 M H₂SO₄?

(a) 1.8 mL   (b) 2.7 mL   (c) 3.3 mL   (d) 4.0 mL   (e) 4.6 mL

c 83. A laboratory stock solution is 1.50 M NaOH. Calculate the volume of this stock solution that would be needed to prepare 300. mL of 0.200 M NaOH.

(a) 2.25 mL   (b) 10.0 mL   (c) 40.0 mL   (d) 1.00 mL   (e) 0.100 mL

a 84. Calculate the molarity of the resulting solution if 25.0 mL of 2.40 M HCl solution is diluted to 300. mL.

(a) 0.200 M   (b) 29.0 M   (c) 2.00 M   (d) 0.400 M   (e) 0.0400 M

e 85. Calculate the molarity of the resulting solution if enough water is added to 50.0 mL of 4.20 M NaCl solution to make a solution with a volume of 2.80 L.

(a) 75.0 M   (b) 0.043 M   (c) 33.1 M   (d) 0.067 M   (e) 0.0750 M

e *86. Calculate the resulting molarity of a solution prepared by mixing 25.0 mL of 0.160 M NaBr and 55.0 mL of 0.0320 M NaBr.

(a) 0.522 M   (b) 0.272 M   (c) 0.230 M   (d) 0.0658 M   (e) 0.0720 M

c *87. Calculate the molarity of the resulting solution prepared by diluting 25.0 mL of 18.0% ammonium chloride, NH_4Cl, (density = 1.05 g/mL) to a final volume of 80.0 mL.

(a) 0.292 M   (b) 0.059 M   (c) 1.10 M   (d) 0.0536 M   (e) 0.00105 M

d *88. A sample of commercial perchloric acid is 70.0% $HClO_4$ by mass; its density is 1.664 g/mL. How many milliliters of this concentrated $HClO_4$ would be required to prepare 500. mL of 1.50 M $HClO_4$ solution?

(a) 33.0 mL   (b) 45.3 mL   (c) 54.1 mL   (d) 64.7 mL   (e) 78.6 mL

Using Solutions in Chemical Reactions

b 89. How many grams of KOH are contained in 400. mL of 0.250 M KOH solution?

(a) 12.4 g   (b) 5.61 g   (c) 89.8 g   (d) 35.1 g   (e) 8.98 g

c 90. Silver nitrate, $AgNO_3$, reacts with sodium chloride as indicated by the following equation. What mass of NaCl would be required to react with 200. mL of 0.200 M $AgNO_3$ solution?

$$AgNO_3 + NaCl \rightarrow AgCl + NaNO_3$$

(a) 0.117 g   (b) 1.17 g   (c) 2.34 g   (d) 4.68 g   (e) 3.06 g

b 91. What mass of calcium carbonate, $CaCO_3$, is required to react with 100. mL of 2.00 M HCl solution?

$$CaCO_3 + 2HCl \rightarrow CaCl_2 + CO_2 + H_2O$$

(a) 5.00 g   (b) 10.0 g   (c) 15.0 g   (d) 20.0 g   (e) 23.0 g

d 92. What volume of 3.00 molar sulfuric acid, H_2SO_4, is required to react with 250. grams of calcium carbonate, $CaCO_3$?

$$CaCO_3 + H_2SO_4 \rightarrow CaSO_4 + CO_2 + H_2O$$

(a) 208 mL   (b) 462 mL   (c) 767 mL   (d) 833 mL   (e) 946 mL

a 93. What volume of 0.130 M HCl solution will just react with 0.424 gram of $Ba(OH)_2$?

$$2HCl + Ba(OH)_2 \rightarrow BaCl_2 + 2H_2O$$

(a) 38.1 mL   (b) 32.6 mL   (c) 24.1 mL   (d) 18.6 mL   (e) 96.7 mL

e 94. If 40.0 mL of H_2SO_4 solution reacts with 0.212 gram of Na_2CO_3, what is the molarity of the H_2SO_4 solution?

$$Na_2CO_3 + H_2SO_4 \rightarrow Na_2SO_4 + CO_2 + H_2O$$

(a) 0.50 M   (b) 0.10 M   (c) 0.20 M   (d) 0.40 M   (e) 0.050 M

a 95. Some antacids use magnesium hydroxide to neutralize (react with) the excess hydrochloric acid in a hyperacidic stomach. What mass of magnesium hydroxide would a student need to consume if his "upset" stomach held 350. mL of 0.0325 M HCl?

$$Mg(OH)_2 + 2HCl \rightarrow MgCl_2 + H_2O$$

(a) 0.33 g (b) 0.66 g (c) 16 g (d) 0.98 g (e) 23 g

c 96. What volume of 0.132 M KOH solution would react completely with 25.8 mL of 0.198 M HCl according to the following equation?

$$HCl + KOH \rightarrow KCl + H_2O$$

(a) 86.4 mL (b) 25.7 mL (c) 38.7 mL (d) 17.2 mL (e) 3.86 mL

d 97. What volume of 0.385 molar nitric acid, HNO_3, is required to react with 48.0 mL of 0.0770 M calcium hydroxide, $Ca(OH)_2$, according to the following equation?

$$2HNO_3 + Ca(OH)_2 \rightarrow Ca(NO_3)_2 + 2H_2O$$

(a) 24.8 mL (b) 9.62 mL (c) 38.4 mL (d) 19.2 mL (e) 45.0 mL

a 98. What volume of 0.150 M $AgNO_3$ solution is required to react with 80.0 mL of 0.0660 M $CaCl_2$ solution according to the following reaction?

$$2AgNO_3 + CaCl_2 \rightarrow 2AgCl + Ca(NO_3)_2$$

(a) 70.4 mL (b) 140 mL (c) 26.8 mL (d) 35.2 mL (e) 60.2 mL

b 99. Witherite is a mineral that contains barium carbonate. If a 1.68-g sample of witherite were to react completely with 24.6 mL of 0.2558 M HCl, what would be the percent of barium carbonate in the witherite sample? (Barium carbonate is the only compound present that reacts with the hydrochloric acid.)

$$BaCO_3 + 2HCl \rightarrow BaCl_2 + CO_2 + H_2O$$

(a) 74.2% (b) 37.0% (c) 62.1% (d) 23.4% (e) 13.5%

4 Some Types of Chemical Reactions

The Periodic Table: Metals, Nonmetals, and Metalloids

a 1. Which of the following responses contains all the true statements and no others?

   I. The elements at the far right of the periodic table, except the noble gases, have the greatest tendency to form anions.
   II. The elements with the least tendency to form ions are those at the far left of the periodic table.
   III. Bonds in compounds consisting of two adjacent elements in the periodic table are likely to be covalent.
   IV. The elements at the far left of the periodic table possess poor electrical conductivity.

   (a) I and III (b) I, II, and IV (c) II and IV
   (d) I, II, and III (e) IV

b 2. The chemical behavior of a group of elements is determined by the _____ of the atoms in the group.

   (a) mass numbers (b) atomic numbers (c) atomic weights
   (d) atomic mass units (e) Avogadro numbers

d 3. Which response includes all the following that are properties of **most** metals, and no other properties?

   I. They tend to form cations.
   II. They are good heat insulators.
   III. They have outer electronic shells that contain more than four electrons.
   IV. They tend to form ionic compounds when they combine with the elements of Group VIIA.

   (a) I, II, and III (b) II, III, and IV (c) III and IV
   (d) I and IV (e) I and III

c 4. Which response includes all of the following that are properties of **most** nonmetals, and no other properties?

   I. Outer electronic shells contain four or more electrons.
   II. They tend to form positive ions.
   III. They are more likely to form ionic compounds than covalent compounds when they combine with other nonmetals.
   IV They have high electronegativities.

   (a) I, II, and III (b) II and III (c) I and IV
   (d) II and IV (e) none of these answers

40

d 5. Which of the following is a metalloid?

(a) Cr (b) K (c) U (d) Si (e) Pb

c 6. Which of the following is an alkaline earth metal?

(a) Zr (b) Sn (c) Ca (d) Li (e) Th

b 7. Which one of the following is an alkaline earth metal?

(a) potassium, K (b) magnesium, Mg (c) iron, Fe
(d) tin, Sn (e) bismuth, Bi

d 8. The "eka-silicon" of Mendeleev is now known as _____.

(a) phosphorus, P (b) zirconium, Zr (c) tin, Sn
(d) germanium, Ge (e) gallium, Ga

c 9. Which element is **not** correctly classified?

	Element	Classification
(a)	S	nonmetal
(b)	Ge	metalloid
(c)	Se	metalloid
(d)	Li	metal
(e)	In	metal

c 10. Which element and group are **not** correctly matched?

	Element	Periodic Group or Classification
(a)	Sb	metalloid
(b)	Kr	noble gas
(c)	Al	alkali metal
(d)	F	halogen
(e)	Ca	alkaline earth metal

Aqueous Solutions—An Introduction

b 11. Which one of the following compounds is **not** a salt?

(a) Ag_2SO_4 (b) $Ca(OH)_2$ (c) KCl (d) $Ca(NO_3)_2$ (e) $NaCH_3COO$

c 12. Which one of the following compounds is **not** a salt?

(a) LiI (b) $Al(ClO_4)_3$ (c) HI
(d) $Fe(ClO_3)_3$ (e) NH_4Br

d 13. Which one of the following compounds is a salt?

(a) $HClO_2$ (b) CH_3COOH (c) HCN
(d) NH_4CH_3COO (e) KOH

b 14. Which one of the following is a strong acid?

(a) HF
(b) HNO_3
(c) CH_3COOH
(d) H_2SO_3
(e) H_2CO_3

e 15. Which one of the following is a strong acid?

(a) HClO
(b) $(COOH)_2$
(c) H_3PO_4
(d) HSO_3^-
(e) H_2SO_4

a 16. Which one of the following is a strong acid?

(a) HI
(b) HF
(c) HNO_2
(d) HCN
(e) HBrO

e 17. Which one of the following is a strong acid?

(a) HNO_2
(b) H_2SO_3
(c) HF
(d) HClO
(e) $HClO_3$

e 18. Which one of the following is a weak acid?

(a) $HClO_4$
(b) HCl
(c) HBr
(d) HI
(e) CH_3COOH

e 19. Which one of the following is a weak acid?

(a) HBr
(b) HNO_3
(c) $HClO_4$
(d) HCl
(e) $HClO_2$

a 20. Consider the following reaction.

$$NH_3(g) + H_2O(\ell) \rightleftharpoons NH_4^+(aq) + OH^-(aq)$$

Which one of the following statements is **false**?

(a) The double arrows indicate that ammonia, NH_3, is only very slightly soluble in water.
(b) The reaction is reversible.
(c) When ammonia is added to water, NH_4^+ and OH^- ions are produced in a 1:1 ratio.
(d) When solutions of NH_4Cl and NaOH are mixed, some ammonia is produced.
(e) Ammonia is considered a weak base.

d 21. Which one of the following ionic hydroxides is a soluble base?

(a) $Cu(OH)_2$
(b) $Fe(OH)_2$
(c) $Fe(OH)_3$
(d) $Sr(OH)_2$
(e) $Al(OH)_3$

c 22. Which one of the following is a strong and soluble base?

(a) $Ni(OH)_2$
(b) $Mg(OH)_2$
(c) RbOH
(d) $Cr(OH)_3$
(e) $Fe(OH)_3$

c 23. Which one of the following is **not** a soluble base?

(a) KOH (b) CsOH (c) Fe(OH)$_2$
(d) Ca(OH)$_2$ (e) Ba(OH)$_2$

c 24. Which one of the following is an **insoluble** base?

(a) RbOH (b) Ba(OH)$_2$ (c) Al(OH)$_3$
(d) LiOH (e) KOH

d 25. Which one of the following compounds is **incorrectly** identified as to type of compound?

	Substance	Type of Compound
(a)	KOH	strong base
(b)	HClO$_4$	strong acid
(c)	HNO$_2$	weak acid
(d)	NH$_3$	insoluble base
(e)	H$_2$SO$_3$	weak acid

c 26. Which one of the following substances is **insoluble** in water?

(a) RbOH (b) KSCN (c) BaCO$_3$ (d) LiBr (e) Na$_3$PO$_4$

c 27. Which one of the following substances is **insoluble** in water?

(a) MgS (b) Cr(CH$_3$COO)$_3$ (c) Hg$_2$Cl$_2$
(d) RbClO$_4$ (e) Mn(NO$_3$)$_3$

e *28. Which one of the following salts is **insoluble** in water?

(a) NaF (b) NH$_4$ClO$_3$ (c) CuCl$_2$
(d) CaSO$_4$ (e) Ba$_3$(AsO$_4$)$_2$

d 29. Which one of the following salts is **insoluble** in water?

(a) Na$_2$SO$_4$ (b) K$_2$CO$_3$ (c) Li$_3$PO$_4$ (d) BaSO$_4$ (e) (NH$_4$)$_3$AsO$_4$

e 30. Which one of the following salts is **insoluble** in water?

(a) FeCl$_3$ (b) Al$_2$(SO$_4$)$_3$ (c) Cr(NO$_3$)$_3$
(d) (NH$_4$)$_2$CO$_3$ (e) AgCl

a 31. Which one of the following salts is **insoluble** in water?

(a) Mg$_3$(PO$_4$)$_2$ (b) AgNO$_3$ (c) CrCl$_3$ (d) (NH$_4$)$_2$S (e) KHCO$_3$

a 32. Which one of the following salts is soluble in water?

(a) KClO$_3$ (b) BaSO$_4$ (c) Ag$_3$PO$_4$ (d) CuS (e) FeCO$_3$

c 33. Which one of the following salts is **insoluble** in water?

(a) MgSO$_4$ (b) KNO$_3$ (c) AgBr (d) FeCl$_3$ (e) NaBr

c 34. Which one of the following salts is soluble in water?

(a) FeS (b) Ag$_2$CO$_3$ (c) NH$_4$F (d) CrPO$_4$ (e) Hg$_2$Cl$_2$

e 35. Which one of the following salts is **insoluble** in water?

(a) K$_2$S (b) NaClO (c) Mg(NO$_3$)$_2$
(d) Rb$_2$CO$_3$ (e) Ba$_3$(PO$_4$)$_2$

a 36. Which one of the following salts is **insoluble** in water?

(a) Ca$_3$(PO$_4$)$_2$ (b) Pb(CH$_3$COO)$_2$ (c) Al(NO$_3$)$_3$
(d) FeBr$_3$ (e) Na$_2$CO$_3$

e 37. Which one of the following salts is **insoluble** in water?

(a) Rb$_2$SO$_4$ (b) (NH$_4$)$_2$S (c) KI
(d) Cr(NO$_3$)$_3$ (e) Ni(CO$_3$)$_2$

d 38. Which one of the following salts is **insoluble** in water?

(a) NH$_4$CN (b) Ni(NO$_3$)$_2$·6H$_2$O (c) LiNO$_3$
(d) MnS (e) Ba(CH$_3$COO)$_2$

c 39. Which response includes all of the following salts that are **insoluble** in water, and no others?

I. KI II. AgBr III. (NH$_4$)$_2$CO$_3$
IV. Pb(CH$_3$COO)$_2$ V. PbSO$_4$

(a) II, III, and IV (b) I (c) II and V
(d) III, IV, and V (e) II and IV

a 40. Which response includes all of the following compounds that are **insoluble** in water, and no others?

I. Mg$_3$(PO$_4$)$_2$ II. MgSO$_4$ III. NaCN
IV. Ni(ClO$_4$)$_2$ V. HCN

(a) I (b) I, II, and IV (c) III, IV, and V
(d) IV and V (e) I, II, and V

d 41. Which one of the following compounds is **not** a strong electrolyte?

(a) RbF (b) Ni(ClO$_3$)$_2$ (c) Mg(NO$_3$)$_2$
(d) HF (e) HNO$_3$

44

a 42. Which one of the following is **not** a strong electrolyte?

(a) BaSO$_4$ (b) MgS (c) Na$_2$S (d) HI (e) Sr(OH)$_2$

c 43. Which one of the following is a weak electrolyte?

(a) KBr (b) H$_2$SO$_4$ (c) H$_2$CO$_3$ (d) NH$_4$ClO (e) Ca(OH)$_2$

d 44. Which one of the following is a weak electrolyte?

(a) HClO$_3$ (b) HCl (c) Ca(OH)$_2$ (d) HClO (e) NH$_4$NO$_3$

b 45. Which one of the following is **not** a strong electrolyte?

(a) HBr (b) Sn(OH)$_2$ (c) KMnO$_4$
(d) Ba(OH)$_2$ (e) NaBrO$_3$

c 46. Which response includes all of the following substances that are strong electrolytes, and no others?

I. CH$_3$COOH II. NH$_4$Cl III. Cr(OH)$_3$ IV. KOH

(a) I and II (b) II and III (c) II and IV
(d) I and IV (e) II, III, and IV

a 47. Which response includes all of the following substances that are strong electrolytes, and no others?

I. K$_2$CO$_3$ II. HNO$_3$ III. H$_2$SO$_3$ IV. CuCO$_3$

(a) I and II (b) I and IV (c) II, III, and IV
(d) III and IV (e) I, II, and III

e 48. Which response includes all of the following substances that are strong electrolytes, and no others?

I. LiOH II. Ca(CH$_3$COO)$_2$ III. HClO$_4$ IV. H$_3$PO$_4$

(a) I (b) II and III (c) II, III, and IV
(d) I and III (e) I, II, and III

d 49. Which response includes all of the following substances that are strong electrolytes, and no others?

I. Na$_2$SO$_4$ II. HCN III. KClO$_3$ IV. Mn(OH)$_3$

(a) II, III, and IV (b) I and II (c) II and IV
(d) I and III (e) I, III, and IV

d 50. Which response includes all of the following substances that are strong electrolytes, and no others?

I. Na_2CrO_4 II. HI   III. $(COOH)_2$ IV. $(NH_4)_2S$

(a) I and IV
(b) II and III
(c) I, II, and III
(d) I, II, and IV
(e) I and II

Reactions in Aqueous Solutions

b 51. What is the **total ionic** equation for the following formula unit equation?

$$BaCl_2(aq) + Na_2SO_4(aq) \rightarrow BaSO_4(s) + 2NaCl(aq)$$

(a) $[Ba^{2+}(aq)+Cl^-(aq)] + [Na^+(aq)+SO_4^{2-}(aq)] \rightarrow BaSO_4(s) + [Na^+(aq)+Cl^-(aq)]$
(b) $[Ba^{2+}(aq)+2Cl^-(aq)] + [2Na^+(aq)+SO_4^{2-}(aq)] \rightarrow BaSO_4(s) + 2[Na^+(aq)+Cl^-(aq)]$
(c) $[Ba^{2+}(aq)+2Cl^-(aq)] + 2[Na^+(aq)+SO_4^{2-}(aq)] \rightarrow BaSO_4(s) + 2[Na^+(aq)+Cl^-(aq)]$
(d) $[Ba^{2+}(aq)+2Cl^-(aq)] + [Na^+(aq)+SO_4^{2-}(aq)] \rightarrow BaSO_4(s) + [Na^+(aq)+Cl^-(aq)]$
(e) $Ba^{2+}(aq) + SO_4^{2-}(aq) \rightarrow BaSO_4(s)$

c 52. What is the **total ionic** equation for the following formula unit equation?

$$HF(aq) + KOH(aq) \rightarrow KF(aq) + H_2O(\ell)$$

(a) $[H^+(aq)+F^-(aq)] + [K^+(aq)+OH^-(aq)] \rightarrow [K^+(aq)+F^-(aq)] + [2H^+(\ell)+O^{2-}(\ell)]$
(b) $[H^+(aq)+F^-(aq)] + [K^+(aq)+OH^-(aq)] \rightarrow [K^+(aq)+F^-(aq)] + H_2O(aq)$
(c) $HF(aq) + [K^+(aq)+OH^-(aq)] \rightarrow [K^+(aq)+F^-(aq)] + H_2O(\ell)$
(d) $HF(aq) + [K^+(aq)+OH^-(aq)] \rightarrow [K^+(aq)+F^-(aq)] + [2H^+(aq)+OH^-(aq)]$
(e) $HF(aq) + OH^-(aq) \rightarrow F^-(aq) + H_2O(\ell)$

d 53. What is the **net ionic** equation for the following formula unit equation?

$$CH_3COOH(aq) + NaOH(aq) \rightarrow NaCH_3COO(aq) + H_2O(\ell)$$

(a) $CH_3COO^-(aq) + H^+(aq) + Na^+(aq) + OH^-(aq) \rightarrow NaCH_3COO(aq) + H_2O(\ell)$
(b) $CH_3COO^-(aq) + H^+(aq) + Na^+(aq) + OH^-(aq) \rightarrow Na^+CH_3COO^-(aq) + H_2O(\ell)$
(c) $CH_3COOH(aq) + Na^+(aq) + OH^-(aq) \rightarrow Na^+(aq) + CH_3COO^-(aq) + H_2O(\ell)$
(d) $CH_3COOH(aq) + OH^-(aq) \rightarrow CH_3COO^-(aq) + H_2O(\ell)$
(e) $CH_3COOH(aq) + Na^+(aq) + OH^-(aq) \rightarrow NaCH_3COO(aq) + H_2O(\ell)$

a 54. What is the **net ionic** equation for the following formula unit equation?

$$Cu(NO_3)_2(aq) + H_2S(aq) \rightarrow CuS(s) + 2HNO_3(aq)$$

(a) $Cu^{2+}(aq) + H_2S(aq) \rightarrow CuS(s) + 2H^+(aq)$
(b) $[Cu^{2+}(aq)+2NO_3^-(aq)] + H_2S(aq) \rightarrow CuS(s) + 2[H^+(aq)+2NO_3^-(aq)]$
(c) $Cu^{2+}(aq) + 2H^+(aq) + S^{2-}(aq) \rightarrow CuS(s) + 2H^+(aq)$
(d) $Cu^{2+}(aq) + S^{2-}(aq) \rightarrow CuS(s)$
(e) $Cu^{2+}(aq) + 2NO_3^-(aq) + 2H^+(aq) + S^{2-}(aq) \rightarrow CuS(s) + 2H^+(aq) + 2NO_3^-(aq)$

a *55. Which of the following equations could **not** be a formula unit equation for the net ionic equation: $H^+(aq) + OH^-(aq) \rightarrow H_2O(\ell)$?

(a) $HCN(aq) + NaOH(aq) \rightarrow NaCN(aq) + H_2O(\ell)$
(b) $HNO_3(aq) + NaOH(aq) \rightarrow NaNO_3(aq) + H_2O(\ell)$
(c) $HCl(aq) + KOH(aq) \rightarrow KCl(aq) + H_2O(\ell)$
(d) $HClO_4(aq) + LiOH(aq) \rightarrow LiClO_4(aq) + H_2O(\ell)$
(e) $Ba(OH)_2(aq) + 2HCl(aq) \rightarrow BaCl_2(aq) + 2H_2O(\ell)$

Oxidation Numbers

d 56. Determine the oxidation number of the underlined element in $K_2\underline{C}O_3$.

(a) –1   (b) –2   (c) – 4   (d) +4   (e) +5

d 57. Determine the oxidation number of the underlined element in $\underline{S}O_3^{2-}$.

(a) +2   (b) –2   (c) +3   (d) +4   (e) –3

d 58. Determine the oxidation number of the underlined element in $Na\underline{Mn}O_4$.

(a) +1   (b) +6   (c) +3   (d) +7   (e) +5

e 59. Determine the oxidation number of the underlined element in $Na\underline{N}O_3$.

(a) +1   (b) +2   (c) +3   (d) +4   (e) +5

d 60. Determine the oxidation number of the underlined element in $(NH_4)_2\underline{C}O_3$.

(a) +1   (b) +2   (c) +3   (d) +4   (e) +6

e 61. Determine the oxidation number of the underlined element in $H_2\underline{Cr}O_4$.

(a) +1   (b) +2   (c) +3   (d) +4   (e) +6

b 62. Determine the oxidation number of the underlined element in $H_2\underline{S}O_4$.

(a) +7   (b) +6   (c) +3   (d) +4   (e) +5

a 63. Determine the oxidation number of the underlined element in $Li_2\underline{S}O_4$.

(a) +6   (b) +2   (c) +3   (d) +4   (e) +5

c 64. Determine the oxidation number of the underlined element in $H\underline{As}Cl_4$.

(a) +1   (b) –3   (c) +3   (d) –5   (e) +5

a 65. Determine the oxidation number of the underlined element in $H_2\underline{P}O_2^-$.

(a) +1   (b) +2   (c) +3   (d) +4   (e) +5

d 66. Determine the oxidation number of the underlined element in $H_2\underline{Mn}O_3$.

(a) +1   (b) +2   (c) +3   (d) +4   (e) +5

e 67. Determine the oxidation number of the underlined element in $\underline{N}_2H_5^+$.

(a) +1   (b) +2   (c) +3   (d) –1   (e) –2

a 68. Determine the oxidation number of the underlined element in $Na_2\underline{Se}O_4$.

(a) +6   (b) +2   (c) +3   (d) +4   (e) +5

d 69. Determine the oxidation number of the underlined element in $\underline{Se}O_3^{2-}$.

(a) +6   (b) +8   (c) +3   (d) +4   (e) +5

a 70. Determine the oxidation number of the underlined element in $H_2\underline{P}_2O_7$.

(a) +6   (b) +7   (c) +3   (d) +4   (e) +5

a 71. Determine the oxidation number of the underlined element in $K\underline{Mn}O_4$.

(a) +7   (b) +6   (c) +3   (d) +4   (e) +5

d 72. What are the oxidation numbers (oxidation states) of the elements in $KClO_2$?

(a) K = +3, Cl = +3, O = –2
(b) K = –1, Cl = –5, O = +2
(c) K = +1, Cl = +5, O = –2
(d) K = +1, Cl = +3, O = –2
(e) K = +1, Cl = +2, O = –3

c 73. What are the oxidation numbers (oxidation states) of the elements in $K_2Cr_2O_7$?

(a) K = +1, Cr = +7, O = −2
(b) K = +1, Cr = +12, O = −2
(c) K = +1, Cr = +6, O = −2
(d) K = +1, Cr = +8, O = −2
(e) K = +2, Cr = +6, O = −2

d 74. What are the oxidation numbers (oxidation states) of the elements in H_3PO_4?

(a) H = +1, P = +7, O = −2
(b) H = +1, P = +11, O = −2
(c) H = +3, P = +5, O = −2
(d) H = +1, P = +5, O = −2
(e) H = +3, P = +4, O = −8

b 75. What are the oxidation numbers (oxidation states) of the elements in $NaClO_4$?

(a) Na = +1, Cl = +9, O = −2
(b) Na = +1, Cl = +7, O = −2
(c) Na = +1, Cl = −1, O = 0
(d) Na = +2, Cl = +6, O = −2
(e) Na = +1, Cl = +5, O = −2

e 76. What are the oxidation numbers (oxidation states) of the elements in HCO_3^-?

(a) H = +1, C = +5, O = −2
(b) H = +1, C = +3, O = −2
(c) H = +1, C = +2, O = −2
(d) H = +2, C = +2, O = −2
(e) H = +1, C = +4, O = −2

Oxidation–Reduction Reactions—An Introduction

b 77. Which response contains all of the following that are oxidation-reduction reactions and no others?

I. $PCl_3(\ell) + 3H_2O(\ell) \rightarrow 3HCl(aq) + H_3PO_3(aq)$
II. $Fe_2O_3(s) + 3CO(g) \rightarrow 2Fe(s) + 3CO_2(g)$
III. $CaCO_3(s) + 2HClO_3(aq) \rightarrow Ca(ClO_3)_2(aq) + CO_2(g) + H_2O(\ell)$

(a) I (b) II (c) III (d) II and III (e) I and II

c 78. Which response includes all the following that are oxidation-reduction reactions, and no others?

I. $BaSO_3(s) \rightarrow BaO(s) + SO_2(g)$
II. $2K(s) + Br_2(\ell) \rightarrow 2KBr(s)$
III. $H_2CO_3(aq) + Ca(OH)_2(aq) \rightarrow CaCO_3(s) + 2H_2O(\ell)$
IV. $SnS_2(s) + 6HCl(aq) \rightarrow H_2SnCl_6(s) + 2H_2S(aq)$
V. $3Cl_2(g) + 6KOH(aq) \rightarrow 5KCl(aq) + KClO_3(aq) + 3H_2O(\ell)$

(a) II, III, and IV
(b) I and III
(c) II and V
(d) I and IV
(e) another one or another combination

b 79. What is the oxidizing agent in the following reaction?

$$16H^+(aq) + 2MnO_4^-(aq) + 10SO_4^{2-}(aq) \rightarrow 2Mn^{2+}(aq) + 5S_2O_8^{2-}(aq) + 8H_2O(\ell)$$

(a) H^+ (b) MnO_4^- (c) SO_4^{2-} (d) Mn^{2+} (e) $S_2O_8^{2-}$

b *80. What is the oxidizing agent in the following reaction?

$$6KOH(aq) + 3Cl_2(g) \rightarrow KClO_3(aq) + 5KCl(aq) + 3H_2O(\ell)$$

(a) KOH   (b) Cl_2 (c) $KClO_3$ (d) KCl   (e) H_2O

a 81. What is the reducing agent in the following reaction?

$$Cu(s) + 4H^+(aq) + SO_4^{2-}(aq) \rightarrow Cu^{2+}(aq) + 2H_2O(\ell) + SO_2(g)$$

(a) Cu   (b) H^+ (c) SO_4^{2-} (d) Cu^{2+} (e) SO_2

b *82. What is the reducing agent in the following reaction?

$$6KOH(aq) + 3Cl_2(g) \rightarrow KClO_3(aq) + 5KCl(aq) + 3H_2O(\ell)$$

(a) KOH   (b) Cl_2 (c) $KClO_3$ (d) KCl   (e) H_2O

c 83. The oxidizing agent is _____, and it is _____ in the following reaction.

$$2MnO_4^-(aq) + 3S^{2-}(aq) + 4H_2O(\ell) \rightarrow 2MnO_2(s) + 3S(s) + 8OH^-(aq)$$

(a) MnO_4^-, oxidized   (b) S^{2-}, oxidized   (c) MnO_4^-, reduced
(d) S^{2-}, reduced   (e) H_2O, reduced

b 84. In the following reaction $NaClO_3$ is _____.

$$5NaClO_3(aq) + 3H_2O(\ell) + 3I_2(s) \rightarrow 6HIO_3(aq) + 5NaCl(aq)$$

(a) the oxidizing agent and is oxidized   (b) the oxidizing agent and is reduced
(c) the reducing agent and is oxidized   (d) the reducing agent and is reduced
(e) neither an oxidizing agent nor a reducing agent

c 85. In the following reaction CO is _____.

$$Fe_2O_3(s) + 3CO(g) \rightarrow 2Fe(s) + 3CO_2(g)$$

(a) the oxidizing agent and is oxidized   (b) the oxidizing agent and is reduced
(c) the reducing agent and is oxidized   (d) the reducing agent and is reduced
(e) neither an oxidizing agent nor a reducing agent

e 86. In the following reaction H_2O is _____.

$$5NaClO_3(aq) + 3H_2O(\ell) + 3I_2(s) \rightarrow 6HIO_3(aq) + 5NaCl(aq)$$

(a) the oxidizing agent and is oxidized   (b) the oxidizing agent and is reduced
(c) the reducing agent and is oxidized   (d) the reducing agent and is reduced
(e) neither an oxidizing agent nor a reducing agent

Combination Reactions

b 87. Which of the following reactions is a combination reaction?

(a) $AgNO_3(aq) + HCl(aq) \rightarrow AgCl(s) + HNO_3(aq)$
(b) $Na_2O(s) + CO_2(g) \rightarrow Na_2CO_3(s)$
(c) $C_3H_8(g) + 5O_2(g) \rightarrow 3CO_2(g) + 4H_2O(\ell)$
(d) $2H_2O(\ell) \rightarrow 2H_2(g) + O_2(g)$
(e) $KOH(aq) + HCl(aq) \rightarrow KCl(aq) + H_2O(\ell)$

e 88. Which of the following reactions is **not** a combination reaction?

(a) $2H_2(g) + O_2(g) \rightarrow 2H_2O(\ell)$
(b) $2Li(s) + Cl_2(g) \rightarrow 2LiCl(s)$
(c) $SO_2(g) + H_2O(\ell) \rightarrow H_2SO_3(\ell)$
(d) $2PbO(s) + O_2(g) \rightarrow 2PbO_2(s)$
(e) $Ca(s) + 2HCl(aq) \rightarrow CaCl_2(aq) + H_2(g)$

c 89. Which of the following reactions is **not** a combination reaction?

(a) $2CO(g) + O_2(g) \rightarrow 2CO_2(g)$
(b) $PBr_3(\ell) + Br_2(\ell) \rightarrow PBr_5(s)$
(c) $AgNO_3(aq) + HCl(aq) \rightarrow AgCl(s) + HNO_3(aq)$
(d) $2Ca(s) + O_2(g) \rightarrow 2CaO(s)$
(e) $CO_2(g) + H_2O(\ell) \rightarrow H_2CO_3(aq)$

Decomposition Reactions

c 90. Which of the following statements about a decomposition reaction is **not** true?

(a) It may or may not also be an oxidation-reduction reaction.
(b) It may produce two elements as products.
(c) It may also be a combination reaction.
(d) It may produce an element and a compound as products..
(e) It may produce two different compounds as products.

e 91. Which of the following reactions is a decomposition reaction?

(a) $2H_2(g) + O_2(g) \rightarrow 2H_2O(\ell)$
(b) $Fe_2O_3(s) + 3CO(g) \rightarrow 2Fe(s) + 3CO_2(g)$
(c) $C_3H_8(g) + 5O_2(g) \rightarrow 3CO_2(g) + 4H_2O(\ell)$
(d) $2AgNO_3(aq) + Zn(s) \rightarrow 2Ag(s) + Zn(NO_3)_2(aq)$
(e) $2KClO_3(s) \rightarrow 2KCl(s) + 3O_2(g)$

Chapter 4 51

b 92. Which of the following reactions is **not** a decomposition reaction?

(a) $2HgO(s) \rightarrow 2Hg(\ell) + O_2(g)$
(b) $H_2(g) + Cl_2(g) \rightarrow 2HCl(g)$
(c) $NH_4NO_3(s) \rightarrow N_2O(g) + 2H_2O(g)$
(d) $Mg(OH)_2(s) \rightarrow MgO(s) + H_2O(g)$
(e) $2H_2O_2(aq) \rightarrow 2H_2O(\ell) + O_2(g)$

a 93. Which of the following reactions is **not** a decomposition reaction?

(a) $C_4H_8(g) + 6O_2(g) \rightarrow 4CO_2(g) + 4H_2O(\ell)$
(b) $Na_2CO_3(s) \rightarrow Na_2O(s) + CO_2(g)$
(c) $2KClO_3(s) \rightarrow 2KCl(s) + 3O_2(g)$
(d) $2H_2O(\ell) \rightarrow 2H_2(g) + O_2(g)$
(e) $BaSO_3(s) \rightarrow BaO(s) + SO_2(g)$

Displacement Reactions

b 94. Which response includes all of the following that are displacement reactions, and no other reactions?

I. $P_4O_{10}(s) + 6Na_2O(s) \rightarrow 4Na_3PO_4(s)$
II. $2AgNO_3(aq) + Zn(s) \rightarrow 2Ag(s) + Zn(NO_3)_2(aq)$
III. $Ca(s) + 2HCl(aq) \rightarrow CaCl_2(aq) + H_2(g)$
IV. $Fe(OH)_2(s) + 2HCl(aq) \rightarrow FeCl_2(aq) + 2H_2O(\ell)$

(a) I and II   (b) II and III   (c) II and IV
(d) I and III   (e) I, III, and IV

e 95. Which response includes all of the following that are displacement reactions, and no other reactions?

I. $Li_2O(s) + CO_2(g) \rightarrow Li_2CO_3(s)$
II. $Cl_2(g) + 2NaBr(aq) \rightarrow Br_2(\ell) + 2NaCl(aq)$
III. $HNO_3(aq) + KOH(aq) \rightarrow KNO_3(aq) + H_2O(\ell)$
IV. $SO_3(\ell) + H_2O(\ell) \rightarrow H_2SO_4(\ell)$

(a) I and II   (b) I and IV   (c) I, III, and IV
(d) I, II, and IV   (e) II

Harcourt, Inc

e 96. Which response includes all of the following that are displacement reactions, and no other reactions?

    I. $MgSO_3(s) \xrightarrow{heat} MgO(s) + SO_2(g)$
    II. $2H_2O_2(\ell) \xrightarrow{electricity} 2H_2O(\ell) + O_2(g)$
    III. $2K(s) + H_2(g) \rightarrow 2KH(s)$
    IV. $Mg(s) + 2HBr(aq) \rightarrow MgBr_2(aq) + H_2(g)$

(a) I and IV   (b) I and II   (c) II and IV
(d) III and IV   (e) IV

d 97. Which response includes all of the following that are displacement reactions, and no other reactions?

    I. $N_2O_5(s) + H_2O(\ell) \rightarrow 2HNO_3(\ell)$
    II. $WO_3(s) + 3H_2(g) \xrightarrow{heat} W(s) + 3H_2O(\ell)$
    III. $2CdS(s) + 3O_2(g) \xrightarrow{heat} 2CdO(s) + 2SO_2(g)$
    IV. $2Cr(s) + 6HClO_4(aq) \rightarrow 2Cr(ClO_4)_3(aq) + 3H_2(g)$

(a) I and IV   (b) II and III   (c) II, III, and IV
(d) II and IV   (e) I, II, and IV

b 98. Which response includes all of the following that are displacement reactions, and no other reactions?

    I. $BaO(s) + SO_2(g) \rightarrow BaSO_3(s)$
    II. $2Rb(s) + 2H_2O(\ell) \rightarrow 2RbOH(aq) + H_2(g)$
    III. $2HI(aq) + Ca(OH)_2(aq) \rightarrow CaI_2(aq) + 2H_2O(\ell)$
    IV. $2HgO(s) \xrightarrow{heat} 2Hg(\ell) + O_2(g)$

(a) I and IV   (b) II   (c) II, III, and IV
(d) II and IV   (e) I, II, and IV

a 99. Which response includes all of the following that are displacement reactions, and no other reactions?

    I. $2KBr(aq) + F_2(g) \rightarrow 2KF(aq) + Br_2(\ell)$
    II. $N_2O_3(g) \xrightarrow{heat} NO(g) + NO_2(g)$
    III. $PF_3(g) + F_2(g) \rightarrow PF_5(g)$
    IV. $2Na(s) + 2H_2O(\ell) \xrightarrow{heat} 2NaOH(aq) + H_2(g)$

(a) I and IV   (b) I and II   (c) II, III, and IV
(d) II and IV   (e) I, II, and IV

e 100. Which one of the following acids does **not** produce $H_2(g)$ as a major product of reaction with a moderately active metal like zinc?

(a) H_2SO_4 (b) HCl   (c) $HClO_4$ (d) $HClO_3$ (e) HNO_3

c 101. Which one of the following metals is **least** active?

(a) Ca (b) K (c) Pt (d) Li (e) Al

b 102. Which one of the following ions would be displaced from aqueous solution by magnesium?

(a) Ca^{2+} (b) Cu^{2+} (c) Li^+ (d) K^+ (e) Na^+

a 103. Which one of the following ions would **not** be displaced from aqueous solution by magnesium?

(a) Na^+ (b) Cu^{2+} (c) Hg^{2+} (d) Ag^+ (e) Au^+

c 104. Which one of the following elements reacts with water at room temperature to produce hydrogen?

(a) Ti (b) Zn (c) Ca (d) Mg (e) Al

b 105. Which response includes all of the following reactions that **will occur** in aqueous solution, and no others?

I. $2NaF(aq) + Cl_2(g) \rightarrow 2NaCl(aq) + F_2(g)$
II. $2NaCl(aq) + I_2(s) \rightarrow 2NaI(aq) + Cl_2(g)$
III. $2NaBr(aq) + Cl_2(g) \rightarrow 2NaCl(aq) + Br_2(\ell)$
IV. $2NaI(aq) + Br_2(\ell) \rightarrow 2NaBr(aq) + I_2(s)$
V. $2NaBr(aq) + I_2(s) \rightarrow 2NaI(aq) + Br_2(\ell)$

(a) I, II, and III (b) III and IV (c) II, IV, and V
(d) I and III (e) III, IV, and V

c 106. Which response lists all the reactions below that **will occur** in aqueous solution, and no other reactions?

I. $Br_2(\ell) + 2F^-(aq) \rightarrow 2Br^-(aq) + F_2(g)$
II. $Cl_2(g) + 2I^-(aq) \rightarrow 2Cl^-(aq) + I_2(s)$
III. $Br_2(\ell) + 2I^-(aq) \rightarrow 2Br^-(aq) + I_2(s)$
IV. $Cl_2(g) + 2Br^-(aq) \rightarrow 2Cl^-(aq) + Br_2(\ell)$

(a) II and III (b) I, II, and IV (c) II, III, and IV
(d) I and IV (e) I, II, III, and IV

Metathesis (Acid—Base) Reactions

a 107. Which one of the following represents the **net ionic** equation for the reaction of nitric acid with aluminum hydroxide?

(a) $3H^+(aq) + Al(OH)_3(s) \rightarrow Al^{3+}(aq) + 3H_2O(\ell)$
(b) $3HNO_3(aq) + Al(OH)_3(s) \rightarrow 3Al(NO_3)_3(aq) + 3H_2O(\ell)$
(c) $2H^+(aq) + Al(OH)_2(s) \rightarrow Al^{2+}(aq) + 2H_2O(\ell)$
(d) $H^+(aq) + OH^-(aq) \rightarrow H_2O(\ell)$
(e) $3NO_3^-(aq) + Al^{3+}(aq) \rightarrow Al(NO_3)_3(aq)$

b 108. Write the balanced **formula unit** equation for the complete reaction of barium hydroxide with perchloric acid. What is the coefficient of H_2O?

(a) 1   (b) 2   (c) 3   (d) 4   (e) 5

a 109. Write the **formula unit** equation for the complete reaction of hydrobromic acid with calcium hydroxide. What is the sum of the coefficients? (Do not forget coefficients of one.)

(a) 6   (b) 7   (c) 3   (d) 4   (e) 5

a 110. Write the balanced **formula unit** equation for the complete neutralization of dilute H_2SO_4 with KOH in aqueous solution. What is the sum of the coefficients? (Do not forget coefficients of one.)

(a) 6   (b) 8   (c) 9   (d) 4   (e) 5

e 111. Write the **formula unit** equation for the complete neutralization of $Ca(OH)_2$ with dilute H_2SO_4. What is the sum of the coefficients? (Do not forget coefficients of one.)

(a) 6   (b) 7   (c) 8   (d) 4   (e) 5

c 112. Write the balanced **formula unit** equation for the reaction of nitric acid with aqueous ammonia. What is the sum of the coefficients?

(a) 7   (b) 8   (c) 3   (d) 4   (e) 6

a 113. Write the **formula unit** equation for the complete neutralization of iron(II) hydroxide with hydrochloric acid. What is the sum of the coefficients? (Do not forget coefficients of one.)

(a) 6   (b) 7   (c) 8   (d) 4   (e) 5

d 114. Write the **formula unit** equation for the reaction of nitrous acid with potassium hydroxide. What is the sum of the coefficients? (Do not forget coefficients of one.)

(a) 6   (b) 8   (c) 3   (d) 4   (e) 5

e 115. Write the balanced **formula unit** equation for the complete reaction of sulfurous acid with aqueous sodium hydroxide. What is the sum of the coefficients? (Do not forget coefficients of one.)

(a) 8 (b) 10 (c) 12 (d) 4 (e) 6

e 116. Write the **formula unit** equation for the complete reaction of $Ca(OH)_2$ with H_2S. What is the sum of the coefficients? (Do not forget coefficients of one.)

(a) 6 (b) 7 (c) 3 (d) 4 (e) 5

a *117. Write the **formula unit** equation for the complete reaction of LiOH with oxalic acid, $(COOH)_2$. What is the sum of the coefficients? (Do not forget coefficients of one.)

(a) 6 (b) 7 (c) 8 (d) 4 (e) 5

b 118. Write the balanced **formula unit** equation for the complete neutralization of dilute H_3PO_4 with $Ca(OH)_2$ in aqueous solution. What is the sum of the coefficients? (Do not forget coefficients of one.)

(a) 10 (b) 12 (c) 14 (d) 16 (e) 18

e 119. Write the **formula unit** equation for the reaction of the appropriate acid and base to produce the salt $Mg(CH_3COO)_2$. What is the sum of all coefficients? (Do not forget coefficients of one.)

(a) 8 (b) 10 (c) 12 (d) 4 (e) 6

e 120. Write the **formula unit** equation for the reaction by which calcium carbonate is produced from an acid and a base. What is the sum of the coefficients? (Do not forget coefficients of one.)

(a) 6 (b) 7 (c) 3 (d) 4 (e) 5

a 121. Write the balanced **total ionic** equation for the reaction of nitric acid with aqueous ammonia. What is the coefficient preceding NH_3?

(a) 1 (b) 2 (c) 3 (d) 4 (e) 5

a 122. Write the balanced **total ionic** equation for the complete reaction of phosphoric acid and calcium hydroxide. What is the coefficient preceding water in this equation?

(a) 6 (b) 8 (c) 12 (d) 4 (e) 5

c 123. Write the **net ionic** equation for the complete neutralization of HBr by $Ca(OH)_2$. Use H^+ rather than H_3O^+. What is the sum of the coefficients? (Do not forget coefficients of one.)

(a) 6 (b) 7 (c) 3 (d) 4 (e) 5

a 124. Write the **net ionic** equation for the complete reaction of barium hydroxide and hydrochloric acid. Use H^+ rather than H_3O^+. What is the sum of the coefficients? (Do not forget coefficients of one.)

(a) 3   (b) 7   (c) 8   (d) 4   (e) 5

c 125. Write the balanced **net ionic** equation for the complete neutralization of dilute sulfuric acid with potassium hydroxide in aqueous solution. Use H^+ rather than H_3O^+. What is the sum of the coefficients? (Do not forget coefficients of one.)

(a) 6   (b) 8   (c) 3   (d) 4   (e) 5

d 126. Write the **net ionic** equation for the complete neutralization of calcium hydroxide with dilute sulfuric acid. Use H^+ rather than H_3O^+. What is the sum of the coefficients? (Do not forget coefficients of one.)

(a) 6   (b) 7   (c) 4   (d) 3   (e) 5

c 127. Write the balanced **net ionic** equation for the reaction of HNO_3 with aqueous NH_3. Use H^+ rather than H_3O^+. What is the sum of the coefficients? (Do not forget coefficients of one.)

(a) 7   (b) 8   (c) 3   (d) 4   (e) 5

c 128. Write the **net ionic** equation for the complete neutralization of iron(II) hydroxide with HCl acid. Use H^+ rather than H_3O^+. What is the sum of the coefficients? (Do not forget coefficients of one.)

(a) 8   (b) 7   (c) 6   (d) 4   (e) 5

b 129. Write the balanced **net ionic** equation for the complete neutralization of chromium(III) hydroxide with sulfuric acid. Use H^+ rather than H_3O^+. What is the sum of the coefficients? (Do not forget coefficients of one.)

(a) 11   (b) 8   (c) 16   (d) 18   (e) 22

d 130. Write the **net ionic** equation for the neutralization of HNO_2 by KOH. What is the sum of the coefficients? (Do not forget coefficients of one.)

(a) 6   (b) 8   (c) 3   (d) 4   (e) 5

d 131. Write the **net ionic** equation for the complete reaction of carbonic acid with potassium hydroxide. What is the total charge on the ions on the right side of the equation?

(a) 1+   (b) 2+   (c) 3+   (d) 2–   (e) 1–

b 132. Write the balanced **net ionic** equation for the complete neutralization of H_2SO_3 with aqueous NaOH. What is the total charge on the right side of the balanced equation?

(a) 1–   (b) 2–   (c) 6–   (d) 4–   (e) 8–

Chapter 4   57

a*133. Write the **net ionic** equation for the complete reaction of lithium hydroxide with oxalic acid, $(COOH)_2$. What is the sum of the coefficients? (Do not forget coefficients of one.)

(a) 6   (b) 7   (c) 8   (d) 4   (e) 5

c 134. Write the balanced **net ionic** equation for the complete reaction of phosphoric acid and calcium hydroxide. What is the sum of the coefficients? (Do not forget coefficients of one.)

(a) 10   (b) 12   (c) 18   (d) 4   (e) 19

b 135. Write the **net ionic** equation for the reaction of the acid and base needed to produce calcium carbonate. What is the sum of the coefficients? (Do not forget coefficients of one.)

(a) 6   (b) 7   (c) 3   (d) 4   (e) 5

Metathesis (Precipitation) Reactions

b 136. Will a precipitate form when 0.1 M aqueous solutions of $AgNO_3$ and $NaCl$ are mixed? If a precipitate does form, **identify** the precipitate and give the **net ionic** equation for the reaction.

(a) No precipitate forms.
(b) AgCl precipitates. $Ag^+(aq) + Cl^-(aq) \rightarrow AgCl(s)$
(c) Ag_3N precipitates. $6Ag^+(aq) + 2NO_3^-(aq) \rightarrow 2Ag_3N(s) + 3O_2(g)$
(d) AgCl precipitates. $Ag^+(aq) + NaCl(aq) \rightarrow AgCl(s) + Na^+(aq)$
(e) $NaNO_3$ precipitates. $NO_3^-(aq) + Na^+(aq) \rightarrow NaNO_3(s)$

a 137. Will a precipitate form when 0.1 M aqueous solutions of HBr and $Pb(CH_3COO)_2$ are mixed? If a precipitate does form, **identify** the precipitate and give the **net ionic** equation for the reaction.

(a) $PbBr_2$ precipitates. $2[H^+(aq)+Br^-(aq)] + [Pb^{2+}(aq)+2CH_3COO^-(aq)] \rightarrow$
                                $PbBr_2(s) + 2CH_3COOH(aq)$
(b) CH_3COOH precipitates $H^+(aq) + CH_3COO^-(aq) \rightarrow CH_3COOH(s)$
(c) $PbBr_2$ precipitates. $Pb^{2+}(aq) + 2Br^-(aq) \rightarrow PbBr_2(s)$
(d) $PbBr_2$ precipitates. $Pb(CH_3COO)_2(aq) + 2Br^-(aq) \rightarrow$
                                $PbBr_2(s) + 2CH_3COO^-(aq)$
(e) No precipitate forms.

d 138. Will a precipitate form when 0.1 M aqueous solutions of $Ba(NO_3)_2$ and H_2CO_3 are mixed? If a precipitate does form, **identify** the precipitate and give the **net ionic** equation for the reaction.

(a) No precipitate forms.
(b) $BaCO_3$ precipitates. $Ba^{2+}(aq) + CO_3^{2-}(aq) \rightarrow BaCO_3(s)$
(c) $BaCO_3$ precipitates. $Ba^{2+}(aq) + H_2CO_3(aq) \rightarrow BaCO_3(s) + H_2(g)$
(d) $BaCO_3$ precipitates. $Ba^{2+}(aq) + H_2CO_3(aq) \rightarrow BaCO_3(s) + 2H^+(aq)$
(e) $H_2(NO_3)_2$ precipitates. $2H^+(aq) + 2NO_3^-(aq) \rightarrow H_2(NO_3)_2(s)$

c 139. Will a precipitate form when 0.1 M aqueous solutions of $CaCl_2$ and K_2CO_3 are mixed? If a precipitate does form, **identify** the precipitate and give the **net ionic** equation for the reaction.

(a) No precipitate forms.
(b) CaC_2 precipitates. $Ca^{2+}(aq) + 2CO_3^{2-}(aq) \rightarrow CaC_2(s) + 3O_2(g)$
(c) $CaCO_3$ precipitates. $Ca^{2+}(aq) + CO_3^{2-}(aq) \rightarrow CaCO_3(s)$
(d) $CaCO_3$ precipitates. $Ca^{2+}(aq) + K_2CO_3(aq) \rightarrow CaCO_3(s) + 2K^+(aq)$
(e) KCl precipitates. $K^+(aq) + Cl^-(aq) \rightarrow KCl(s)$

a 140. Will a precipitate form when 0.1 M aqueous solutions of NH_4NO_3 and $NaBr$ are mixed? If it does form, **identify** the precipitate and give the **net ionic** equation for the reaction.

(a) No precipitate forms.
(b) $NaNO_3$ precipitates. $Na^+(aq) + NO_3^-(aq) \rightarrow NaNO_3(s)$
(c) NH_4BrO_3 precipitates. $NH_4^+(aq) + NO_3^-(aq) + Br^-(aq) \rightarrow NH_4BrO_3(s) + N_2$
(d) NH_4N precipitates. $2NH_4^+(aq) + 2NO_3^-(aq) \rightarrow 2NH_4N(s) + 3O_2(g)$
(e) NH_4Br precipitates. $NH_4^+(aq) + Br^-(aq) \rightarrow NH_4Br(s)$

Metathesis (Gas-Formation) Reactions

b 141. The equation, $2H^+(aq) + CO_3^{2-}(aq) \rightarrow H_2O(\ell) + CO_2(g)$, is the **net ionic** equation for the reaction of an aqueous mixture of

(a) $CaCO_3$ and HCl.   (b) Na_2CO_3 and HCl.   (c) H_2CO_3 and NaOH.
(d) $BaCO_3$ and H_2SO_4.   (e) $(COOH)_2$ and KOH.

c 142. The equation, $2H^+(aq) + SO_3^{2-}(aq) \rightarrow H_2O(\ell) + SO_2(g)$, is the **net ionic** equation for the reaction of an aqueous mixture of

(a) $BaSO_3$ and H_2SO_4.   (b) H_2SO_4 and $CaSO_3$.   (c) K_2SO_3 and HCl.
(d) $PbSO_3$ and HBr.   (e) K_2SO_4 and HCl.

d 143. Which of the following equations is the **net ionic** equation for the reaction of iron(II) sulfide and hydrochloric acid?

(a) $S^{2-}(aq) + 2H^+(aq) \rightarrow 2H_2S(g)$
(b) $Fe_2S_3(s) + 6H^+(aq) \rightarrow 3H_2S(g) + 2Fe^{3+}(aq)$
(c) $Fe^{2+}(aq) + 2Cl^- \rightarrow FeCl_2(s)$
(d) $FeS(s) + 2H^+(aq) \rightarrow Fe^{2+}(aq) + H_2S(g)$
(e) $Fe_2S(s) + 2H^+(aq) \rightarrow Fe_2^{2+} + H_2S(g)$

Summary of Reaction Types

e 144. The *driving force* of a metathesis reaction is the removal of ions from the solution. Which of the following statements is **not** true for a metathesis reaction?

(a) No change in oxidation numbers occurs during a metathesis reaction.
(b) The *driving force* may be the formation of an insoluble solid compound.
(c) The *driving force* may be the formation of an insoluble or slightly soluble gas.
(d) The *driving force* may be the formation of weak or nonelectrolyte such as water.
(e) The *driving force* may be the reduction of an ion into an insoluble element.

d 145. Which of the following reactions is (are) metathesis reactions?

I. $2NaCl(aq) + Hg_2(NO_3)_2(aq) \rightarrow Hg_2Cl_2(s) + 2NaNO_3(aq)$
II. $H_2SO_4(aq) + 2NaOH(aq) \rightarrow Na_2SO_4(aq) + 2H_2O(\ell)$
III. $K_2CO_3(aq) + 2HCl(aq) \rightarrow 2KCl(aq) + H_2O(\ell) + CO_2(g)$

(a) only I   (b) only II   (c) only III   (d) I, II and III   (e) only I and II

e 146. Which of the following reactions is (are) **not** metathesis reactions?

I. $Cl_2(g) + 2KBr(aq) \rightarrow Br_2(\ell) + 2KCl(aq)$
II. $BaCl_2(aq) + H_2SO_4(aq) \rightarrow BaSO_4(s) + 2HCl(aq)$
III. $CH_4(g) + 2O_2(g) \rightarrow CO_2(g) + 2H_2O(g)$

(a) only I   (b) only II   (c) only III   (d) only I and II   (e) only I and III

a 147. Classify the following reaction by giving the reaction type that applies.
$$2NiS(s) + 3O_2(g) \xrightarrow{heat} 2NiO(s) + 2SO_2(g)$$

(a) redox   (b) combination   (c) decomposition
(d) displacement   (e) metathesis

e 148. Classify the following reaction by giving the reaction type that applies.
$$2HI(aq) + Ba(OH)_2(aq) \rightarrow 2H_2O(\ell) + BaI_2(aq)$$

(a) redox   (b) combination   (c) decomposition
(d) displacement   (e) metathesis

c 149. Classify the following reaction by giving **all** of these reaction type(s) that apply.
 I. redox II. combination III. decomposition
 IV. displacement V. metathesis

$$2PbO(s) + O_2(g) \rightarrow 2PbO_2(s)$$

(a) I and V (b) only II (c) I and II (d) only III (e) only V

a 150. Classify the following reaction by giving **all** of these reaction type(s) that apply.
 I. redox II. combination III. decomposition
 IV. displacement V. metathesis

$$2H_2O_2(aq) \rightarrow 2H_2O(\ell) + O_2(g)$$

(a) I and III (b) only II (c) I and II (d) only IV (e) only V

a 151. Classify the following reaction by giving **all** of these reaction type(s) that apply.
 I. redox II. combination III. decomposition
 IV. displacement V. metathesis

$$SnCl_2(aq) + 2HgCl_2(aq) \rightarrow SnCl_4(aq) + Hg_2Cl_2(s)$$

(a) only I (b) only II (c) only III (d) only IV (e) only V

c 152. Classify the following reaction by giving **all** of these reaction type(s) that apply.
 I. redox II. combination III. decomposition
 IV. displacement V. metathesis

$$CaCO_3(s) \xrightarrow{heat} CaO(s) + CO_2(g)$$

(a) only I (b) only II (c) only III (d) only IV (e) I and V

c 153. Classify the following reaction by giving **all** of these reaction type(s) that apply.
 I. redox II. combination III. decomposition
 IV. displacement V. metathesis

$$Ba(OH)_2(aq) + Na_2CO_3(aq) \rightarrow BaCO_3(s) + 2NaOH(aq)$$

(a) only I (b) only II (c) only V (d) only IV (e) II and III

b 154. Classify the following reaction by giving **all** of these reaction type(s) that apply.
 I. redox II. combination III. decomposition
 IV. displacement V. metathesis

$$BaO(s) + H_2O(\ell) \rightarrow Ba(OH)_2(aq)$$

(a) only I (b) only II (c) only III (d) only IV (e) II and III

e 155. Classify the following reaction by giving **all** of these reaction type(s) that apply.
 I. redox II. combination III. decomposition
 IV. displacement V. metathesis

$$Zn(s) + 2HCl(aq) \rightarrow ZnCl_2(aq) + H_2(g)$$

(a) only I (b) only II (c) only III (d) only V (e) I and IV

Naming Binary Compounds

d 156. Which of the following matched pairs of name and formula has an error?

	Formula	Name
(a)	CaF_2	calcium fluoride
(b)	HCN	hydrogen cyanide
(c)	KBr	potassium bromide
(d)	Na_2S	sodium sulfate
(e)	$Ba(SCN)_2$	barium thiocyanate

b 157. Which of the following matched pairs of name and formula has an error?

	Formula	Name
(a)	Cl_2O_7	dichlorine heptoxide
(b)	As_4O_6	tetraarsenic oxide
(c)	NO	nitrogen oxide
(d)	SO_3	sulfur trioxide
(e)	N_2O_5	dinitrogen pentoxide

c 158. Which of the following matched pairs of name and formula has an error?

	Formula	Name
(a)	As_2O_5	diarsenic pentoxide
(b)	I_2O_7	diiodine heptoxide
(c)	N_2O_4	dinitrogen oxide
(d)	P_4O_{10}	tetraphosphorus decoxide
(e)	S_4N_4	tetrasulfur tetranitride

a 159. Which of the following matched pairs of name and formula has an error?

	Formula	Name
(a)	Mn_2O_3	manganese(II) oxide
(b)	$FeCl_3$	ferric chloride
(c)	$SnBr_2$	stannous bromide
(d)	CuS	copper(II) sulfide
(e)	CrO	chromium(II) oxide

b 160. Which of the following matched pairs of name and formula has an error?

	Formula	Name
(a)	N_2O_4	dinitrogen tetroxide
(b)	P_4O_6	phosphorus hexoxide
(c)	OF_2	oxygen difluoride
(d)	NO	nitrogen oxide
(e)	Mn_2O_7	manganese(VII) oxide

d 161. Which of the following matched pairs of name and formula has an error?

	Formula	Name
(a)	FeF$_3$	ferric fluoride
(b)	SnO	stannous oxide
(c)	FeI$_2$	ferrous iodide
(d)	Cu$_2$S	copper(II) sulfide
(e)	CuO	cupric oxide

Naming Ternary Acids and Their Salts

c 162. Which of the following is the formula for potassium bisulfate?

(a) K$_2$SO$_3$ (b) K(SO$_3$)$_2$ (c) KHSO$_4$ (d) KHSO$_3$ (e) K(HSO$_4$)$_2$

d 163. Which of the following matched pairs of name and formula has an error?

	Formula	Name
(a)	H$_2$CO$_3$	carbonic acid
(b)	H$_2$SO$_3$	sulfurous acid
(c)	HNO$_3$	nitric acid
(d)	HClO$_2$	hypochlorous acid
(e)	HBrO$_3$	bromic acid

c 164. Which of the following matched pairs of name and formula has an error?

	Formula	Name
(a)	KBrO	potassium hypobromite
(b)	K$_2$SO$_4$	potassium sulfate
(c)	K$_2$BrO$_4$	potassium perbromate
(d)	KNO$_2$	potassium nitrite
(e)	K$_2$CO$_3$	potassium carbonate

d 165. Which of the following matched pairs of name and formula has an error?

	Formula	Name
(a)	NaClO$_3$	sodium chlorate
(b)	NaNO$_2$	sodium nitrite
(c)	NaHSO$_4$	sodium hydrogen sulfate
(d)	Na$_2$SeO$_4$	sodium selenite
(e)	Na$_3$PO$_4$	sodium phosphate

b 166. Which of the following matched pairs of name and formula has an error?

	Formula	Name
(a)	LiClO$_2$	lithium chlorite
(b)	HIO$_3$	periodic acid
(c)	HClO$_2$	chlorous acid
(d)	HBrO	hypobromous acid
(e)	Sr(ClO$_4$)$_2$	strontium perchlorate

e 167. Which of the following matched pairs of name and formula has an error?

	Formula	Name
(a)	H_2SO_3	sulfurous acid
(b)	$HClO_3$	chloric acid
(c)	H_2CO_3	carbonic acid
(d)	$HBrO_4$	perbromic acid
(e)	HIO_2	hypoiodous acid

a 168. Which of the following matched pairs of name and formula has an error?

	Formula	Name
(a)	K_2CrO_4	potassium tetroxochromate
(b)	Na_2CO_3	sodium carbonate
(c)	LiClO	lithium hypochlorite
(d)	$NaClO_4$	sodium perchlorate
(e)	$Ca(MnO_4)_2$	calcium permanganate

Naming Binary and Ternary Compounds

d 169. Which of the following matched pairs of name and formula has an error?

	Formula	Name
(a)	$SnCl_4$	stannic chloride
(b)	$SiCl_4$	silicon tetrachloride
(c)	Cl_2O_7	dichlorine heptoxide
(d)	$Mg(HCO_3)_2$	magnesium carbonite
(e)	$NaClO_4$	sodium perchlorate

b 170. Which of the following matched pairs of name and formula has an error?

	Formula	Name
(a)	$(NH_4)_2CO_3$	ammonium carbonate
(b)	Cs_2PO_4	cesium phosphate
(c)	SnF_2	stannous fluoride
(d)	$Cd(MnO_4)_2$	cadmium permanganate
(e)	$NaNO_2$	sodium nitrite

d 171. Which of the following matched pairs of name and formula has an error?

	Formula	Name
(a)	$SnCl_4$	tin(IV) chloride
(b)	$Fe_2(CrO_4)_3$	iron(III) chromate
(c)	SnO	tin(II) oxide
(d)	Cu_2O	copper(II) oxide
(e)	$MnCO_3$	manganese(II) carbonate

d 172. Which of the following matched pairs of name and formula has an error?

	Formula	Name
(a)	CuBr	cuprous bromide
(b)	Fe$_2$O$_3$	iron(III) oxide
(c)	Cu(NO$_3$)$_2$	copper(II) nitrate
(d)	SnO$_2$	tin(II) oxide
(e)	SnCl$_2$	stannous chloride

c 173. Which of the following matched pairs of name and formula has an error?

	Formula	Name
(a)	Na$_2$S	sodium sulfide
(b)	CaH$_2$	calcium hydride
(c)	KSO$_4$	potassium sulfate
(d)	MgBr$_2$	magnesium bromide
(e)	Li$_2$O	lithium oxide

a 174. Which of the following matched pairs of name and formula has an error?

	Formula	Name
(a)	Al(NO$_2$)$_3$	aluminum nitrate
(b)	BaSO$_3$	barium sulfite
(c)	Cr(SCN)$_3$	chromium(III) thiocyanate
(d)	N$_2$O$_3$	dinitrogen trioxide
(e)	Sb$_4$O$_6$	tetraantimony hexoxide

5 The Structure of Atoms

Some of the following values may be useful for solving some of problems 57-81.
speed of light = 3.00 x 10⁸ m/s   1 joule = 1 kg·m²/s²
Planck's constant = 6.63 x 10⁻³⁴ J·s   1 Ångstrom = 1 x 10⁻¹⁰ m

****

Fundamental Particles and the Nuclear Atom

d 1. Which of the following has a negative charge?

(a) nucleus   (b) neutron   (c) proton   (d) electron   (e) alpha particle

d 2. In interpreting the results of his "oil drop" experiment in 1909, Robert Millikan was able to determine _____.

(a) the charge on a proton
(b) that electrically neutral particles (neutrons) are present in the nuclei of atoms
(c) that the masses of protons and neutrons are nearly identical
(d) the charge on an electron
(e) the extremely dense nature of the nuclei of atoms

a 3. Goldstein's "canal rays" are characterized by _____.

(a) positive charge   (b) negative charge   (c) no charge
(d) no apparent mass   (e) neither charge nor mass

c 4. Which statement regarding the "gold foil" experiment is **false**?

(a) It was performed by Rutherford and his research group early in the 20th century.
(b) Most of the alpha particles passed through the foil undeflected.
(c) The alpha particles were repelled by electrons.
(d) It suggested the nuclear model of the atom.
(e) It suggested that atoms are mostly empty space.

b 5. In the Rutherford gold foil experiment, the fact that most of the alpha particles were **not** deflected as they passed through the gold foil indicates that

(a) the nucleus is positively charged.
(b) the atom is mostly empty space.
(c) atoms are solid spheres touching each other in the solid state.
(d) gold is very dense.
(e) none of the above is correct.

a 6. Ernest Rutherford's model of the atom did not specifically include the _____.

(a) neutron   (b) nucleus   (c) proton
(d) electron   (e) electron or the proton

65
Harcourt, Inc

c 7. The Rutherford "gold foil experiment" suggested _____.

(a) that electrons have negative charges
(b) that protons have charges equal in magnitude but opposite in sign to those of electrons
(c) that atoms have a tiny, positively charged, massive center
(d) the ratio of the mass of an electron to the charge of the electron
(e) the existence of canal rays

a 8. Which of the following particles has the smallest mass?

(a) an electron (b) a proton (c) a neutron
(d) a hydrogen atom (e) a hydrogen nucleus

b 9. The number of electrons in a neutral atom of an element is always equal to the _____ of the element.

(a) mass number (b) atomic number (c) atomic mass unit
(d) isotope number (e) Avogadro's number

e 10. What is the correct chronological order of the establishment of the existence of subatomic particles?

	discovered first	discovered second	discovered last
(a)	neutron	electron	proton
(b)	electron	neutron	proton
(c)	neutron	proton	electron
(d)	proton	neutron	electron
(e)	electron	proton	neutron

b 11. Some of the following list match scientists with their contributions to our understanding of the structure of matter. Others do not. Which possible response includes all the correct "matches", and no others?

I. Stoney - determined that the atomic number of an element is directly proportional to the square root of the reciprocal of the wavelength of a particular x-ray spectral line.
II. Davy - determined that elements are held together by electrical forces
III. Thomson - determined ratio of charge to mass for electrons
IV. Millikan - determined the ratio of the mass of a proton to the mass of an electron
V. Chadwick - discovered the neutron

(a) I, II, and IV (b) II, III, and V (c) I, IV, and V
(d) II, IV, and V (e) I, II, and III

e 12. Which statement is **false**?

(a) Ordinary chemical reactions do not involve changes in nuclei.
(b) Atomic nuclei are very dense.
(c) Nuclei are positively charged.
(d) Electrons contribute only little to the mass of an atom.
(e) The nucleus occupies nearly all the volume of an atom.

c 13. What is a possible explanation for why the neutron was the last of the three fundamental subatomic particles to be discovered?

(a) The atoms of very few elements contain neutrons in their nuclei.
(b) Its existence was not suspected until Rutherford's gold foil experiment.
(c) It was difficult to detect because it has no charge.
(d) Because its mass is similar to a proton's mass, both are effected similarly by an electric field.
(e) It is difficult to detect because it is located outside the nucleus with the more numerous, negatively charged electrons.

b 14. What was Moseley's conclusion on the basis of the analysis of his X-ray experiments?

(a) Each element has at least one isotope.
(b) Each element differs from its preceding element (in the periodic table) by having one more proton in its nucleus.
(c) Each element has at least one neutron in its atoms.
(d) An atom of an element has the same number of electrons and neutrons.
(e) He was able to calculate the charge of an electron.

d 15. A graduate student wished to recreate Millikan's "oil drop" experiment. In order to perfect his technique and test his apparatus, he practices performing the experiments many times. If one of his early attempts yielded the following data for the charges on four equal-sized oil drops, what charge per electron would he have calculated and how many electrons were on the drop with the largest number of electrons?

6.60×10^{-19} C 3.30×10^{-19} C 9.90×10^{-19} C 8.25×10^{-19} C

(a) 3.30×10^{-19} C, 3e⁻ (b) 3.30×10^{-19} C, 1e⁻ (c) 1.10×10^{-19} C, 9e⁻
(d) 1.65×10^{-19} C, 6e⁻ (e) 1.98×10^{-19} C, 5e⁻

Mass Number and Isotopes

e 16. The atomic number of an element gives the number of _____ and _____ in the atom while the mass number gives the total number of _____ and _____.

(a) neutrons, protons   neutrons, electrons
(b) neutrons, electrons   protons, electrons
(c) neutrons, electrons   neutrons, protons
(d) protons, electrons   neutrons, electrons
(e) protons, electrons   neutrons, protons

d 17. The mass number of an atom is the number of _____ in the atom.

(a) protons
(b) neutrons
(c) protons plus the number of electrons
(d) protons plus the number of neutrons
(e) electrons plus the number of neutrons

e 18. The difference between the mass number of an atom and the atomic number of the atom is always equal to _____.

(a) 6.02 x 10²³
(b) the atomic number of the element
(c) the atomic mass unit
(d) the number of protons in the nucleus
(e) the number of neutrons in the nucleus

d 19. Isotopes are atoms of the same element that

(a) have different numbers of electrons.
(b) have different numbers of protons.
(c) have different atomic numbers.
(d) have different numbers of neutrons.
(e) have different nuclear charges.

c 20. The atomic number of a certain element is 19, and its atomic weight is 39. An atom of the element contains _____ protons, _____ neutrons, and the chemical symbol for the element is _____.

(a) 19, 19, F
(b) 19, 20, F
(c) 19, 20, K
(d) 20, 19, K
(e) none of these

b 21. Which response includes all the true statements and none of the false ones?

I. The number 23 in the symbol $^{23}_{11}$Na represents the sum of the number of protons and electrons in the nucleus of a sodium atom.
II. If we know that an element is sulfur, then its atomic number must be 16.
III. All neutral atoms of a particular element are identical in every respect.

(a) I
(b) II
(c) III
(d) I and II
(e) II and III

d 22. Which pair of species has the most nearly identical properties?

(a) $^{23}_{11}$Na and $^{23}_{11}$Na⁺
(b) $^{35}_{17}$Cl and $^{80}_{35}$Br
(c) $^{7}_{3}$Li and $^{9}_{4}$Be
(d) $^{12}_{6}$C and $^{13}_{6}$C
(e) $^{19}_{9}$F and $^{19}_{9}$F⁻

c 23. Which pair of species has the most nearly identical chemical properties?

(a) Cl⁻ and F⁻
(b) K⁺ and Na⁺
(c) ¹⁶O and ¹⁸O
(d) O and S
(e) Rb⁺ and Sr²⁺

a 24. Give the number of protons, neutrons, and electrons in an atom of the $^{22}_{10}$Ne isotope.

(a) 10 p, 12 n, 10 e
(b) 10 p, 10 n, 12 e
(c) 10 p, 12 n, 12 e
(d) 12 p, 10 n, 12 e
(e) 10 p, 22 n, 10 e

c 25. Give the number of protons, neutrons, and electrons in an atom of the ^{41}K isotope.

    (a) 19 p, 22 n, 22 e   (b) 41 p, 19 n, 41 e   (c) 19 p, 22 n, 19 e
    (d) 19 p, 16 n, 19 e   (e) 15 p, 26 n, 15 e

b 26. Give the number of protons, neutrons, and electrons in the $^{41}_{21}$Sc^{3+} ion.

    (a) 21 p, 20 n, 21 e   (b) 21 p, 20 n, 18 e   (c) 21 p, 20 n, 24 e
    (d) 20 p, 21 n, 17 e   (e) 21 p, 41 n, 18 e

d 27. Give the number of protons, neutrons, and electrons in the $^{34}_{16}$S^{2-} ion.

    (a) 16 p, 18 n, 16 e   (b) 16 p, 18 n, 14 e   (c) 16 p, 16 n, 19 e
    (d) 16 p, 18 n, 18 e   (e) 34 p, 16 n, 18 e

c 28. What is the symbol for an atom composed of 17 protons, 20 neutrons, and 17 electrons?

    (a) $^{37}_{20}$Ca   (b) $^{35}_{17}$Cl   (c) $^{37}_{17}$Cl   (d) $^{54}_{37}$Rb   (e) $^{36}_{17}$Cl

e 29. What is the symbol for an atom composed of 33 protons, 42 neutrons, and 33 electrons?

    (a) $^{42}_{33}$As   (b) $^{75}_{42}$Mo   (c) $^{65}_{33}$As   (d) $^{42}_{42}$Mo   (e) $^{75}_{33}$As

e 30. What is the symbol for a species composed of 35 protons, 44 neutrons, and 36 electrons?

    (a) $^{79}_{35}$Br   (b) $^{79}_{35}$Br$^+$ (c) $^{79}_{36}$Br$^-$ (d) $^{80}_{36}$Kr   (e) $^{79}_{35}$Br$^-$

e 31. What is the symbol for a species composed of 20 protons, 20 neutrons, and 18 electrons?

    (a) ^{40}Ca   (b) ^{40}Ar   (c) ^{34}S^{2-} (d) ^{39}K$^+$ (e) ^{40}Ca^{2+}

d 32. How many neutrons does the phosphorus isotope $^{32}_{15}$P contain?

    (a) 8   (b) 2   (c) 15   (d) 17   (e) 32

b 33. Consider the $^{35}_{17}$Cl isotope. An atom of this isotope contains _____ neutrons.

    (a) 17   (b) 18   (c) 30   (d) 35   (e) none of these

c 34. An atom of the ^{40}K isotope contains _____ neutrons.

    (a) 19   (b) 20   (c) 21   (d) 40   (e) none of these

a 35. An atom of the ^{40}Ar isotope contains _____ protons.

    (a) 18   (b) 20   (c) 22   (d) 40   (e) 19

b 36. Consider the ^{40}Ar isotope. An atom of this isotope contains _____ neutrons.

    (a) 18   (b) 22   (c) 32   (d) 40   (e) none of these c 37. An atom of the ^{23}Na isotope contains _____ electrons.

(a) 23 (b) 12 (c) 11 (d) 10 (e) 13

c 38. These three species ^{80}Se, ^{81}Br and ^{82}Kr have

(a) the same atomic mass.
(b) the same number of protons.
(c) the same number of neutrons.
(d) the same mass number.
(e) the same number of electrons.

d 39. Which response includes all the following statements that are true, and no false statements?

I. Isotopes of an element differ only in the number of protons.
II. The number of protons in an atom is its atomic number.
III. The mass number of an atom is the sum of the number of protons plus electrons in the atom.
IV. The volume occupied by the nucleus of an atom represents a large percentage of the total volume of the atom.

(a) I and III (b) II, III, and IV (c) II and IV
(d) II (e) I and IV

d 40. Which statement is false?

(a) Cathode rays are emitted by the negative electrode of an operating cathode ray tube.
(b) If two electrodes are placed perpendicular to the line of flow of cathode rays, the rays are repelled by the negative electrode.
(c) Isotopes differ only in mass number.
(d) Canal rays are unaffected by magnetic fields but are affected by electric fields.
(e) Canal rays are positive rays.

Mass Spectrometry and Isotopic Abundance

b 41. Why is the atomic weight of chlorine 35.4527 instead of exactly 35?

(a) Each chlorine atom has an odd number of protons.
(b) There are two naturally occurring isotopes of chlorine.
(c) Each isotope of chlorine has a different number of protons.
(d) Every atom of chlorine has 18 neutrons.
(e) Chlorine is a diatomic molecule.

b 42. Which of the following statements is **false**?

(a) Mass spectrometers measure the charge-to-mass ratio of charged particles.
(b) Mass spectrometers create negative ions by bombarding a gas sample with low energy electrons causing the gas molecules to absorb electrons.
(c) Mass spectrometers can be used to measure the masses of isotopes.
(d) In nature some elements exist in only one isotopic form.
(e) Mass spectrometers can be used to determine isotopic abundance.

d 43. In a mass spectrometry positive ions are produced and then accelerated by an electric field toward a magnetic field. The magnetic field deflects the ions from their straight-line path. The extent to which the beam of ions is deflected depends upon four factors. Which of the following is **not** one of these factors?

(a) charges on the particles
(b) magnetic field strength
(c) masses of the particles
(d) diameter of the particles
(e) magnitude of accelerating voltage (electric field strength)

b 44. Which response lists all the true statements and no false one?

I. Cathode rays are deflected away from a positive plate in a cathode ray tube.
II. The "canal rays" observed by Goldstein in modified cathode ray tubes were positively charged.
III. In a mass spectrometer, a given magnetic field will alter the path of a $^4_2$He⁺ ion more than it will a $^3_2$He⁺ ion.
IV. In nature all elements exist in at least two isotopic forms.

(a) I   (b) II   (c) III   (d) I, II, and IV   (e) IV

The Atomic Weight Scale and Atomic Weights

d 45. What is the atomic weight of a **hypothetical** element consisting of two isotopes, one with mass = 64.23 amu (26.00%), and one with mass = 65.32 amu?

(a) 65.16 amu   (b) 64.37 amu   (c) 64.96 amu
(d) 65.04 amu   (e) 64.80 amu

b 46. If an element consisted of three isotopes in the following relative abundance, what would the atomic weight of the element be? This is a **hypothetical** example.

30.00%   37.00 amu
50.00%   38.00 amu
20.00%   40.00 amu

(a) 38.00 amu   (b) 38.10 amu   (c) 38.20 amu
(d) 39.98 amu   (e) none of these

d 47. A **hypothetical** element consists of the following naturally occurring isotopes. What is the atomic weight of the element?

Isotopes	Mass	Abundance
1	46.041 amu	26.00%
2	47.038 amu	58.00%
3	49.034 amu	16.00%

(a) 46.78 amu   (b) 46.89 amu   (c) 47.02 amu
(d) 47.10 amu   (e) 47.24 amu

a 48. Suppose a **hypothetical** element consists of the following four isotopes. Calculate its atomic weight.

Mass	Natural Abundance
62.982 amu	14.260%
63.978 amu	31.660%
64.973 amu	8.180%
65.968 amu	45.900%

(a) 64.831 amu (b) 64.822 amu (c) 64.816 amu
(d) 64.808 amu (e) 64.802 amu

e 49. A **hypothetical** element consists of four isotopes having the following percentage natural abundance and isotopic masses. What is its atomic weight? isotope 1: 16.06% (52.100 amu); isotope 2: 28.36% (53.097 amu); isotope 3: 42.60% (54.093 amu); isotope 4: 12.98% (55.090 amu).

(a) 53.182 amu (b) 53.262 amu (c) 53.314 amu
(d) 53.487 amu (e) 53.620 amu

d 50. Naturally occurring lithium exists as two isotopes, $^6_3$Li (mass = 6.0151 amu) and $^7_3$Li (mass = 7.0160 amu). The atomic weight of lithium is 6.941 amu. What is the percent abundance of $^7_3$Li?

(a) 4.3% (b) 7.8% (c) 95.7% (d) 92.5% (e) 89.2%

c 51. The atomic weight of silver is 107.868 amu. Naturally occurring silver consists of two isotopes. ^{107}Ag (mass = 106.904 amu) and ^{109}Ag (mass = 108.905 amu). What percentage of naturally occurring silver is the heavier isotope?

(a) 52.6% (b) 45.4% (c) 48.2% (d) 51.7% (e) 62.7%

e *52. A hypothetical element has an atomic weight of 48.68 amu. It consists of three isotopes having masses of 47.00 amu, 48.00 amu, and 49.00 amu. The lightest-weight isotope has a natural abundance of 10.0%. What is the percent abundance of the heaviest isotope?

(a) 66.0% (b) 12.0% (c) 18.0% (d) 72.0% (e) 78.0%

b 53. The atomic weight of antimony is 121.76 amu. There are two naturally occurring isotopes of antimony. ^{121}Sb has an isotopic mass of 120.9038 amu and has a natural abundance of 57.40%. What is the isotopic mass of the other isotope, ^{123}Sb?

(a) 122.4 amu (b) 122.9 amu (c) 122.2 amu
(d) 123.2 amu (e) 123.1 amu

e 54. The atomic weight of rubidium is 85.4678 amu. Rubidium consists of two isotopes, ^{85}Rb (72.15%) and ^{87}Rb (27.85%). The mass of an atom of ^{85}Rb is 84.9117 amu. What is the mass of an atom of ^{87}Rb?

(a) 86.7271 amu (b) 86.8013 amu (c) 86.8220 amu
(d) 86.8621 amu (e) 86.9085 amu

e *55. The five isotopes of nickel occur in the following percentages with the following isotopic masses. The atomic weight is 58.69 amu. What is the isotopic mass of ^{60}Ni? (Choose the closest answer.)

Isotope	Mass (amu)	%
^{58}Ni	57.935	68.27
^{60}Ni	?	26.10
^{61}Ni	60.931	1.13
^{62}Ni	61.928	3.59
^{64}Ni	63.928	0.91

(a) 59.31 amu (b) 58.24 amu (c) 58.62 amu
(d) 60.57 amu (e) 59.93 amu

Electromagnetic Radiation

a 56. Which of the responses contains all the true statements and no others regarding electromagnetic radiation (light)?

I. As wavelength increases frequency decreases.
II. As energy increases frequency decreases.
III. As wavelength increases energy decreases.
IV. The product of wavelength and frequency is constant.

(a) I, III, and IV (b) I and II (c) I, II, and IV
(d) II III, and IV (e) III and IV

e 57. What is the frequency of light having a wavelength of 4.50 x 10^{-6} cm?

(a) 2.84 x 10^{-12} s^{-1} (b) 2.1 x 10^4 s^{-1} (c) 4.29 x 10^{14} s^{-1}
(d) 1.06 x 10^{22} s^{-1} (e) 6.67 x 10^{15} s^{-1}

e 58. What is the frequency of light of wavelength 7000 Å?

(a) 8.41 x 10^{15} s^{-1} (b) 4.72 x 10^{14} s^{-1} (c) 2.48 x 10^{13} s^{-1}
(d) 6.67 x 10^{15} s^{-1} (e) 4.29 x 10^{14} s^{-1}

d 59. What is the wavelength in meters of radiation with a frequency = 4.80 x 10^{13} s^{-1}?

(a) 3.48 x 10^{-8} m (b) 1.60 x 10^1 m (c) 1.44 x 10^{22} m
(d) 6.25 x 10^{-6} m (e) 7.10 x 10^{-7} m

d 60. What is the wavelength of yellow light having a frequency of 5.17 x 10^{14} Hz?

(a) 3.60 x 10^{-10} m (b) 1.55 x 10^{23} m (c) 6.45 x 10^{-28} m
(d) 5.80 x 10^{-7} m (e) 2.72 x 10^{-6} m

a 61. What is the wavelength in Ångstroms of radiation that has a frequency of 4.00 x 10^{14} s^{-1}?

(a) 7.50 x 10^3 Å (b) 7.96 x 10^{-20} Å (c) 6.28 x 10^{-1} Å
(d) 1.33 x 10^2 Å (e) 4.06 x 10^2 Å

c 62. What is the wavelength of green light having a frequency of 6.10×10^{14} Hz?

(a) 1.67×10 Å
(b) 1.07×10^{17} Å
(c) 4.92×10^{3} Å
(d) 2.38×10^{-13} Å
(e) 6.61×10^{-8} Å

e 63. What is the energy, in J/photon, of ultraviolet light with a frequency of 2.70×10^{16} Hz?

(a) 6.00×10^{8} J/photon
(b) 8.00×10^{6} J/photon
(c) 2.46×10^{-18} J/photon
(d) 4.07×10^{-19} J/photon
(e) 1.79×10^{-17} J/photon

a 64. Calculate the energy (J/photon) of a photon of wavelength 4.50×10^{-8} m.

(a) 4.42×10^{-18} J
(b) 2.98×10^{-39} J
(c) 4.72×10^{-24} J
(d) 3.02×10^{-22} J
(e) none of these

c 65. What is the energy, in J/photon, of blue light having a wavelength of 4.240×10^{-7} m?

(a) 2.81×10^{-38} J/photon
(b) 6.42×10^{-17} J/photon
(c) 4.69×10^{-19} J/photon
(d) 7.07×10^{-20} J/photon
(e) 6.28×10^{-11} J/photon

b 66. What is the energy in joules of a photon of light of wavelength 3.75×10^{3} Å?

(a) 3.30×10^{-13} J
(b) 5.30×10^{-19} J
(c) 1.10×10^{-17} J
(d) 1.38×10^{-14} J
(e) 2.22×10^{-11} J

b 67. The energy of a photon of light is 1.5×10^{-20} J. What is its wavelength?

(a) 4.67×10^{4} Å
(b) 1.33×10^{5} Å
(c) 5.12×10^{4} Å
(d) 6.25×10^{5} Å
(e) 7.60×10^{3} Å

The Photoelectric Effect

e 68. Which statement regarding the photoelectric effect is false?

(a) Electrons can be ejected only if the light is of sufficiently short wavelength.
(b) The current increases with increasing intensity of the light.
(c) Electrons can be ejected only if the light is of sufficiently high energy.
(d) The current does not depend on the color of the light as long as the wavelength is short enough.
(e) The wavelength limit sufficient for the ejection of electrons is the same for all metals.

Atomic Spectra and the Bohr Atom

a 69. The Rydberg equation is an empirical equation that describes mathematically _____.

(a) the lines in the emission spectrum of hydrogen
(b) the results of the oil drop experiment
(c) the results of the cathode ray experiments
(d) the Bohr model of the atom
(e) the possible paths of two isotopes of the same element in a mass spectrometer

b 70. Which of the responses contains all the statements that are consistent with the Bohr theory of the atom (and no others)?

    I.   An electron can remain in a particular orbit as long as it continually absorbs radiation of a definite frequency.
    II.   The **lowest** energy orbits are those closest to the nucleus.
    III.   An electron can jump from the K shell to the M shell by **emitting** radiation of a definite frequency.

(a) II and III   (b) II   (c) I and II   (d) III   (e) I, II, and III

a 71. Which one of the following statements is **not** consistent with the Bohr theory?

(a)   An atom has a number of discrete energy levels (orbits) in which an electron can exist as long as it continually emits radiation of a definite energy.
(b)   An electron may move to a lower energy orbit by emitting radiation of frequency proportional to the energy difference between the two orbits.
(c)   An electron may move to a higher energy orbit by absorbing radiation of frequency proportional to the energy difference between the two orbits.
(d)   An electron moves in a circular orbit around the nucleus.
(e)   The energy of an electron is quantized.

c 72. One of the spectral lines in the emission spectrum of mercury has a wavelength of 6.234×10^{-7} m. What is the frequency of the line?

(a) 4.81×10^6 s^{-1} (b) 1.87×10^{14} s^{-1} (c) 4.81×10^{14} s^{-1}
(d) 6.45×10^{11} s^{-1} (e) 1.87×10^6 s^{-1}

a 73. The emission spectrum of gold shows a line of wavelength 2.676×10^{-7} m. How much energy is emitted as the excited electron falls to a lower energy level?

(a) 7.43×10^{-19} J/atom   (b) 1.07×10^{-20} J/atom   (c) 6.05×10^{-19} J/atom
(d) 3.60×10^{-20} J/atom   (e) 5.16×10^{-20} J/atom

c 74. When sodium compounds are heated in a bunsen burner flame, they emit light at a wavelength of 5890 Å. If 1.0×10^{-4} mole of sodium atoms each emit a photon of this wavelength, how many kilojoules of energy are emitted?

(a) 1.21 kJ   (b) 2.03×10^{-3} kJ   (c) 2.03×10^{-2} kJ
(d) 8.08×10^{-3} kJ   (e) 6.20×10^{-2} kJ

c 75. One of the spectral lines in the emission spectrum of mercury has a wavelength of 6.234×10^{-7} m. How much energy is emitted if 1.00 mole of mercury atoms emits light of 6.234×10^{-7} m? Express your answer in kJ/mol.

(a) 127 kJ/mol   (b) 485 kJ/mol   (c) 192 kJ/mol
(d) 5.56×10^{-1} kJ/mol   (e) 1.74×10^{-1} kJ/mol

d*76. When an electron of an excited hydrogen atom falls from level $n = 2$ to level $n = 1$, what wavelength of light is emitted? $R = 1.097 \times 10^7$ m^{-1}

(a) 18.2 Å   (b) 970 Å   (c) 4800 Å   (d) 1215 Å   (e) 1820 Å

The Wave Nature of the Electron

e 77. What is the de Broglie wavelength of a proton of mass 1.67 x 10^{-24} g moving at a velocity of 1.25 x 10^6 m/s?

(a) 5.00 x 10^{-1} m (b) 3.14 x 10^{12} m (c) 3.18 x 10^{-16} m
(d) 4.96 x 10^{-4} m (e) 3.18 x 10^{-13} m

d 78. An electron of mass 9.11 x 10^{-28} g is traveling at 2.50 x 10^6 m/s. Calculate its de Broglie wavelength (in Å).

(a) 0.029 Å (b) 345 Å (c) 0.14 Å (d) 2.91 Å (e) 2.90 x 10^{-3} Å

a *79. An alpha particle of mass 4.0026 amu has a velocity of 10.0% of the speed of light. What is its de Broglie wavelength (in m)?

(a) 3.32 x 10^{-15} m (b) 3.30 x 10^{-18} m (c) 1.22 x 10^{-15} m
(d) 3.70 x 10^{-16} m (e) 3.50 x 10^{-21} m

b 80. What is the velocity of a neutron of mass 1.67 x 10^{-27} kg that has a de Broglie wavelength of 2.05 Å?

(a) 19 m/s (b) 1.94 x 10^3 m/s (c) 8.13 x 10^{-7} m/s
(d) 1.93 x 10^4 m/s (e) 5.16 x 10^{-4} m/s

c 81. What is the de Broglie wavelength of a 16.0 lb shotput moving at a velocity of 7.26 m/s?

(a) 1.30 x 10^{-38} m (b) 1.85 x 10^{-30} m (c) 1.26 x 10^{-35} m
(d) 2.60 x 10^{-36} m (e) 6.63 x 10^{-31} m

c 82. Which match is **incorrect**?

(a) de Broglie   predicted that electrons have wave-like properties
(b) Rutherford   showed that most of the mass of an atom is in its nucleus
(c) Thomson   demonstrated the existence of the neutron
(d) Faraday   proposed that the elements of a chemical compound are held together by electrical forces
(e) Millikan   determined the magnitude of the charge on an electron

The Quantum Mechanical Picture of the Atom

c 83. The Heisenberg Uncertainty Principle states that _____.

(a) no two electrons in the same atom can have the same set of four quantum numbers
(b) two atoms of the same element must have the same number of protons
(c) it is impossible to determine accurately both the position and momentum of an electron simultaneously
(d) electrons of atoms in their ground states enter energetically equivalent sets of orbitals singly before they pair up in any orbital of the set
(e) charged atoms (ions) must generate a magnetic field when they are in motion

e 84. Which response includes all the following statements that are true, and no others?

   I. When an electron falls to a lower energy level in an atom, it emits electromagnetic radiation.
   II. The energy of electromagnetic radiation is directly proportional to its frequency.
   III. The product of wavelength and the speed of light is frequency.
   IV. Atoms can exist only in certain energy states.

   (a) I (b) I and III (c) II (d) II and III (e) I, II, and IV

Quantum Numbers

b 85. A(An) _____ is a region of space in which there is a high probability of finding an electron in an atom.

   (a) shell (b) atomic orbital (c) core
   (d) major energy level (e) nucleus

b 86. Which response lists all the true statements about the four quantum numbers?

   I. n = principal quantum number, $n = 1,2,3, \ldots$
   II. ℓ = angular momentum quantum number, $\ell = 0,1,2,3, \ldots, (n+1)$
   III. m_ℓ = magnetic quantum number, $m_\ell = (-\ell), \ldots, 0, \ldots, (+\ell)$
   IV. m_s = spin quantum number, $m_s = \pm \frac{1}{2}$

   (a) I, II, and III (b) I, III, and IV (c) I, II, and IV
   (d) I and II (e) I and IV

a 87. Which response lists all the true statements about the four quantum numbers?

   I. n = principal quantum number, $n = 1,2,3, \ldots$
   II. ℓ = angular momentum quantum number, $\ell = 0,1,2,3, \ldots, (n-1)$
   III. m_ℓ = magnetic quantum number, $m_\ell = 0, 1, \ldots, \ell$
   IV. m_s = spin quantum number, $m_s = \pm \frac{1}{2}$

   (a) I, II, and IV (b) I, II, and III (c) I and III
   (d) II and III (e) II, III, and IV

b 88. Which quantum number is often designated by the letters s, p, d and f?

   (a) n (b) ℓ (c) m_ℓ (d) m_s (e) ψ

b 89. How many orbitals are there in a sublevel if the angular momentum quantum number for electrons in that sublevel is three?

   (a) one (b) seven (c) three (d) nine (e) five

a 90. The principal quantum number of an orbital is $n = 1$. This must be a(n) ___ orbital.

   (a) s (b) p (c) d (d) f (e) g

c 91. The orientation in space of an orbital is designated by which quantum number?

(a) n (b) ℓ (c) m_ℓ (d) m_s (e) ψ

Atomic Orbitals

b 92. Which statement is **false**?

(a) The third shell (or major energy level) with $n = 3$ has no f orbitals.
(b) There are 10 d orbitals in a set.
(c) A set of p orbitals can accommodate a maximum of 6 electrons.
(d) None of the p orbitals are spherically symmetric.
(e) The fifth shell (or major energy level) has one set of f orbitals.

b 93. Which response includes all the following statements that are true, and no others?

I.   An s orbital can accommodate a maximum of two electrons.
II.   A set of d orbitals can accommodate a maximum of ten electrons.
III.   Each d orbital within a set consists of two lobes, 180° apart.
IV.   There are nine f orbitals in a set of f orbitals.

(a) I, III, and IV   (b) I and II   (c) II and IV
(d) II, III, and IV   (e) I and IV

c 94. Which response includes all the following statements that are true, and no false statements?

I.   Each set of d orbitals contains 7 orbitals.
II.   Each set of d orbitals can hold a maximum of 14 electrons.
III.   The first energy level contains only s and p orbitals.
IV.   The s orbital in any shell is always spherically symmetrical.

(a) I and II   (b) I, III, and IV   (c) IV
(d) II and IV   (e) III

c 95. Choose the response that contains all the true statements and no others regarding atomic orbitals.

I.   Each shell, except the first, contains three mutually perpendicular, equal arm, dumbbell–shaped orbitals known as p orbitals.
II.   Each set of d orbitals can accommodate a maximum of 12 electrons.
III.   Each shell, except the first, contains a spherically shaped s orbital.
IV.   No known elements have electrons occupying f orbitals in their ground states.

(a) I, III, and IV   (b) I, II, and IV   (c) I
(d) I, II, and III   (e) II

e 96. The maximum number of electrons that can occupy an energy level or shell (n = principle quantum number) is _____.

(a) n (b) $2n$ (c) $n + 1$ (d) $n - 1$ (e) $2n^2$

d 97. The third energy level or shell of an atom can hold a maximum of _____ electrons.

(a) 8   (b) 2   (c) 16   (d) 18   (e) 25

d 98. The total number of electrons that can be accommodated in the 4th shell is _____.

(a) 8   (b) 2   (c) 18   (d) 32   (e) 50

b 99. Identify the following orbital.

(a) $d_{x^2-y^2}$ (b) d_{z^2} (c) p_z (d) d_{xy} (e) f_z

d 100. What is the value of the angular momentum quantum number, ℓ, for the following orbital?

(a) $\frac{1}{2}$ (b) –1   (c) 0   (d) 1   (e) 2

e 101. What is the value of the angular momentum quantum number, ℓ, for the following orbital?

(a) $\frac{1}{2}$ (b) –1   (c) 0   (d) 1   (e) 2

Electron Configurations

b 102. No two electrons in the same atom can have the same set of four quantum numbers is a statement of _____.

(a) the Aufbau Principle
(b) the Pauli Exclusion Principle
(c) Dalton's Theory
(d) Hund's Rule
(e) the Heisenberg Uncertainty Principle

e 103. All orbitals of a given degenerate set must be singly occupied before pairing begins in that set is a statement of _____.

(a) the Heisenberg Uncertainty Principle
(b) the Bohr Theory
(c) the Aufbau Principle
(d) Planck's Theory
(e) Hund's Rule

d 104. Which element has the electron configuration below?

$$1s^2 2s^2 2p^6 3s^2 3p^6 3d^{10} 4s^2 4p^3$$

(a) V   (b) Ca   (c) P   (d) As   (e) Se

c 105. The electron configuration $1s^2 2s^2 2p^6 3s^2 3p^6 4s^2 3d^6$ represents the element _____.

(a) Mn   (b) Se   (c) Fe   (d) Co   (e) Kr

e 106. If an element has the following electron configuration, what is the symbol for the element?
$$1s^2 2s^2 2p_x^2 2p_y^2 2p_z^2 3s^2 3p_x^2 3p_y^2 3p_z^1$$

(a) Al (b) Si (c) P (d) S (e) Cl

a 107. What is the electron configuration of silicon, Si?

(a) $1s^2 2s^2 2p^6 3s^2 3p^2$
(b) $1s^2 1p^6 2s^2 2p^4$
(c) $1s^2 1p^6 2s^2 2p^2$
(d) $1s^2 2s^2 2p^6 2d^4$
(e) $1s^2 2s^2 2p^6 3s^2 3p^4$

d 108. Which of the following is the electron configuration of sulfur in its ground state?

(a) $1s^2 1p^6 2s^2 2p^6$
(b) $1s^2 2s^2 2p^3 3p^3 3d^4$
(c) $1s^2 2s^2 2p^3 3s^2 3p^6 3d^1$
(d) $1s^2 2s^2 2p^6 3s^2 3p^4$
(e) $1s^2 2s^2 2p^4 3s^2 3p^6$

c 109. What is the electron configuration of tellurium, $_{52}$Te?

(a) $1s^2 2s^2 2p^6 3s^2 3p^6 4s^2 3d^{10} 4p^6 5s^2 4d^{10} 4f^2$
(b) $1s^2 2s^2 2p^6 3s^2 3p^6 4s^2 3d^{10} 4p^6 4f^{14}$
(c) $1s^2 2s^2 2p^6 3s^2 3p^6 4s^2 3d^{10} 4p^6 5s^2 4d^{10} 5p^4$
(d) $1s^2 2s^2 2p^6 3s^2 3p^6 4s^2 3d^{10} 4p^6 5s^2 4d^{10} 5p^6 6s^2 5d^4$
(e) $1s^2 2s^2 2p^6 2d^{10} 3s^2 3p^6 4s^2 3d^{10} 4p^6 5s^2 5p^2$

d 110. What is the electron configuration of tin, Sn?

(a) [Kr]$5s^2 3d^{10} 3f^{14} 5p^4$
(b) [Kr]$5s^2 3d^{10} 4d^{14} 5p^4$
(c) [Kr]$5s^2 3d^{10} 4f^{14} 5p^2$
(d) [Kr]$5s^2 4d^{10} 5p^2$
(e) [Xe]$5s^2 4d^{10} 5p^2$

c 111. What is the electron configuration of In?

(a) $1s^2 2s^2 2p^6 3s^2 3p^6 3d^{10} 4s^2 4p^6 4d^{10} 5s^2$
(b) $1s^2 2s^2 2p^6 3s^2 2d^{10} 3p^6 4s^2 3d^{10} 4p^6 4d^{10} 5s^2 5p^1$
(c) $1s^2 2s^2 2p^6 3s^2 3p^6 3d^{10} 4s^2 4p^6 4d^{10} 5s^2 5p^1$
(d) $1s^2 2s^2 2p^6 3s^2 2d^{10} 3p^6 4s^2 3d^{10} 4p^6 5s^2 4d^1$
(e) $1s^2 2s^2 2p^6 3s^2 3p^6 3d^{10} 4s^2 4p^6 4d^{11} 5s^2$

b 112. Which one of the following electron configurations is **incorrect**?

(a) $_9$F $1s^2 2s^2 2p^5$ (b) $_{12}$Mg   [Ne]$3s^2$
(c) $_{17}$Cl [Ne] $3s^2 3p^5$ (d) $_6$C   [He]$2s^2 2p^4$
(e) $_{19}$K $1s^2 2s^2 2p^6 3s^2 3p^6 4s^1$

b*113. Which one of the following ground state electron configurations is **incorrect**?

(a) $_{20}$Ca $1s^2 2s^2 2p^6 3s^2 3p^6 4s^2$
(b) $_{25}$Mn [Ar]$4s^2 4d^5$
(c) $_{29}$Cu [Ar]$3d^{10} 4s^1$
(d) $_{50}$Sn [Kr]$4d^{10} 5s^2 5p^2$
(e) $_{54}$Xe [Kr]$4d^{10} 5s^2 5p^6$

e 114. Which element has the following electron configuration?

	1s	2s	2p	3s	3p
	↑↓	↑↓	↑↓ ↑↓ ↑↓	↑↓	↑ ↑ ↑

(a) Na (b) Mg (c) Cl (d) Br (e) P

b 115. The electron configuration

	1s	2s	2p	3s	3p	4s
	↑↓	↑↓	↑↓ ↑↓ ↑↓	↑↓	↑↓ ↑↓ ↑↓	↑↓

represents the element _____.

(a) Rb (b) Ca (c) Ge (d) Ti (e) Sr

c 116. What is the electron configuration of phosphorus, P?

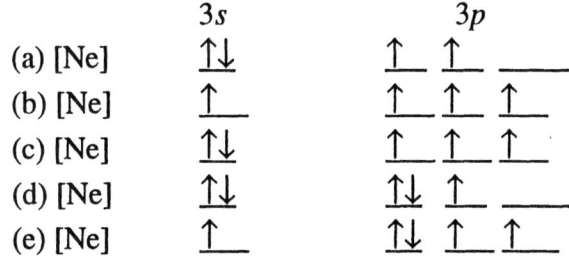

		3s	3p
(a)	[Ne]	↑↓	↑ ↑ ___
(b)	[Ne]	↑	↑ ↑ ↑
(c)	[Ne]	↑↓	↑ ↑ ↑
(d)	[Ne]	↑↓	↑↓ ↑ ___
(e)	[Ne]	↑	↑↓ ↑ ↑

e 117. What is the electron configuration of oxygen, O?

	1s	2s	2p
(a)	↑↓	↑↓	↑ ↑ ↑
(b)	↑↓	↑	↑↓ ↑ ↑
(c)	↑↓	___	↑↓ ↑↓ ↑↓
(d)	↑↓	↑↓	↑↓ ↑↓ ___
(e)	↑↓	↑↓	↑↓ ↑ ↑

c 118. Paramagnetism is characteristic of systems containing _____.

(a) no unpaired electrons
(b) only p electrons as valence electrons
(c) one or more unpaired electrons
(d) only d electrons as valence electrons
(e) only s electrons as valence electrons

d 119. If the following set of quantum numbers represents the "last" electron added to complete the ground state electron configuration of an element according to the Aufbau Principle, which one of the following **could** be the symbol for the element?

$$n = 3, \ell = 1, m_\ell = 0, m_s = \pm\tfrac{1}{2}$$

(a) Na (b) V (c) Zn (d) Si (e) Th

b 120. Which of the following sets of quantum numbers **could** represent the "last" electron added to complete the electron configuration for a ground state atom of Br according to the Aufbau Principle.

	n	ℓ	m_ℓ	m_s
(a)	4	0	0	$-\frac{1}{2}$
(b)	4	1	1	$-\frac{1}{2}$
(c)	3	1	1	$-\frac{1}{2}$
(d)	4	1	2	$+\frac{1}{2}$
(e)	4	2	1	$-\frac{1}{2}$

b 121. Which statement is **false**?

- (a) If an electron has the quantum number $n = 3$, the electron could be in a d sublevel.
- (b) If an electron has the quantum number $\ell = 2$, the only possible values of m_ℓ are 0 and 1.
- (c) If an electron has $m_\ell = -1$, it might be in a p, d, or f sublevel but not in an s sublevel.
- (d) An electron that has $n = 3$ cannot be in an f sublevel.
- (e) An electron that has $n = 5$ could be in an s, p, d, or f sublevel.

c 122. Which one of the following statements is **false**?

- (a) If an electron has the quantum number $n = 2$, it may be in a p sublevel.
- (b) If an electron has $\ell = 1$, it must be in a p sublevel.
- (c) Two electrons in the same atom may have quantum numbers, n, l, m_ℓ, m_s of 2, 1, –1, 1/2, and 2, –1, –1, 1/2.
- (d) Two electrons in the same atom may **not** have quantum numbers of 2, 1, –1, –1/2 and 2, 1, –1, –1/2.
- (e) If an electron has $n = 1$, it must be in an s orbital.

b*123. What statement is **false**?

- (a) A set of p orbitals in a specific energy level always fills before the d orbitals in the same energy level are occupied.
- (b) There are five d orbitals per major energy level (shell) beginning with the fourth energy level.
- (c) A sample of chromium is attracted toward a magnetic field.
- (d) A sample of beryllium is weakly repelled by a magnetic field.
- (e) A set of d orbitals can accommodate as many as ten electrons.

The Periodic Table and Electron Configurations

a 124. The number of electrons present in the *s* orbitals in the outermost electron shell of the Group IA metals is _____.

(a) one (b) two (c) three (d) four (e) five

b 125. The number of electrons present in the *s* orbitals in the outermost electron shell of the alkaline earth (Group IIA) metals is _____.

(a) one (b) two (c) three (d) four (e) five

b 126. The number of electrons in *p* orbitals in the highest energy levels of the Group IVA elements is _____.

(a) one (b) two (c) three (d) four (e) six

c 127. The number of electrons present in *p* orbitals in the outermost electron shell of the Group VA elements is:

(a) one (b) two (c) three (d) four (e) five

c 128. The **total** number of electrons in *s* orbitals in a germanium atom (Z = 32) is _____.

(a) 18 (b) 15 (c) 8 (d) 20 (e) 6

c 129. The **total** number of electrons in *p* orbitals in a palladium atom (atomic number = 46) in its ground state is _____.

(a) 6 (b) 12 (c) 18 (d) 24 (e) 30

e 130. An element with the outermost electron configuration ns^2np^3 would be in Group

(a) VIIIA (b) IIA (c) IIIA (d) VIIA (e) VA

c 131. An element has the outermost electron configuration ns^2np^5. The element could be _____.

(a) Si (b) P (c) Br (d) Ar (e) Mn

e 132. The outer electron configuration ns^2np^4 corresponds to which one of the following elements in its ground state?

(a) As (b) Ca (c) Cr (d) Br (e) S

c*133. An element has the following outer electron configuration in its ground state, where *n* represents the highest occupied energy level: $(n-1)d^{10}ns^2np^4$. Which of the elements listed below could it be?

(a) Si (b) S (c) Se (d) Ge (e) none of these

b*134. An element has the following outer electron configuration in its ground state, where n represents the highest occupied energy level: $(n-1)d^{10}ns^1$. Which of the elements listed below could it be?

(a) K   (b) Ag   (c) Ge   (d) Ga   (e) Cd

d 135. An element with the outermost electron configuration

    ↑↓   ↑↓ ↑ ↑
    ns   np

could be _____.

(a) $_{82}$Pb   (b) $_{40}$Zr   (c) $_{42}$Mo   (d) $_{34}$Se   (e) $_{19}$K

c 136. How many unpaired electrons are there in a neutral arsenic atom (element 33)?

(a) one   (b) two   (c) three   (d) four   (e) five

a 137. Which of the following atoms has the greatest number of unpaired electrons in its ground state?

(a) N   (b) Cl   (c) S   (d) Ti   (e) Cu

d 138. Which one of the following elements is diamagnetic in its ground state?

(a) K   (b) Sc   (c) Cr   (d) Zn   (e) Ge

b 139. Which one of the following elements is paramagnetic in its ground state?

(a) He   (b) Se   (c) Kr   (d) Hg   (e) Mg

c 140. An element that has four electrons in its outer shell in its ground state is _____.

(a) Nb   (b) Cr   (c) Sn   (d) Ti   (e) O

e 141. The number of electrons present in the p orbitals in the outermost electron shell (major energy level, n) of the halogen atoms is _____.

(a) one   (b) two   (c) six   (d) seven   (e) five

6 Chemical Periodicity

More About the Periodic Table

e 1. Accordingly to the periodic law the properties of elements repeat at regular intervals when the elements are arranged in order of

(a) their increasing atomic mass.
(b) their increasing atomic size.
(c) their increasing number of neutrons in the nucleus.
(d) their increasing number of isotopes.
(e) their increasing number of protons in the nucleus.

b 2. Which periodic group or series of elements is **not** correctly matched with its common family name?

(a) alkaline earth metals   IIA
(b) alkali metals   IIIA
(c) lanthanides $_{58}Ce - _{71}Lu$
(d) halogens   VIIA
(e) noble gases   VIIIA

b 3. Of the following, which element does **not** match its designation?

(a) $_{38}Sr$ representative metal
(b) $_{49}In$ representative nonmetal
(c) $_{14}Si$ metalloid
(d) $_{74}W$ d-transition metal
(e) $_{90}Th$ f-transition metal

c 4. _____ is a representative element.

(a) $_{22}Ti$ (b) $_{42}Mo$ (c) $_{81}Tl$ (d) $_{24}Cr$ (e) $_{92}U$

d 5. _____ is a noble gas.

(a) $_{22}Ti$ (b) $_{42}Mo$ (c) $_{81}Tl$ (d) $_{36}Kr$ (e) $_{92}U$

e 6. _____ is an actinide.

(a) $_{22}Ti$ (b) $_{42}Mo$ (c) $_{81}Tl$ (d) $_{36}Kr$ (e) $_{92}U$

b 7. Which one of the following is an inner transition (f-transition) element?

(a) Rb   (b) Ho   (c) Co   (d) Ru   (e) Bi

c 8. Which of the following is **not** a representative element?

(a) Cl   (b) Sr   (c) Ni   (d) K   (e) N

a 9. Choose the response that includes all of the listed elements that are *d*-transition elements, and no others.

I. $_{22}$Ti   II. $_{42}$Mo   III. $_{81}$Tl   IV. $_{36}$Kr   V. $_{92}$U

(a) I and II (b) II and III (c) III and IV
(d) IV and V (e) I, III, and V

c 10. What would be the outer electron configuration of group VIA (O, S, Se, . . .)?

(a) ns^2np^6 (b) ns^2np^2 (c) ns^2np^4 (d) np^6 (e) ns^0np^6

e 11. What would be the outer electron configuration of group IVA (C, Si, Ge, . . .)?

(a) $ns^2nd^2np^0$ (b) ns^2np^4 (c) ns^0np^4
(d) ns^1np^3 (e) ns^2np^2

e 12. What would be the outer electron configuration of alkaline earth metals?

(a) ns^2np^2 (b) np^2 (c) ns^0np^2 (d) nd^2 (e) ns^2

b 13. What would be the outer electron configuration of halogens?

(a) ns^2np^6 (b) ns^2np^5 (c) ns^2np^7 (d) ns^2np^4 (e) $ns^2nd^5np^0$

b 14. Choose the term that best describes **all** members of this series of elements:

V, Cr, Fe, Co, Ni

(a) metalloids   (b) *d*–transition elements   (c) alkaline earth metals
(d) alkali metals   (e) representative elements

e 15. Choose the term that best describes **all** members of this series of elements:

K, Ca, Ba, Cl, N

(a) metalloids   (b) *d*–transition elements   (c) alkaline earth metals
(d) alkali metals   (e) representative elements

Atomic Radii

e 16. Which of the following statements is **false**?

(a) The effective nuclear charge experienced by an electron in an outer shell is less than the actual nuclear charge.
(b) Within a family (vertical group in the periodic table) of representative elements atomic radii increase from top to bottom.
(c) Electrons in inner shells screen, or shield, electrons in outer shells from the full effect of the nuclear charge.
(d) The atomic radii of representative elements decrease from left to right across a period (horizontal row in the periodic table).
(e) Transition elements have larger atomic radii than the preceding IA and IIA elements in the same period because transition elements have electrons in their *d* orbitals.

Harcourt, Inc

a 17. Which element has the **largest** atomic radius?

(a) Al (b) Si (c) P (d) S (e) Cl

e 18. Which element has the **largest** atomic radius?

(a) B (b) Al (c) Ga (d) In (e) Tl

a 19. Which element has the **largest** atomic radius?

(a) Cs (b) Ba (c) Tl (d) Pb (e) Bi

b 20. Which element has the **largest** atomic radius?

(a) Ga (b) In (c) Ge (d) P (e) O

c 21. Which element has the **largest** atomic radius?

(a) $_3$Li (b) $_{11}$Na (c) $_{37}$Rb (d) $_9$F (e) $_{53}$I

c 22. Which element has the **largest** atomic radius?

(a) Mo (b) Mg (c) Ba (d) Cl (e) At

e 23. Which element has the **largest** atomic radius?

(a) $_9$F (b) $_{16}$S (c) $_{11}$Na (d) $_{28}$Ni (e) $_{37}$Rb

e 24. Which element has the **smallest** radius?

(a) Na (b) Mg (c) Al (d) Si (e) P

a 25. Which element has the **smallest** radius?

(a) F (b) Cl (c) Br (d) I (e) At

e 26. Which element has the **smallest** radius?

(a) Ca (b) Ga (c) Ge (d) As (e) Se

e 27. Which element has the **smallest** radius?

(a) K (b) Na (c) Rb (d) Mg (e) Cl

e 28. Arrange the following elements in order of **increasing** atomic radii.

Sr, Rb, Sb, I, In

(a) Rb < Sr < In < Sb < I
(b) I < Sb < In < Rb < Sr
(c) In < Sb < I < Sr < Rb
(d) Sb < I < In < Sr < Rb
(e) I < Sb < In < Sr < Rb

b 29. Arrange the following elements in order of **increasing** atomic radii.

K, Na, Mg, Cs, Cl

(a) Na < Mg < Cl < K < Cs
(b) Cl < Mg < Na < K < Cs
(c) Cs < K < Cl < Mg < Na
(d) Cl < Mg < Cs < K < Na
(e) Cl < Mg < Na < Cs < K

c 30. Arrange the following elements in order of **decreasing** atomic radii.

Pb, P, Cl, F, Si

(a) Cl > F > Pb > Si > P
(b) Pb > Si > P > F > Cl
(c) Pb > Si > P > Cl > F
(d) Pb > Cl > P > Si > F
(e) Pb > Cl > P > Si > F

Ionization Energy

b 31. The first ionization energy of sulfur is less than that of phosphorus. A reasonable explanation for this fact involves

(a) the stability of the half-filled subshell in atomic sulfur.
(b) pairing of two electrons in one 3p orbital in sulfur atoms.
(c) the smaller size of sulfur atoms relative to phosphorus atoms.
(d) the ease with which phosphorus attains a noble gas electronic configuration.
(e) the higher electronegativity of sulfur relative to phosphorus.

e 32. Which element has the **lowest** first ionization energy?

(a) F   (b) Cl   (c) Br   (d) I   (e) At

a 33. Which element has the **lowest** first ionization energy?

(a) Al   (b) Si   (c) P   (d) S   (e) Cl

e 34. Which element has the **lowest** first ionization energy?

(a) F   (b) B   (c) O   (d) S   (e) Sr

b*35. Which element has the **lowest** first ionization energy?

(a) Be   (b) B   (c) C   (d) N   (e) O

a 36. Which element has the **highest** first ionization energy?

(a) B   (b) Al   (c) Ga   (d) In   (e) Tl

e 37. Which element has the **highest** first ionization energy?

(a) Na   (b) Mg   (c) Al   (d) Si   (e) P

e 38. Which element has the **highest** first ionization energy?

    (a) Li   (b) Cs   (c) Cl   (d) I   (e) Ar

d 39. Which element has the **highest** first ionization energy?

    (a) Be   (b) B   (c) C   (d) N   (e) O

a*40. Arrange the following elements in order of **increasing** first ionization energy.

    Mg, Al, Si, P, S

    (a) Al < Mg < Si < S < P   (b) Mg < Al < Si < P < S
    (c) Al < Mg < Si < P < S   (d) Mg < Al < Si < S < P
    (e) Al < Mg < P < Si < S

d 41. Arrange the following elements in order of **decreasing** first ionization energy.

    Rb, In, Sn, Sb, As

    (a) Sb > Sn > In > As > Rb   (b) As > In > Sn > Sb > Rb
    (c) Rb > As > Sb > Sn > In   (d) As > Sb > Sn > In > Rb
    (e) As > Sn > Sb > In > Rb

b 42. Arrange the following elements in order of **decreasing** first ionization energy.

    Be, Ca, Cs, Mg, K

    (a) Mg > Be > Ca > K > Cs   (b) Be > Mg > Ca > K > Cs
    (c) Cs > K > Ca > Be > Mg   (d) Ca > Mg > Be > Cs > K
    (e) Ca > Mg > Be > K > Cs

Electron Affinity

c 43. The amount of energy absorbed in the process in which an electron is added to a neutral gaseous atom is defined as _____.

    (a) shielding effect   (b) electronegativity   (c) electron affinity
    (d) first ionization energy   (e) standard reduction potential

c 44. Which of the following elements has the most negative electron affinity?

    (a) Si   (b) P   (c) S   (d) Se   (e) Te

c 45. Which of the following elements has the most negative electron affinity?

    (a) Ge   (b) I   (c) Br   (d) Se   (e) K

d 46. The general electron configuration for the element group that would have the largest negative value for the electron affinity for its atoms is _____?

    (a) ns^2np^6 (b) ns^2np^4 (c) ns^1 (d) ns^2np^5 (e) ns^2np^3

e 47. Arrange the following elements in order of **increasing** values of electron affinity, i.e., from most negative to least negative. (Note: None of these elements is an exception to the general trends of electron affinities.)

$$Cl, Se, S, Cs, Rb, Te$$

(a) Cl < S < Se < Rb < Te < Cs
(b) Cl > Te > Se > S > Rb > Cs
(c) Cl > Se > S > Te > Rb > Cs
(d) Cl < S < Se < Te < Cs < Rb
(e) Cl < S < Se < Te < Rb < Cs

Ionic Radii

d 48. Which one of the following species is **not** isoelectronic with neon?

(a) Mg^{2+} (b) Na^+ (c) O^{2-} (d) Cl^- (e) Al^{3+}

e 49. Which one of the following pairs contains isoelectronic species?

(a) Na, Na^+
(b) S, Se
(c) S^{2-}, Se^{2-}
(d) F_2, Cl_2
(e) Na^+, O^{2-}

b 50. Which of the following ions is **not** isoelectronic with a noble gas?

(a) Mg^{2+} (b) P^{2-} (c) Cs^+ (d) Se^{2-} (e) Ba^{2+}

d 51. Which response includes all of the following that are isoelectronic with Kr, and no other species?

I. S^{2-} II. Ar   III. K^+ IV. Sr^{2+} V. Br^-

(a) I, II, and III
(b) II and IV
(c) III and IV
(d) IV and V
(e) I, IV, and V

c*52. Consider the group of ions that are isoelectronic with krypton. Which response contains all the true statements and no others?

I. The ion with the highest positive charge is the largest ion.
II. The ion with the highest atomic number bears the highest positive charge.
III. The ion with the lowest atomic number bears the least negative charge.
IV. The ion with a 1– charge is obtained by adding one electron to a Group VIIA element.
V. All the ions have a noble gas electronic configuration.

(a) II and III
(b) I, II, and V
(c) II, IV, and V
(d) I and IV
(e) II, III, IV, and V

d 53. Consider the following atoms and ions: Na, Na^+, Mg, Mg^{2+}, O, O^{2-}, F, F^-. Which one of the following groups includes only isoelectronic species?

(a) Na, Mg, O, F
(b) Na, Na^+, Mg, Mg^{2+}
(c) O, O^{2-}, F, F^-
(d) Na^+, Mg^{2+}, O^{2-}, F^-
(e) none of these

e 54. Which ion has the **largest** radius?

(a) O^{2-} (b) S^{2-} (c) Se^{2-} (d) Te^{2-} (e) Po^{2-}

b 55. Which ion has the **largest** radius?

(a) Li^+ (b) Na^+ (c) Be^{2+} (d) Mg^{2+} (e) Al^{3+}

a 56. Which ion has the **largest** radius?

(a) P^{3-} (b) S^{2-} (c) Cl^- (d) N^{3-} (e) F^-

a 57. Which ion has the **smallest** radius?

(a) F^- (b) Cl^- (c) Br^- (d) I^- (e) At^-

e 58. Which ion has the **smallest** radius?

(a) As^{3-} (b) Se^{2-} (c) Br^- (d) Rb^+ (e) Sr^{2+}

e*59. Which ion has the **smallest** radius?

(a) K^+ (b) Cs^+ (c) Sr^{2+} (d) Ba^{2+} (e) Tl^{3+}

c 60. Which ion has the **smallest** radius?

(a) K^+ (b) Rb^+ (c) Ca^{2+} (d) Sr^{2+} (e) Ba^{2+}

d 61. Which ion or atom has the **largest** radius?

(a) S (b) S^{2-} (c) Se (d) Se^{2-} (e) Br

e 62. Which ion or atom has the **largest** radius?

(a) Sr^{2+} (b) Rb^+ (c) Kr (d) Br^- (e) Se^{2-}

b 63. Arrange the following set of ions in order of **increasing** ionic radii.

$$Ca^{2+}, Cl^-, K^+, P^{3-}, S^{2-}$$

(a) $Ca^{2+} < K^+ < P^{3-} < S^{2-} < Cl^-$ (b) $Ca^{2+} < K^+ < Cl^- < S^{2-} < P^{3-}$
(c) $K^+ < Cl^- < Ca^{2+} < S^{2-} < P^{3-}$ (d) $Cl^- < S^{2-} < P^{3-} < Ca^{2+} < K^+$
(e) $P^{3-} < S^{2-} < Cl^- < K^+ < Ca^{2+}$

e 64. Arrange the following set of ions in order of **decreasing** ionic radii.

$$Al^{3+}, Ga^{3+}, Ca^{2+}, Rb^+, K^+$$

(a) $Rb^+ > Ga^{3+} > Ca^{2+} > K^+ > Al^{3+}$ (b) $Rb^+ > K^+ > Ca^{2+} > Al^{3+} > Ga^{3+}$
(c) $Ga^{3+} > Al^{3+} > Ca^{2+} > Rb^+ > K^+$ (d) $Rb^+ > Ca^{2+} > K^+ > Ga^{3+} > Al^{3+}$
(e) $Rb^+ > K^+ > Ca^{2+} > Ga^{3+} > Al^{3+}$

d 65. Which of the lists of increasing ionic radii is **not** correct?

(a) Cl⁻ < Br⁻ < I⁻
(b) Al³⁺ < Mg²⁺ < Na⁺
(c) Ca²⁺ < Sr²⁺ < Ba²⁺
(d) S²⁻ < Cl⁻ < K⁺
(e) Na⁺ < O²⁻ < N³⁻

c 66. Which of these comparisons according to radius is (are) correct?

I. Na⁺ > Mg²⁺   II. In³⁺ > Sr²⁺   III. Cl⁻ > K⁺   IV. Cl > K

(a) I and II   (b) II and IV   (c) I and III   (d) II and III   (e) III and IV

Electronegativity

e 67. Which one of the following properties is based on the attraction of an atom for electrons in a chemical bond?

(a) binding energy
(b) mass defect
(c) electron affinity
(d) ionization energy
(e) electronegativity

e 68. Which element has the **lowest** electronegativity?

(a) H   (b) Li   (c) Na   (d) K   (e) Cs

a 69. Which element has the **lowest** electronegativity?

(a) K   (b) Ca   (c) Ga   (d) Ge   (e) As

c 70. Which element has the **lowest** electronegativity?

(a) P   (b) As   (c) Sb   (d) Te   (e) I

b 71. Which element has the **lowest** electronegativity?

(a) Be   (b) Ba   (c) Ga   (d) Tl   (e) I

a 72. Which element has the **highest** electronegativity?

(a) C   (b) Si   (c) Ge   (d) Sn   (e) Pb

b 73. Which element has the **highest** electronegativity?

(a) ₃Li   (b) ₇N   (c) ₁₉K   (d) ₃₃As   (e) ₅₆Ba

d 74. Which element has the **highest** electronegativity?

(a) B   (b) Ge   (c) Ca   (d) O   (e) At

c 75. Arrange the following elements in order of **increasing** electronegativities.

At, Bi, Cl, F, I

(a) At < Bi < Cl < F < I
(b) F < Cl < Bi < I < At
(c) Bi < At < I < Cl < F
(d) F < Cl < I < At < Bi
(e) At < Bi < I < Cl < F

a 76. Arrange the following elements in order of **decreasing** electronegativities.

Al, Cs, Mg, Na, P

(a) P > Al > Mg > Na > Cs
(b) Cs > Na > Mg > Al > P
(c) Al > Mg > Na > Cs > P
(d) P > Al > Mg > Cs > Na
(e) P > Cs > Na > Mg > Al

b 77. Which comparison of electronegativities is **not** correct?

(a) Br > Se   (b) K > Mg   (c) O > S   (d) N > Be   (e) I > Ba

c 78. Which of these comparisons of electronegativities is (are) correct?

I. O > N   II. S > P   III. Rb > Li   IV. Br > Cl

(a) I and IV   (b) I and III   (c) I and II   (d) II and III   (e) III and IV

d 79. Which of the these elements has the greatest attraction for electrons in a covalent bond?

(a) Ge   (b) As   (c) Se   (d) Br   (e) Kr

c 80. Which of the following pairs of elements would be expected to form an ionic compound?

(a) S, F   (b) H, C   (c) Rb, Cl   (d) As, Br   (e) C, I

b 81. Which pair of elements below would be **least likely** to form an ionic bond between them?

(a) Na and S
(b) C and N
(c) Al and F
(d) Mg and Br
(e) Cs and O

e 82. Chlorine is **most likely** to form an ionic compound with _____.

(a) F   (b) O   (c) C   (d) N   (e) Li

Hydrogen and the Hydrides

c 83. Which one of the following reactions does not produce $H_2(g)$?

(a) electrolysis of H_2O
(b) zinc with HCl(aq)
(c) combustion of ethane, C_2H_6
(d) iron with steam
(e) steam cracking of hydrocarbons

a 84. Which one of the following reactions is the "water gas" reaction?

(a) $C(s) + H_2O(g) \xrightarrow{1500°C} CO(g) + H_2(g)$
(b) $2CH_4(g) + 3O_2(g) \xrightarrow{heat} 2CO(g) + 4H_2O(g)$
(c) $CH_4(g) \xrightarrow{catalyst} C(s) + 2H_2(g)$
(d) $2Na(s) + 2H_2O(\ell) \rightarrow 2NaOH(aq) + H_2(g)$
(e) $C_4H_{10}(g) \xrightarrow{catalyst} 2C_2H_2(g) + 3H_2(g)$

d 85. Which pair of gases is the product of the "water gas reaction"?

(a) $H_2O + H_2$ (b) $CO + H_2O$ (c) $O_2 + H_2O$
(d) $CO + H_2$ (e) $H_2 + O_2$

c 86. Which one of the following (pure) compounds of hydrogen is ionic?

(a) HF   (b) HCl   (c) NaH   (d) H_2O (e) H_2Se

d 87. Which one of the following is an ionic hydride?

(a) CH_4 (b) H_2Se (c) AsH_3 (d) BaH_2 (e) B_2H_6

d 88. Which one of the following is a molecular hydride?

(a) RbH   (b) SrH_2 (c) NaH   (d) PH_3 (e) CaH_2

b *89. Which of the following is **incorrectly** classified?

(a) KH – ionic hydride   (b) CH_4 – ionic hydride   (c) HI – molecular hydride
(d) CsH – ionic hydride   (e) H_2S – molecular hydride

d 90. Which response includes all the elements below whose compounds with hydrogen would be expected to be molecular hydrides, and no others?

Br, K, P, B

(a) K, B   (b) Br, P   (c) K, Br   (d) Br, P, B   (e) Br

c 91. Which one of the following hydrides is basic?

(a) H_2Te (b) B_2H_6 (c) CaH_2 (d) HI   (e) CH_4

c *92. Which one of the following hydrides is acidic?

(a) RbH   (b) SrH_2 (c) H_2S (d) NaH   (e) NH_3

d *93. What would be the general balanced equation for the reaction of the ionic hydride MH_2 with excess water?

(a) $MH_2(s) + H_2O(\ell) \rightarrow MOH(aq) + H_2(g)$
(b) $MH_2(s) + H_2O(\ell) \rightarrow MO(s) + 2H_2(g)$
(c) $MH_2(s) + H_2O(\ell) \rightarrow MH_2(s) + H_2O(g)$
(d) $MH_2(s) + 2H_2O(\ell) \rightarrow M(OH)_2(aq) + 2H_2(g)$
(e) $MH_2(s) + 3H_2O(\ell) \rightarrow MO(s) + 4H_2(g) + O_2(g)$

a 94. Write the **balanced formula unit equation** for the reaction of nitrogen with hydrogen at high temperature and pressure. What is the sum of the coefficients? Don't forget coefficients of one. Use the smallest whole number coefficients.

(a) 6   (b) 7   (c) 8   (d) 4   (e) 5

c 95. Write the **balanced formula unit equation** for the reaction of magnesium with hydrogen at high temperature. What is the sum of the coefficients? Don't forget coefficients of one. Use the smallest whole number coefficients.

(a) 5   (b) 6   (c) 3   (d) 4   (e) none of these

d 96. Write the balanced **formula unit** equation for the reaction of zinc with dilute sulfuric acid, H_2SO_4. What is the sum of the coefficients in the balanced equation? Use the smallest whole numbers possible. Don't forget coefficients of "one".

(a) six   (b) seven   (c) eight   (d) four   (e) five

b 97. Write the balanced **formula unit** equation for the reaction of sodium hydride with water. What is the sum of the coefficients in the balanced equation? Use the smallest whole numbers possible. Don't forget coefficients of "one".

(a) three   (b) four   (c) five   (d) six   (e) eight

b 98. Write the general balanced **formula unit** equation for the reaction of hydrogen with a halogen. What is the sum of the coefficients in the balanced equation? Use the smallest whole numbers possible. Don't forget coefficients of "one".

(a) three   (b) four   (c) five   (d) six   (e) eight

Oxygen and the Oxides

b 99. Oxygen was discovered by Priestley in 1774 when he observed the _____.

(a) electrolysis of water
(b) thermal decomposition of mercury(II) oxide, HgO
(c) reaction of sulfuric acid, H_2SO_4, with sodium
(d) results of fractional distillation of air
(e) the thermal decomposition of potassium chlorate, $KClO_3$

c 100. Which statement does **not** accurately describe ozone?

(a)   Its molecules are angular.
(b)   Its density is about $1\frac{1}{2}$ times that of O_2.
(c)   Its two oxygen – oxygen bond lengths are 1.23 Å and 1.48 Å.
(d)   It is unstable.
(e)   It is a very strong oxidizing agent.

a 101. Which one of the following will **not** react with oxygen to form a peroxide?

(a) Be   (b) Ca   (c) Sr   (d) Ba   (e) all form peroxides c 102. Which one of the following compounds is a superoxide?

(a) Na_2O_2 (b) SrO (c) KO_2 (d) Li_2O (e) Cl_2O_7

c 103. What is the principal product of the reaction of sodium with oxygen?

(a) NaO (b) Na_2O (c) Na_2O_2 (d) NaO_2 (e) Na_2O_3

a 104. Which one of the following does **not** represent correctly the major product formed by the reaction of an alkali metal with oxygen at ordinary temperatures and pressures?

(a) Li_2O_2 (b) Na_2O_2 (c) KO_2 (d) RbO_2 (e) CsO_2

e 105. The principal product of the reaction of iron, Fe, with a limited amount of oxygen is __I__, while with an excess of oxygen it is __II__.

	I	II
(a)	Fe_3O_2	FeO
(b)	Fe_2O_3	FeO
(c)	FeO	Fe_3O_2
(d)	FeO	Fe_3O_4
(e)	FeO	Fe_2O_3

c 106. Which response includes all the oxides below that are ionic and none that are molecular?

I. Na_2O II. CaO III. As_2O_5 IV. SO_3

(a) II and IV (b) I, II, and III (c) I and II
(d) III and IV (e) II and III

e 107. Which response contains all the basic oxides listed below and no others?

I. NO II. CaO III. Li_2O IV. SO_2 V. P_4O_{10}

(a) I and IV (b) II and IV (c) I, III, and IV
(d) IV and V (e) II and III

c 108. Which response includes all of the following oxides that are basic anhydrides and no others?

I. N_2O II. P_4O_6 III. CaO IV. OF_2

(a) I and II (b) II (c) III (d) IV (e) III and IV

e 109. The most common oxidation states of nonmetals do **not** include _____.

(a) zero
(b) the periodic group number of the nonmetal
(c) the periodic group number of the nonmetal minus two
(d) the periodic group number of the nonmetal minus eight
(e) the negative of the periodic group number of the nonmetal

b 110. The most common oxidation states of selenium are +4 and +6 in molecular oxides. What is the formula for the product of the reaction of selenium with a limited amount of oxygen?

(a) SeO (b) SeO$_2$ (c) SeO$_3$ (d) Se$_2$O$_3$ (e) Se$_3$O$_4$

b 111. Which response includes all the oxides below that are molecular and none that are ionic?

I. K$_2$O II. B$_2$O$_3$ III. N$_2$O$_5$ IV. N$_2$O$_3$

(a) I and IV (b) II, III, and IV (c) II and III
(d) III and IV (e) I and II

a 112. Which response includes all of the following oxides that are acid anhydrides?

I. CO$_2$ II. CaO III. ClO$_2$ IV. Tl$_2$O$_3$

(a) I and III (b) II and IV (c) I and IV
(d) II, III, and IV (e) I

e 113. Which of the following oxides does **not** give an acidic solution when dissolved in water?

(a) SO$_2$ (b) CO$_2$ (c) N$_2$O$_5$ (d) P$_4$O$_{10}$ (e) Na$_2$O

b 114. Which one of the following oxides is amphoteric?

(a) Li$_2$O (b) BeO (c) CO$_2$ (d) P$_4$O$_6$ (e) Cl$_2$O

d 115. Which of the following oxides is **not** amphoteric?

(a) Al$_2$O$_3$ (b) BeO (c) SnO$_2$ (d) Br$_2$O$_7$ (e) PbO$_2$

d*116. One of the following oxides is insoluble in water. Which one?

(a) CO$_2$ (b) N$_2$O$_5$ (c) SO$_2$ (d) SiO$_2$ (e) SO$_3$

e 117. Arrange the following in order of increasing ionic character (most ionic at right).

BaO, SiO$_2$, SO$_2$

(a) BaO < SiO$_2$ < SO$_2$ (b) SO$_2$ < BaO < SiO$_2$ (c) SiO$_2$ < SO$_2$ < BaO
(d) BaO < SO$_2$ < SiO$_2$ (e) SO$_2$ < SiO$_2$ < BaO

a 118. Arrange the following in order of **decreasing** basicity.

MgO, Cs$_2$O, Cl$_2$O$_7$, SnO$_2$, P$_4$O$_{10}$

(a) Cs$_2$O > MgO > SnO$_2$ > P$_4$O$_{10}$ > Cl$_2$O$_7$
(b) MgO > Cs$_2$O > P$_4$O$_{10}$ > SnO$_2$ > Cl$_2$O$_7$
(c) Cs$_2$O > SnO$_2$ > MgO > P$_4$O$_{10}$ > Cl$_2$O$_7$
(d) MgO > Cs$_2$O > SnO$_2$ > P$_4$O$_{10}$ > Cl$_2$O$_7$
(e) Cl$_2$O$_7$ > P$_4$O$_{10}$ > SnO$_2$ > Cs$_2$O > MgO

d 119. Arrange the following in order of increasing acidic character (most acidic at the right).

$$Al_2O_3, Na_2O, N_2O_5$$

(a) $Al_2O_3 < Na_2O < N_2O_5$ (b) $N_2O_5 < Al_2O_3 < Na_2O$ (c) $Al_2O_3 < N_2O_5 < Na_2O$
(d) $Na_2O < Al_2O_3 < N_2O_5$ (e) $Na_2O < N_2O_5 < Al_2O_3$

c 120. Which one of the following compounds would be expected to react with oxygen at elevated temperatures to produce both an acidic oxide and a basic oxide?

(a) CH_4 (b) CS_2 (c) CaS   (d) NO   (e) H_2S

d 121. The acid formed by dissolving Cl_2O_7 in water is _____.

(a) HClO   (b) $HClO_2$ (c) $HClO_3$ (d) $HClO_4$ (e) HCl

b 122. The acid for which N_2O_3 can be considered the anhydride is _____.

(a) $H_2N_2O_2$ (b) HNO_2 (c) HNO_3 (d) H_2NO_3 (e) HN_3

d 123. Write the balanced **formula unit** equation for the reaction of sodium with oxygen (use the **major** product). What is the sum of the coefficients?

(a) 2   (b) 7   (c) 3   (d) 4   (e) 5

a 124. Write the balanced **formula unit** equation for the reaction of rubidium, Rb, with oxygen (use the **major** product). What is the sum of the coefficients?

(a) 3   (b) 7   (c) 8   (d) 4   (e) 5

e 125. Write the balanced **formula unit** equation for the reaction of barium with a limited amount (low pressure) of oxygen. What is the sum of the coefficients?

(a) 2   (b) 6   (c) 3   (d) 4   (e) 5

c 126. Write the balanced **formula unit** equation for the reaction of barium with an excess (high pressure) of oxygen. What is the sum of the coefficients?

(a) 2   (b) 8   (c) 3   (d) 4   (e) 5

e 127. Write the balanced **formula unit** equation for the reaction of copper with excess O_2 at elevated temperatures. What is the sum of the coefficients?

(a) 6   (b) 8   (c) 3   (d) 4   (e) 5

e 128. Write the balanced **formula unit** equation for the reaction of arsenic, As_4, heated with a limited amount of oxygen (The molecular formula for the product is twice the empirical formula). What is the sum of the coefficients?

(a) 2   (b) 8   (c) 3   (d) 4   (e) 5

b 129. Write the balanced **formula unit** equation for the reaction of arsenic, As$_4$, heated with an excess of oxygen (The molecular formula for the product is the same as the empirical formula). What is the sum of the coefficients?

(a) 7 (b) 8 (c) 3 (d) 4 (e) 5

c 130. Write the balanced **formula unit** equation for the reaction of calcium oxide with water. What is the sum of the coefficients?

(a) 6 (b) 8 (c) 3 (d) 4 (e) 5

d 131. Write the balanced **formula unit** equation for the reaction of lithium oxide with water. What is the sum of the coefficients?

(a) 2 (b) 6 (c) 3 (d) 4 (e) 5

d 132. Write the balanced **formula unit** equation for the reaction of tetraphosphorus decoxide with excess water. What is the sum of the coefficients?

(a) 10 (b) 7 (c) 9 (d) 11 (e) 5

d 133. Write the balanced **formula unit** equation for the reaction of dichlorine heptoxide with water. What is the sum of the coefficients?

(a) 7 (b) 8 (c) 3 (d) 4 (e) 5

d 134. Write the balanced **formula unit** equation for the reaction of dinitrogen pentoxide with water. What is the sum of the coefficients?

(a) 6 (b) 8 (c) 3 (d) 4 (e) 5

c 135. Write the balanced **formula unit** equation for the reaction of magnesium oxide with sulfur dioxide. What is the sum of the coefficients?

(a) 6 (b) 8 (c) 3 (d) 4 (e) 5

a 136. Write the balanced **formula unit** equation for the reaction of carbon dioxide with sodium oxide. What is the sum of the coefficients?

(a) 3 (b) 6 (c) 7 (d) 4 (e) 5

a 137. Write the balanced **formula unit** equation for the reaction of the complete combustion of ethane, C$_2$H$_6$. What is the sum of the coefficients?

(a) 19 (b) 16 (c) 17 (d) 4 (e) 5

c 138. Write the balanced **formula unit** equation for the reaction of the complete combustion of pentane, C$_5$H$_{12}$. What is the sum of the coefficients?

(a) 16 (b) 19 (c) 20 (d) 21 (e) 22

e 139. Write the balanced **formula unit** equation for the reaction of nitrogen oxide with oxygen. What is the sum of the coefficients?

(a) 3 (b) 6 (c) 8 (d) 4 (e) 5

b 140. Most of the air pollution resulting from combustion of (non-lead-containing) fossil fuels is due to oxides of _____.

(a) N, P, and S
(d) P and C
(b) C, N, and S
(e) P, C, and N
(c) C, P, and S

c 141. Combustion of fossil fuels and the roasting of metal ores often both produce oxides of

(a) P (b) N (c) S (d) Si (e) C

b 142. The brownish color of photochemical smog is due to _____.

(a) CO (b) NO_2 (c) NO (d) SO_2 (e) SO_3

d 143. The two acids that are major contributors to "acid rain" are _____.

(a) H_2CO_3 and HNO_3
(b) H_2SO_4 and H_3PO_4
(c) H_2CO_3 and H_2SO_4
(d) H_2SO_4 and HNO_3
(e) H_3PO_4 and HNO_3

7 Chemical Bonding

Ionic and Covalent Compounds

e 1. Which response includes only the true statements concerning the characteristics of ionic compounds?

    I.   All atoms in the compounds share electrons.
    II.   The compounds are gases at room temperature.
    III.   The compounds have high melting points.
    IV.   Many are soluble in polar solvents.
    V.   Aqueous solutions of these compounds are poor conductors of electricity.

(a) I, II, and IV   (b) I and V   (c) II and III
(d) IV and V   (e) III and IV

a 2. Which response includes only the true statements concerning the characteristics of covalent compounds?

    I.   These compounds can be gases, liquids, or solids with low melting points.
    II.   Most are soluble in polar solvents.
    III.   Liquid and molten compounds do not conduct electricity.
    IV.   Aqueous solutions of these compounds are very good conductors of electricity.

(a) I and III   (b) II and IV   (c) I, III, and IV
(d) I, II, and III   (e) IV

Lewis Dot Formulas of Atoms

b 3. Which Lewis Dot Formula below is **incorrect**?

(a) ·C̈l: (b) B̈   (c) C̈·   (d) ·B̈r: (e) Li·

d 4. Which Lewis Dot Formula is **incorrect**?

(a) Na·   (b) Sr: (c) Ga·   (d) S̈:   (e) ·C̈l:

c *5. Which Lewis Dot Formula below is **incorrect**?

(a) Ca: (b) ·Ï: (c) ·Ċ·   (d) ·N̈·   (e) Äl·

d 6. How many valence electrons does a phosphorus atom have?

(a) 2   (b) 6   (c) 3   (d) 5   (e) 4

b 7. Below are some elements with their number of valence electrons. Which is **incorrect**?

(a) Be – 2   (b) Br – 5   (c) Li – 1   (d) Al – 3   (e) C – 4

c *8 An atom of which element below has the most unpaired electrons?

(a) Ba   (b) Al   (c) P   (d) F   (e) O

Formation of Ionic Compounds

d 9. An element has the following electronic configuration in its outermost shell. In simple ionic compounds the oxidation number of this element would be _____.

$$\underset{ns}{\uparrow\downarrow} \quad \underset{np}{\underline{\quad}\;\underline{\quad}\;\underline{\quad}}$$

(a) –1   (b) –2   (c) –3   (d) +2   (e) +1

c 10. Consider the following electron transfer diagram representing the formation of a binary ionic compound from atoms of its constituent elements. Which response correctly identifies the elements?

2(M [Noble Gas] $\underset{4s}{\uparrow}$ $\underset{4p}{\underline{\quad}\;\underline{\quad}\;\underline{\quad}}$)

X [Noble Gas] $\underset{3s}{\uparrow\downarrow}$ $\underset{3p}{\uparrow\downarrow\;\uparrow\;\uparrow}$

→

2(M⁺ [Noble Gas] $\underset{4s}{\underline{\quad}}$ $\underset{4p}{\underline{\quad}\;\underline{\quad}\;\underline{\quad}}$)

X²⁻ [Noble Gas] $\underset{3s}{\uparrow\downarrow}$ $\underset{3p}{\uparrow\downarrow\;\uparrow\downarrow\;\uparrow\downarrow}$

(a) M = Na, X = O   (b) M = Ca, X = Cl   (c) M = K, X = S
(d) M = Na, X = S   (e) M = Rb, X = O

d 11. Consider the following electron transfer diagram representing the formation of a binary ionic compound from atoms of its constituent elements. Which response correctly identifies the elements?

M [Noble Gas] $\underset{3s}{\uparrow\downarrow}$ $\underset{3p}{\underline{\quad}\;\underline{\quad}\;\underline{\quad}}$

2(X [Noble Gas] $\underset{2s}{\uparrow\downarrow}$ $\underset{2p}{\uparrow\downarrow\;\uparrow\downarrow\;\uparrow}$)

→

M²⁺ [Noble Gas] $\underset{3s}{\underline{\quad}}$ $\underset{3p}{\underline{\quad}\;\underline{\quad}\;\underline{\quad}}$

2(X⁻ [Noble Gas] $\underset{2s}{\uparrow\downarrow}$ $\underset{2p}{\uparrow\downarrow\;\uparrow\downarrow\;\uparrow\downarrow}$)

(a) M = Be, X = F   (b) M = Mg, X = Cl   (c) M = Na, X = O
(d) M = Mg, X = F   (e) M = Ca, X = Cl

e 12. Which Lewis dot notation for atoms and ions is correct for the reaction for the formation of calcium phosphide?

(a) Ca: + ·P̈· → Ca²⁺, [:P̈:]²⁻

(b) 2Ca: + 3·P̈· → 2Ca²⁺, 3[:P̈:]³⁻

(c) Ca: + ·K̈: → Ca²⁺, [:K̈:]²⁻

(d) 3Ca: + 2·K̈: → 3Ca²⁺, 2[:K̈:]²⁻

(e) 3Ca: + 2·P̈· → 3Ca²⁺, 2[:P̈:]³⁻

d 13. Which Lewis dot notation for atoms and ions is correct for the reaction for the formation of aluminum oxide?

(a) 3Äl· + 2·Ö: → 3Al³⁺, 2[:Ö:]²⁻

(b) Äl· + 3·Ö: → Al³⁺, 3[:Ö:]²⁻

(c) 2Al· + 3·Ö: → 2Al³⁺, 3[:Ö:]²⁻

(d) 2Äl· + 3·Ö: → 2Al³⁺, 3[:Ö:]²⁻

(e) Äl + ·Ö: → Al²⁺, [:Ö:]²⁻

c 14. The negative ion F⁻ has the same electronic configuration as the positive ion _____.

(a) Ca²⁺   (b) Li⁺   (c) Mg²⁺   (d) K⁺   (e) Sc³⁺

e 15. Which response includes all of the species listed below that have the electronic configuration $1s^22s^22p^63s^23p^6$?

Cl⁻, Na⁺, K⁺, Ar, P³⁻

(a) Cl⁻ and Na⁺   (b) Cl⁻, Na⁺, and Ar   (c) Na⁺, Ar, and P³⁻
(d) Na⁺, K⁺, and Ar   (e) Cl⁻, K⁺, Ar, and P³⁻

c 16. Which response includes all of the species listed below that have the electron configuration $2s^22p^6$ in their **highest occupied** energy level?

Ca²⁺, Na⁺, Ne, Ar, Cl⁻, O²⁻

(a) Ca²⁺, Ne, and Cl⁻   (b) Na⁺, Ar, and Cl⁻   (c) Na⁺, Ne, and O²⁻
(d) Ca²⁺, Ar, and O²⁻   (e) Na⁺ and Cl⁻

d 17. Of the following, which are isoelectronic with Kr?

S^{2-}, Ar, K^+, Sr^{2+}, Br^-

(a) S^{2-}, Ar, and K^+
(b) Ar and Sr^{2+}
(c) K^+ and Sr^{2+}
(d) Sr^{2+} and Br^-
(e) S^{2-} and Br^-

c 18. Which of the following pairs of species are isoelectronic?

(a) Ne, K^+ (b) Rb, Cs (c) Cl^-, S^{2-} (d) Na, Al^{3+} (e) S^{2-}, Se^{2-}

c 19. Lithium and chlorine react to form LiCl, an ionic compound. The chloride ion, Cl^-, has _____ electrons in its outermost occupied shell.

(a) 6 (b) 2 (c) 8 (d) 18 (e) 7

a 20. Magnesium and nitrogen react to form Mg_3N_2 an ionic compound. The magnesium ion, Mg^{2+}, has _____ electrons in its highest **occupied** energy level.

(a) 8 (b) 2 (c) 10 (d) 4 (e) 5

e 21. What is the charge on the simple (single atom) ion that sulfur forms?

(a) 1+ (b) 2+ (c) 3+ (d) 1– (e) 2–

d 22. Which response includes all of the true statements about Se^{2-}?

I. Se^{2-} has eight electrons in its outermost shell.
II. Se^{2-} has the same electronic configuration as krypton.
III. Se^{2-} has chemical properties similar to those of the noble gases.
IV. Se^{2-} is likely to be a gas at reasonable temperatures and pressures.
V. Se^{2-} has the same number of electrons as Rb^+.
VI. The properties of Se^{2-} are more similar to those of S^{2-} than to those of Br^-.

(a) I, II, and III
(b) II, III, and IV
(c) III and V
(d) I, II, V, and VI
(e) II, III, and IV, and V

e 23. When one mole of calcium, Ca, combines with one-half mole of oxygen, O_2, to form calcium oxide, CaO, _____ mole(s) of electrons are transferred from _____ atoms to _____ atoms.

(a) one-half, oxygen, calcium
(b) one, oxygen, calcium
(c) one, calcium, oxygen
(d) two, oxygen, calcium
(e) two, calcium, oxygen

d 24. Consider the formation of one formula unit of Li_3N from neutral atoms. In the process each lithium atom _____ electron(s) and each nitrogen atom _____ electron(s).

(a) gains three; loses one
(b) loses three; gains one
(c) loses three; gains three
(d) loses one; gains three
(e) gains one; loses three

Chapter 7   105

b 25. What is the formula for the simple ionic compound of barium and sulfur?

(a) BaS$_2$ (b) BaS   (c) Ba$_2$S$_3$ (d) Ba$_2$S   (e) Ba$_3$S$_2$

a 26. What is the formula for the binary ionic compound formed by potassium (element 19) and iodine (element 53)?

(a) KI   (b) K$_2$I   (c) KI$_2$ (d) K$_2$I$_3$ (e) K$_3$I$_2$

e 27. What is the formula for the binary compound of aluminum and nitrogen?

(a) Al$_2$N$_2$ (b) Al$_2$N   (c) AlN$_2$ (d) Al$_2$N$_3$ (e) AlN

d 28. What is the formula for the binary ionic compound of lithium and nitrogen?

(a) LiN   (b) LiN$_2$ (c) Li$_2$N   (d) Li$_3$N   (e) Li$_2$N$_3$

e 29. The formula for the simple ionic compound of calcium and nitrogen is _____.

(a) CaN   (b) Ca$_2$N   (c) Ca$_2$N$_3$ (d) CaN$_2$ (e) Ca$_3$N$_2$

e 30. What is the formula for the binary ionic compound of aluminum and sulfur?

(a) AlS   (b) Al$_2$S   (c) AlS$_2$ (d) Al$_3$S$_2$ (e) Al$_2$S$_3$

b 31. Which one of the formulas below is **incorrect**?

(a) MgCl$_2$ (b) Na$_2$I   (c) InF$_3$ (d) K$_2$S   (e) SrO

d 32. Which one of the formulas below is **incorrect**?

(a) SrCl$_2$ (b) Cs$_2$S   (c) AlCl$_3$ (d) Al$_3$P$_2$ (e) CaSe

d 33. Which of the following formulas is **incorrect**?

(a) CaBr$_2$ (b) NaI   (c) Cs$_2$S   (d) Na$_2$As   (e) Li$_2$O

d 34. Which of the following formulas is **incorrect**?

(a) SrBr$_2$ (b) K$_2$S   (c) MgSe   (d) CsCl$_2$ (e) Al$_2$O$_3$

c 35. Which of the following formulas is **incorrect**?

(a) MgSe   (b) Na$_2$S   (c) Rb$_2$F   (d) AlN   (e) BaCl$_2$

b 36. Which of the following formulas is **incorrect**?

(a) Li$_3$N   (b) Mg$_2$O   (c) KI   (d) NaBr   (e) SrS

Formation of Covalent Bonds

b 37. A chemical bond formed by two atoms sharing one or more pairs of electrons is called a (or an) _____ bond.

(a) ionic
(d) polar
(b) covalent
(e) nonpolar
(c) coordinate covalent

e 38. As two hydrogen atoms approach each other, the electron of each hydrogen atom is attracted by the nucleus of the other hydrogen atom as well as by its own nucleus. The electrons are shared between the two hydrogen atoms, and a hydrogen molecule is formed. The attractive forces between electrons and nuclei are opposed by a repulsive force between the positively charged nucleus of each atom. A plot of the potential energy of the H$_2$ molecule vs. internuclear distance shows a minimum potential energy of –435 kJ/mol at an internuclear distance of 0.74 Å. Which of the following statements is **incorrect**?

(a) The highest electron density for the two electrons of H$_2$ is in the region between the two atoms.
(b) The distance between the two hydrogen nuclei in a stable H$_2$ molecule is 0.74 Å.
(c) As the internuclear distance becomes smaller than 0.74 Å, the repulsive forces of the nuclei increase more rapidly than the attractive forces increase.
(d) As the internuclear distance becomes greater than 0.74 Å, the attractive forces decrease faster than the repulsive forces decrease.
(e) The hydrogen molecule is most stable at an internuclear distance of 0.74 Å because the repulsive forces and the attractive forces each become zero at this distance.

a 39. The Lewis dot formula for Br$_2$ shows

(a) a single covalent bond.
(c) a triple covalent bond.
(e) a total of 8 x 2 = 16 electrons.
(b) a double covalent bond.
(d) a single ionic bond.

e 40. The Lewis dot formula for HBr shows

(a) a single ionic bond.
(c) a triple covalent bond.
(e) a total of 8 electrons.
(b) a double covalent bond.
(d) two ionic bonds.

Lewis Formulas for Molecules and Polyatomic Ions

c 41. Which of the following is the correct Lewis dot formula for H$_2$S?

(a) H:S
     ··
     H

(b) H:S:
     ··
     H

(c) ··
    H:S:
     ··
     H

(d) H:H:S:
         ··

(e) H$^+$ [H:S:]$^-$

b 42. Which of the following is the correct Lewis dash formula for carbon diselenide?

(a) $:\!\ddot{\text{Se}} = \text{C} - \ddot{\text{Se}}\!:$ 　　(b) $:\!\ddot{\text{Se}} = \text{C} = \ddot{\text{Se}}\!:$ 　　(c) $:\!\ddot{\ddot{\text{Se}}} - \ddot{\text{C}} - \ddot{\ddot{\text{Se}}}\!:$

(d) $:\!\ddot{\ddot{\text{Se}}} - \ddot{\text{C}} - \ddot{\ddot{\text{Se}}}\!:$ 　　(e) $:\!\ddot{\text{Se}} - \ddot{\text{C}} - \ddot{\text{Se}}\!:$

e 43. Draw the Lewis dot formula for CO_2. The number of **unshared pairs** of electrons in the outer shell of the central atom is _____.

(a) one 　(b) two 　(c) three 　(d) four 　(e) zero

b 44. The number of **unshared pairs** of electrons in the outer shell of sulfur in H_2S is _____.

(a) one 　(b) two 　(c) three 　(d) four 　(e) zero

e 45. Draw the Lewis dot formula for NH_4^+. How many **unshared pairs** of electrons are in the outer shell of the central nitrogen atom?

(a) 2 　(b) 4 　(c) 6 　(d) 8 　(e) 0

d 46. How many **unshared pairs** of electrons are in the outer shell of the central nitrogen atom of NH_3?

(a) 4 　(b) 3 　(c) 2 　(d) 1 　(e) 0

b 47. How many **shared pairs** of electrons are in the outer shell of the central nitrogen atom of NH_3?

(a) 4 　(b) 3 　(c) 2 　(d) 1 　(e) 0

The Octet Rule

d 48. The total number of valence electrons that must be shown in the dot formula for the $C_4H_5Cl_2FO$ molecule is _____. (It is not necessary to draw the dot formula.)

(a) 42 　(b) 44 　(c) 46 　(d) 48 　(e) 50

d 49. In applying the relationship $S = N - A$ for the PF_3 molecule $N = $ _____, $A = $ _____, and $S = $ _____.

	N	A	S
(a)	34	26	8
(b)	32	24	8
(c)	30	22	8
(d)	32	26	6
(e)	30	24	6

a 50. When writing the relationship among N, A, and S for H_2CO_3, N = _____, A = _____, and S = _____.

	N	A	S
(a)	36	24	12
(b)	36	26	10
(c)	34	22	12
(d)	40	16	14
(e)	34	24	10

d 51. In constructing the dot formula for H_3AsO_4, N = _____, A = _____, and S = _____.

	N	A	S
(a)	44	32	12
(b)	46	30	12
(c)	44	30	14
(d)	46	32	14
(e)	42	30	12

c 52. What is the **correct** Lewis dot formula for H_2CO_3?

(a) 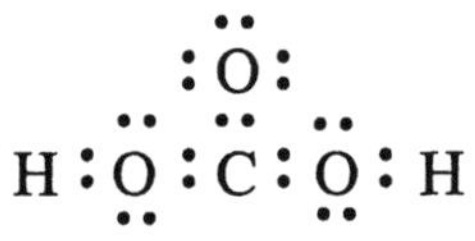 H:Ö:C:Ö:H with :Ö: on top

(b) H:Ö:C:Ö:H with ·Ö· on top

(c) H:Ö:C:Ö:H with ·Ö· on top (different)

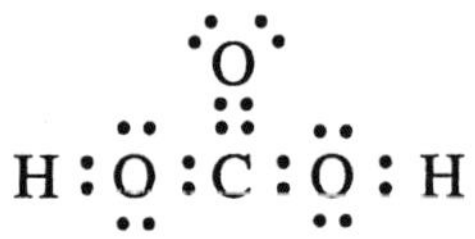

(d) 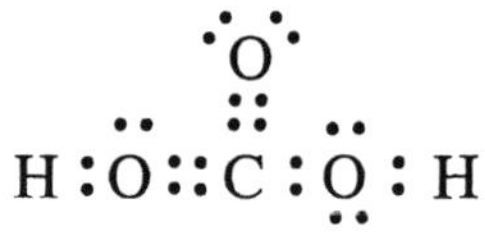 H:Ö::C:Ö:H with ·Ö· on top

(e) 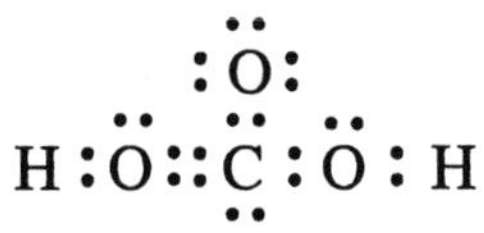 H:Ö::C:Ö:H with :Ö: on top

a 53. Which one of the following Lewis dot formulas is **incorrect**?

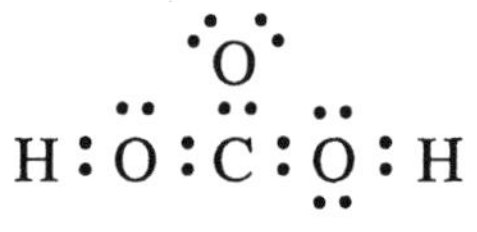

(a) C_2H_4, H₂C:CH₂ structure

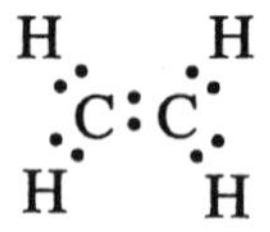

(b) ClO_3^-, [:Ö:Cl:Ö: with :Ö: below]⁻

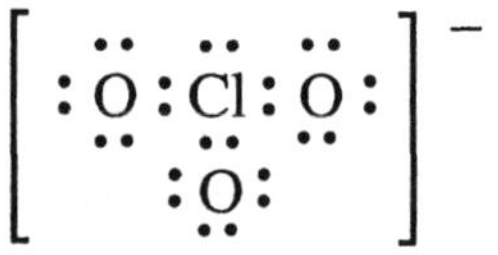

(c) $AsCl_3$, :Cl:As:Cl: with :Cl: below

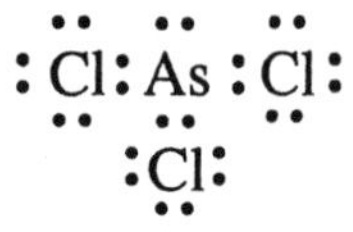

(d) PH_4^+, [H:P:H with H above and H below]⁺

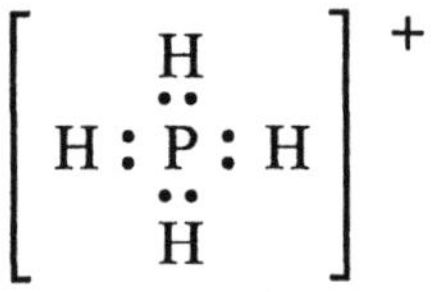

(e) N_2, :N:::N:

d 54. Which one of the following Lewis dot formulas is **incorrect**?

(a) 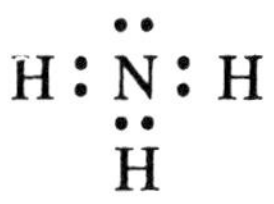 H:N̈:H with H below

(b) 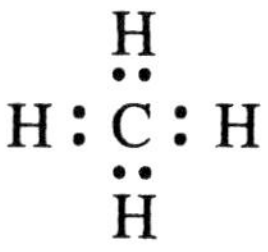 H:C:H with H above and H below

(c) :Ö:H with H below

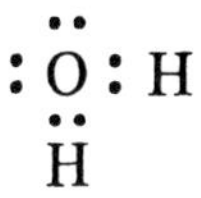

(d)  :C̈l::C̈l:

(e) 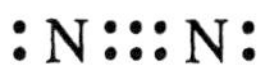 :N:::N:

d 55. Which Lewis dot formula is **incorrect**?

(a) NH_4^+, [H:N̈:H with H above and below]$^+$

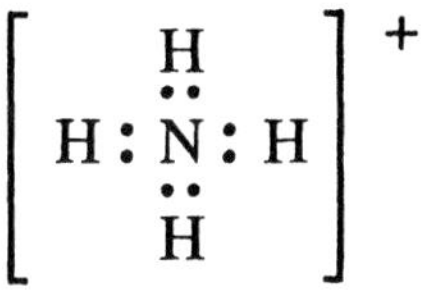

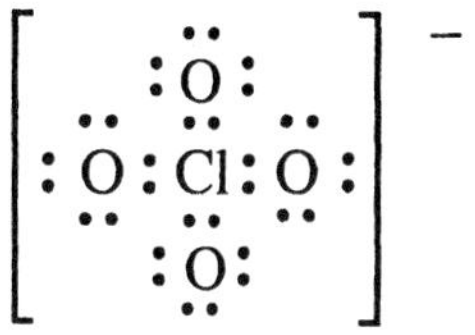

(b) ClO_4^-, [:Ö: above, :Ö:Cl:Ö:, :Ö: below]$^-$

(c) ClO^-, [:C̈l:Ö:]$^-$

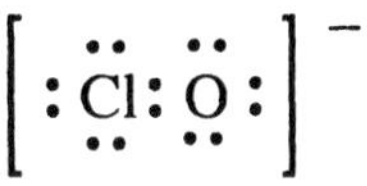

(d) SO_3^{2-}, [:Ö::S:Ö: with :Ö: below]$^{2-}$

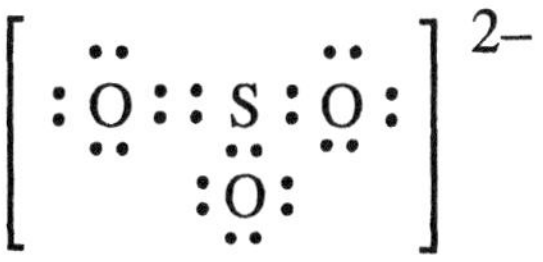

(e) SO_4^{2-}, [:Ö: above, :Ö:S:Ö:, :Ö: below]$^{2-}$

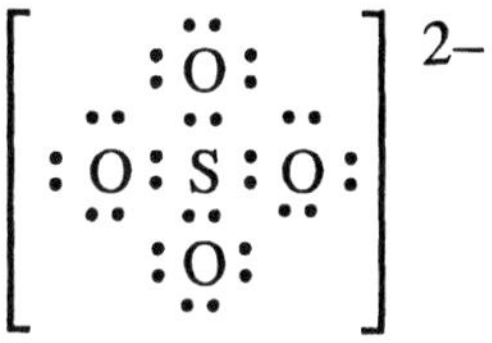

b 56. Which response lists all of the correct Lewis dot formulas and no **incorrect** ones?

I. NH_4^+, [H:N̈:H with H above and below]$^+$

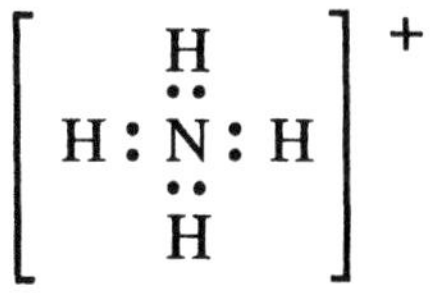

II. ClO_3^-, [:Ö:Cl:Ö: with :Ö: below]$^-$

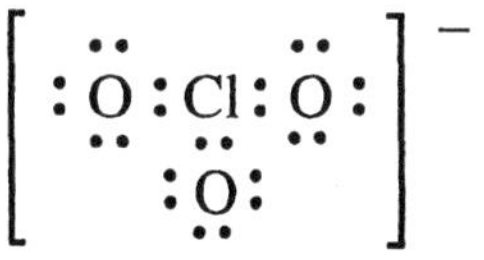

III. SO_4^{2-}, [:Ö: above, :Ö:S:Ö:, :Ö: below]$^{2-}$

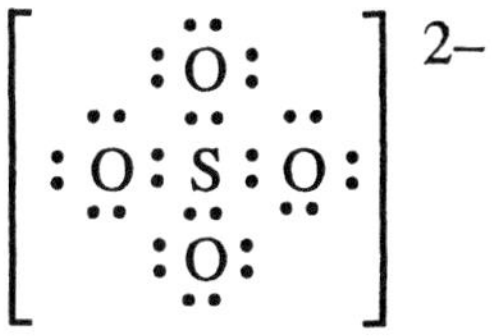

IV. H_3O^+, [H:Ö: with H above and H below]$^+$

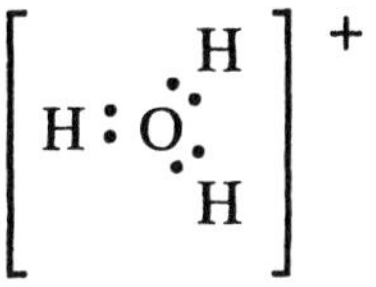

(a) I, II, and IV
(b) I and III
(c) II, III, and IV
(d) II and IV
(e) I, II, and III

b 57. The **total** number of **unshared pairs** of electrons in the N_2 molecule is _____.

(a) one   (b) two   (c) three   (d) four   (e) zero

b 58. The number of **unshared pairs** of electrons in the outer shell of oxygen in Cl_2O is _____.

(a) one (b) two (c) three (d) four (e) zero

a 59. The number of **unshared pairs** of electrons in the outer shell of arsenic in AsF_3 is _____.

(a) one (b) two (c) three (d) four (e) zero

b 60. Draw the dot formula for ethylene, C_2H_4. Each carbon-hydrogen bond is a _____ bond and each carbon-carbon bond is a _____ bond.

(a) single, single (b) single, double (c) single, triple
(d) double, single (e) double, double

c 61. Draw the dot formula for acetylene, C_2H_2. The two carbon atoms are bonded together and each carbon is bonded to one hydrogen. Each carbon-hydrogen bond is a _____ bond and each carbon-carbon bond is a _____ bond.

(a) single, single (b) single, double (c) single, triple
(d) double, single (e) double, double

c 62. The Lewis dot formula for N_2 shows

(a) a single covalent bond. (b) a double covalent bond.
(c) a triple covalent bond. (d) a single ionic bond.
(e) a total of 8 x 2 = 16 electrons.

e 63. The Lewis dot formula for CO_2 shows

(a) two single covalent bonds.
(b) one single covalent bond and one double covalent bond.
(c) one single covalent bond and one triple covalent bond.
(d) a total of 8 x 3 = 24 electrons (dots).
(e) two double covalent bonds.

c 64. Which one of the following molecules contains a triple bond?

(a) AsF_3 (b) NH_3 (c) C_2H_2
(d) H_2CO (Hint: The C is the central atom.)
(e) HOF (Hint: The O is the central atom.)

a 65. The Lewis dot formula for formaldehyde, H_2CO, shows

I. three single covalent bonds. II. one unshared pair of electrons.
III. one double covalent bond. IV. two unshared pairs of electrons.
V. two single covalent bonds.

(a) III, IV and V (b) II, III and V (c) I and II
(d) II and III (e) I and IV

e 66. Which of the following guidelines for drawing Lewis formulas for covalent compounds is **incorrect**?

(a) Carbon always forms 4 bonds.
(b) Hydrogen can never be a central atom.
(c) Representative elements (except hydrogen) usually follow the octet rule.
(d) In neutral species, nitrogen forms 3 bonds and oxygen forms 2 bonds.
(e) One carbon atom in a compound may form both a double bond and a triple bond.

Resonance

b 67. Which molecule exhibits resonance?

(a) BeI_2 (b) O_3 (c) H_2S (d) PF_3 (e) CO_2

b 68. Which numbered response lists all the molecules below that exhibit resonance and none that do not?

I. PF_5 II. HNO_3 III. SO_2 IV. H_2O

(a) I and II   (b) II and III   (c) III and IV   (d) I and III   (e) II, III, and IV

c 69. How many resonance structures does the nitrate ion, NO_3^-, have?

(a) 1   (b) 2   (c) 3   (d) 4   (e) 0

b 70. How many resonance structures does the bicarbonate ion, HCO_3^-, have?

(a) 1   (b) 2   (c) 3   (d) 4   (e) 0

c 71. Assign a formal charge to each atom of $:Cl:As:Cl:$ with $:Cl:$ below.

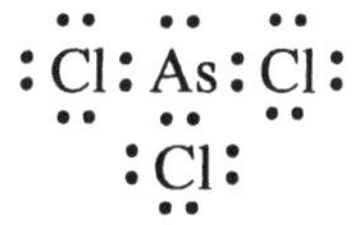

(a) As = 5+, Cl = 1–
(b) As = 5–, Cl = 7+
(c) As = 0, Cl = 0
(d) As = 4+, Cl = 1–
(e) As = 6+, Cl = 2–

c 72. Assign a formal charge to each atom of CSeS, $:Se=C=S:$.

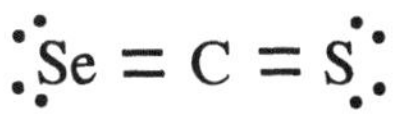

(a) C = 4+, Se = 2–, S = 2–
(b) C = 2+, Se = 1–, S = 1–
(c) C = 0, Se = 0, S = 0
(d) C = 8+, Se = 4–, S = 4–
(e) C = 0, Se = 1–, S = 1+

b 73. Assign a formal charge to each atom of PH_4^+.

(a) P = 0, H = 1+
(b) P = 1+, H = 0
(c) P = 0, H = 0
(d) P = 4+, H = 1–
(e) P = 0, H = $\frac{1}{4}$+

d 74. Assign a formal charge to each atom of $\left[:\overset{..}{\underset{..}{O}}:\overset{..}{Cl}:\overset{..}{\underset{..}{O}}: \atop :\overset{..}{\underset{..}{O}}: \right]^{-}$.

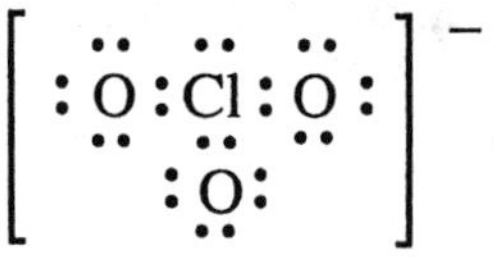

(a) Cl = 1–, O = 0 (b) Cl = 0, O = 0 (c) Cl = 0, O = 1–
(d) Cl = 2+, O = 1– (e) Cl = 2–, O = 1–

Limitations of the Octet Rule for Lewis Formulas

d 75. Which one of the following dot formulas is **incorrect**?

(a) C_2H_4, H₂C::CH₂ structure

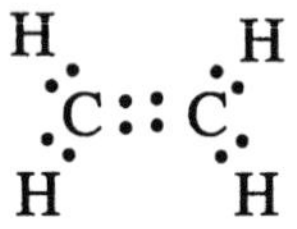

(b) SiF_4, F–Si–F with F above and below

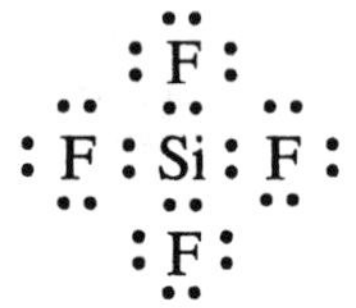

(c) PH_3, H:P:H with H below

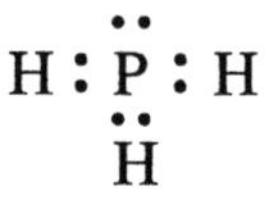

(d) BCl_3, Cl:B:Cl with Cl below

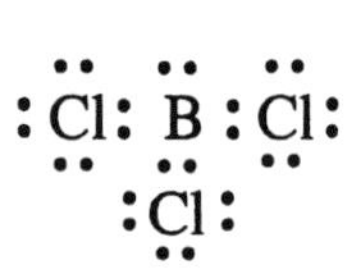

(e) SO_3^{2-}, $\left[:O:S:O: \atop :O: \right]^{2-}$

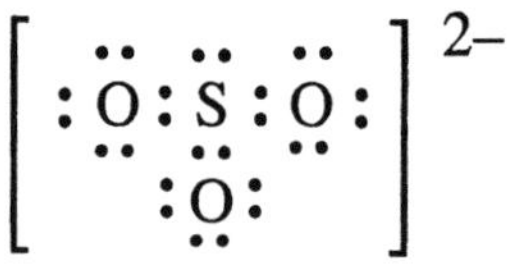

a 76. Which response lists all of the correct Lewis dot formulas and no **incorrect** ones?

I. 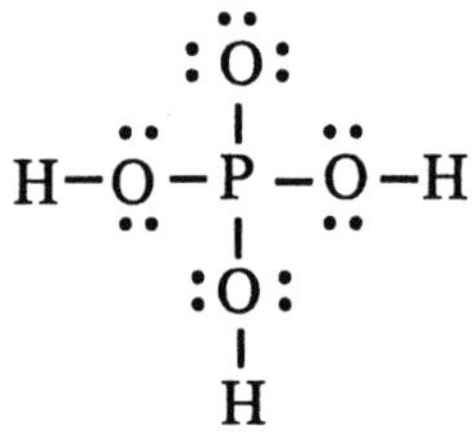 H–O–P(=O)(–O–H)–O–H H_3PO_4

II. 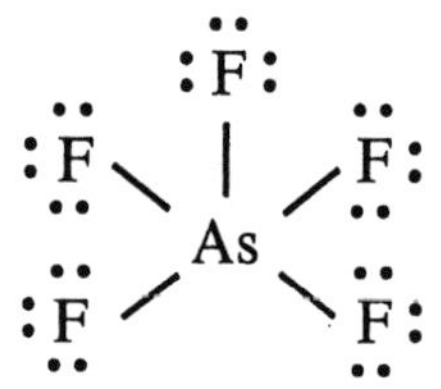 AsF₅ with five F around As

III. 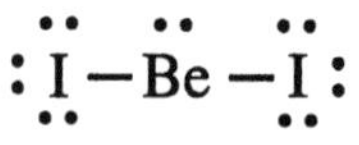 :I–Be–I: BeI_2

IV. 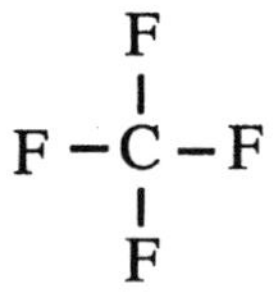 F–C(F)(F)–F CF_4

(a) I and II (b) II, III, and IV (c) III and IV
(d) II and III (e) I, II, and III

b 77. How many **lone pairs** of electrons are there on the Xe atom in the XeF_4 molecule?

(a) one (b) two (c) three (d) four (e) zero

a 78. How many **lone pairs** of electrons are there on the S atom in the SCl_4 molecule?

(a) one (b) two (c) three (d) four (e) zero

c 79. Which molecule below has only **one** unshared pair of electrons in the valence shell of the central atom?

(a) H$_2$S   (b) HF   (c) PH$_3$ (d) BCl$_3$ (e) BeCl$_2$

a 80. Which of the responses includes all of the following molecules that have **no** unshared pairs of electrons about the central atom, and no other molecules?

BCl$_3$, AsF$_5$, NF$_3$, H$_2$S

(a) BCl$_3$ and AsF$_5$
(b) AsF$_5$, NF$_3$, and H$_2$S
(c) NF$_3$ and H$_2$S
(d) BCl$_3$ and H$_2$S
(e) BCl$_3$ and NF$_3$

b 81. Which response lists all the molecules below that have **one** unshared pair of electrons on the central atom, and no other molecules?

H$_2$O, NF$_3$, BF$_3$, OF$_2$

(a) H$_2$O
(b) NF$_3$
(c) NF$_3$ and OF$_2$
(d) H$_2$O, NF$_3$, and OF$_2$
(e) H$_2$O and NF$_3$

e 82. Which one of the following molecules violates the octet rule?

(a) PCl$_3$ (b) CBr$_4$ (c) NF$_3$ (d) OF$_2$ (e) AsF$_5$

d 83. Which one of the following violates the octet rule?

(a) PCl$_4$$^+$ (b) ClF   (c) CCl$_3$$^-$ (d) BCl$_3$ (e) AsCl$_3$

d 84. Which response contains all the molecules below that violate the octet rule, and no others?

BI$_3$, H$_2$Se, SbF$_5$, BeBr$_2$

(a) BI$_3$
(b) H$_2$Se
(c) SbF$_5$ and BeBr$_2$
(d) BI$_3$, SbF$_5$, and BeBr$_2$
(e) BI$_3$ and H$_2$Se

e 85. Which response includes all the molecules below that violate the octet rule and only those?

H$_2$O, SF$_6$, NF$_3$, BeCl$_2$

(a) H$_2$O and SF$_6$
(b) SF$_6$, NF$_3$, and BeCl$_2$
(c) H$_2$O and NF$_3$
(d) SF$_6$
(e) SF$_6$ and BeCl$_2$

e 86. Which response includes all the following covalent molecules in which the central (underlined) atom has an octet of electrons in its highest energy level, and no other molecules?

I. BeI$_2$ II. SiH$_4$ III. BI$_3$ IV. CO$_2$ V. CH$_4$ VI. PH$_3$

(a) I, II, and VI
(b) II, IV, and V
(c) I and III
(d) I, II, and III
(e) II, IV, V, and VI

e 87. Which response includes all of the compounds below that have noble gas configurations around the central atom, and no others?

   I. SF$_6$ II. NF$_3$ III. CF$_4$ IV. BF$_3$ V. PF$_5$

   (a) II, III, and IV   (b) I and IV   (c) I, IV, and V
   (d) II and V   (e) II and III

Polar and Nonpolar Covalent Bonds

e 88. Which would be classified as a **polar covalent** bond?

   (a) C–S bond in CS$_2$ (b) N–Cl bond in NCl$_3$ (c) C–I bond in CI$_4$
   (d) F–F bond in F$_2$ (e) Br–Cl bond in BrCl

e 89. Which of the responses lists only the molecules below that have **polar covalent** bonds?

   HI, I$_2$, S$_8$, KCl

   (a) HI and KCl   (b) I$_2$ (c) S$_8$
   (d) I$_2$ and KCl   (e) HI

e 90. Which response includes all of the following molecules that contain **polar covalent** bonds, and no others?

   NO$_2$, PCl$_3$, N$_2$, H$_2$O, P$_4$, NH$_3$

   (a) NO$_2$, P$_4$, and NH$_3$ (b) PCl$_3$, N$_2$, and H$_2$O   (c) H$_2$O and NH$_3$
   (d) NO$_2$, PCl$_3$, and H$_2$O   (e) NO$_2$, PCl$_3$, H$_2$O, and NH$_3$

b 91. Which response includes all of the molecules that have **nonpolar** bonds, and no others?

   Cl$_2$, BeCl$_2$, I$_2$, BrCl, BCl$_3$

   (a) Cl$_2$, BeCl$_2$, and I$_2$ (b) Cl$_2$ and I$_2$ (c) Cl$_2$, BeCl$_2$, and BrCl
   (d) BeCl$_2$ and BCl$_3$ (e) BrCl

b 92. Which molecule has the **most** polar covalent bond?

   (a) HBr   (b) HF   (c) HI   (d) H$_2$ (e) HCl

d 93. Which molecule has the **least** polar covalent bond?

   (a) HBr   (b) HF   (c) HI   (d) H$_2$ (e) HCl

d 94. The elements of Group VIIA may react with each other to form covalent compounds. Which of the following single covalent bonds in such compounds is the **most polar** bond? The electronegativities of the first four Group VIIA elements are: F = 4.0, Cl = 3.0, Br = 2.8, I = 2.5

   (a) F-F   (b) F-Cl   (c) F-Br   (d) F-I   (e) Cl-I

e 95. Which one of the compounds below has the bonds that are the **most polar**?
Electronegativities: H = 2.1, S = 2.5, P = 2.1, As = 2.1, Cl = 3.0, Si = 1.8, Sb = 1.9

(a) H_2S (b) PH_3 (c) $AsCl_3$ (d) SiH_4 (e) $SbCl_3$

b 96. Which one of the compounds below has the bonds that are **least polar**?
Electronegativities: H = 2.1, S = 2.5, P = 2.1, As = 2.1, Cl = 3.0, Si = 1.8, Sb = 1.9

(a) H_2S (b) PH_3 (c) $AsCl_3$ (d) SiH_4 (e) $SbCl_3$

d 97. Which molecule contains the **least** polar bonds? (Electronegativities: H = 2.1, C = 2.5, F = 4.0, Cl = 3.0, Br = 2.8, I = 2.5)

(a) CF_4 (b) CCl_4 (c) CBr_4 (d) CI_4 (e) CH_4

a 98. Which one of the following molecules contains bonds that are the **most polar**?
(Electronegativities: H = 2.1, Be = 1.5, B = 2.0, N = 3.0, F = 4.0, S = 2.5, Br = 2.8, I = 2.5)

(a) SF_6 (b) BI_3 (c) $BeBr_2$ (d) NH_3 (e) NF_3

Dipole Moments

e 99. Which statement concerning dipole moments is **incorrect**?

(a) Molecular geometry affects the dipole moment of a molecule.
(b) The presence of lone (unshared) pairs of electrons affect the dipole moment of a molecule.
(c) Generally, as electronegativity differences increase in diatomic molecules, the dipole moments increase.
(d) The dipole moment equals the product of charge and distance of separation of the charges.
(e) By careful experimentation, a scientist can measure the dipole moment for each individual bond of a complex molecule.

d 100. Which one of the following molecules has a dipole moment?

(a) Cl_2 (b) H_2 (c) I_2 (d) BrCl (e) N_2

d 101. Which one of the following molecules does **not** have a dipole moment?

(a) BrCl (b) ClF (c) BrF (d) O_2 (e) ICl

b 102. Which molecule would have the strongest dipole moment?

(a) HBr (b) HF (c) HI (d) H_2 (e) HCl

d 103. Which molecule would have the **weakest** dipole moment?

(a) HBr (b) HF (c) HI (d) H_2 (e) HCl

The Continuous Range of Bonding Types

b 104. Which one of the following compounds involves **both** ionic and covalent bonding?

(a) Cl_2 (b) Na_2SO_4 (c) KCl (d) HF (e) HCN

b 105. Which one of the following compounds contains **both** ionic and covalent bonding?

(a) LiF (b) KNO_3 (c) MgO (d) $CaCl_2$ (e) Rb_3N

d 106. Which of the following compounds does **not** contain **both** ionic and covalent bonding?

(a) Na_2SO_4 (b) NH_4NO_3 (c) NH_4Cl (d) KCl (e) $KClO_4$

c 107. Which of the following compounds does **not** contain **both** ionic and covalent bonding?

(a) $NaNO_3$ (b) $BaSO_4$ (c) CBr_4 (d) $NaClO_3$ (e) NH_4Br

e 108. Which response includes all the species listed below that contain **both** ionic and covalent bonds, and no other species?

SiF_4, PCl_3, $LiClO_4$, SF_6, $Al_2(SO_4)_3$

(a) SiF_4, SF_6
(b) PCl_3, $LiClO_4$
(c) $LiClO_4$, SF_6, $Al_2(SO_4)_3$
(d) SiF_4, PCl_3, SF_6
(e) $LiClO_4$, $Al_2(SO_4)_3$

8 Molecular Structure and Covalent Bonding Theories

Valence Shell Electron Pair Repulsion (VSEPR) Theory

d 1. What is the electronic geometry for 5 regions of high electron density on a central atom?

(a) octahedral (b) square planar (c) tetrahedral
(d) trigonal bipyramidal (e) trigonal planar

a 2. What is the electronic geometry for 6 regions of high electron density on a central atom?

(a) octahedral (b) square planar (c) trigonal bipyramidal
(d) tetrahedral (e) trigonal planar

e 3. What angle(s) are associated with a central atom that has octahedral **electronic** geometry?

(a) 109.5° (b) 120° (c) 120° and 180°
(d) 90° and 120° (e) 90° and 180°

a 4. What angle(s) are associated with a central atom that has tetrahedral **electronic** geometry?

(a) 109.5° (b) 120° (c) 120° and 180°
(d) 90° and 120° (e) 90° and 180°

c 5. The selenium hexafluoride molecule is nonpolar and contains no lone (unshared) electron pairs on the selenium atom. What are all of the possible F-Se-F bond angles?

(a) 120° (b) 180° (c) 90° and 180°
(d) 90°, 120°, and 180° (e) 109.5°

d 6. The phosphorus pentachloride molecule is nonpolar and contains no lone (unshared) electron pairs on the phosphorus atom. What are all of the possible Cl-P-Cl bond angles?

(a) 120° (b) 180° (c) 109.5°
(d) 90°, 120°, and 180° (e) 90° and 180°

Polar Molecules: The Influence of Molecular Geometry

d 7. Which of the following statements concerning **polar** molecules is **false**?

(a) There must be at least one polar bond or one unshared pair of electrons on the central atom.
(b) If there are more than one polar bond, they must not be symmetrically arranged so that their polarities cancel.
(c) If there are more than one unshared pair of electrons on the central atom, they must not be symmetrically arranged so that their polarities cancel.
(d) There must be an odd number of polar bonds in order that their polarities not cancel.
(e) A molecule with symmetrically arranged polar bonds can be polar if the central atom is bonded to atoms of different elements.

e 8. Which one of the following molecules is polar?

(a) N_2 (b) P_4 (c) Cl_2 (d) CO_2 (e) H_2O

b 9. Which one of the following molecules is nonpolar?

(a) H_2O (b) CO_2 (c) CSO (d) NF_3 (e) NH_3

Valence Bond (VB) Theory

d *10. Which one of the following statements about compounds or polyatomic ions of the A group elements is **false**?

(a) All compounds in which the central atom is sp^3d^2 hybridized violate the octet rule.
(b) Sulfur hexafluoride is an example of a compound with a central atom that has sp^3d^2 hybridization.
(c) All molecules in which the central element is sp^3d^2 hybridized have octahedral electronic geometry.
(d) All molecules in which the central element is sp^3d^2 hybridized have octahedral molecular geometry.
(e) All ions in which the central atom is sp^3d^2 hybridized have octahedral electronic geometry.

d 11. Which response contains all the following statements that are true, and no others?

I. A set of sp^2 orbitals can be thought of as one s orbital one-third of the time and two p orbitals two-thirds of the time.
II. A set of sp orbitals can accommodate a maximum of six electrons.
III. The orbitals resulting from sp^3d^2 hybridization are directed toward the corners of an octahedron.
IV. A set of sp^3 orbitals results from the mixing of one s orbital and three p orbitals.

(a) II and IV (b) I, II, and IV (c) II, III, and IV
(d) III and IV (e) all are true

b 12. Draw the dot formula for the HNO_2 molecule. The nitrogen atom is the central atom and the hydrogen atom is attached to an oxygen atom. On the basis of the number of regions of high electron density about the nitrogen atom, predict the hybridization at the nitrogen atom.

(a) sp (b) sp^2 (c) sp^3 (d) sp^3d (e) sp^3d^2

c 13. Draw the dot formula for the HNO_2 molecule. The nitrogen atom is the central atom and the hydrogen atom is attached to an oxygen atom. On the basis of the number of regions of high electron density about the nitrogen atom, what is the approximate bond angle about the N atom?

(a) 180° (b) 90° (c) 120° (d) 120° and 90° (e) 109°

b 14. The hybridization associated with the central atom of a molecule in which all the bond angles are 120° is _____.

(a) sp (b) sp^2 (c) sp^3 (d) sp^3d (e) sp^3d^2

a 15. The bond angles associated with sp^3d^2 hybridization are _____.

(a) 90° and 180° (b) 120° (c) 109°
(d) 90° and 120° (e) 109° and 120°

b 16. A neutral molecule having the general formula AB_2 has one lone (unshared) pair of electrons on A. What is the hybridization at A?

(a) sp (b) sp^2 (c) sp^3 (d) sp^3d (e) sp^3d^2

c 17. A neutral molecule of general formula AB_2 is angular. There are two lone (unshared) pairs of electrons on A. What kind of hybrid orbitals are utilized by the central element?

(a) sp (b) sp^2 (c) sp^3 (d) sp^3d (e) sp^3d^2

c 18. A neutral molecule having the general formula AB_3 has one lone (unshared) pair of electrons on A. What is the hybridization of A?

(a) sp (b) sp^2 (c) sp^3 (d) sp^3d (e) sp^3d^2

d 19. A hypothetical AB_3 molecule has two (2) lone (unshared) electron pairs on A. The hybridization at A is _____.

(a) sp (b) sp^2 (c) sp^3 (d) sp^3d (e) sp^3d^2

e 20. A hypothetical AB_4 molecule has two (2) lone (unshared) electron pairs on A. The hybridization at A is _____.

(a) sp (b) sp^2 (c) sp^3 (d) sp^3d (e) sp^3d^2

b 21. The nitrite ion, NO_2^-, is known to be angular. The nitrogen atom is the central atom and the oxygen atoms are equidistant from it. On the basis of these facts, which of the following conclusions may be drawn concerning this ion? Choose the answer that includes all the correct conclusions, and no others.

I. It can be represented by resonance formulas.
II. The dipoles associated with each nitrogen-oxygen bond are equal in magnitude.
III. The dipoles associated with the nitrogen-oxygen bonds are opposite in direction.
IV. The nitrogen atom is sp hybridized.
V. The nitrite ion makes no contribution to the polarity of compounds in which it is contained.

(a) II and III (b) I and II (c) I, III, and V
(d) I, III, and IV (e) II, III, and IV

a 22. The O-N-O bond angles associated with the nitrate ion, NO_3^-, are all 120°. The hybridization at the nitrogen atom is _____.

(a) sp^2 (b) sp^3 (c) sp^3d^2 (d) sp (e) sp^3d

Linear Electronic Geometry

a 23. Which one of the following molecules has sp hybridization at the central atom?

(a) $BeBr_2$ (b) SeF_6 (c) BF_3 (d) PF_5 (e) CF_4

b 24. What are the bond angles in BeI_2 molecules?

(a) 120°   (b) 180°   (c) 90° and 180°
(d) 120° and 90°   (e) 109°

a 25. Cadmium iodide, CdI_2, is a covalent compound. The molecule has the same geometry as a $BeCl_2$ molecule. Which of the following kinds of hybrid orbitals does the cadmium atom likely utilize in CdI_2?

(a) sp (b) sp^2 (c) sp^3 (d) sp^3d (e) sp^3d^2

c 26. Which of the following statements concerning $BeCl_2$ is **false**?

(a) $BeCl_2$ is a linear molecule.
(b) The hybridization of the central Be atom is sp.
(c) The electronegativity difference between Be and Cl is so large (1.5 units) that $BeCl_2$ is an ionic compound.
(d) Because the bond angle is 180° the two bond dipoles cancel to give a nonpolar molecule.
(e) The Be atom in $BeCl_2$ does not satisfy the octet rule.

Trigonal Planar Electronic Geometry

b 27. What kind of hybrid orbitals are utilized by the boron atom in BF_3 molecules?

(a) sp (b) sp^2 (c) sp^3 (d) sp^3d (e) sp^3d^2

c 28. What is the hybridization at the boron atom in gaseous BBr_3?

(a) sp^3d (b) sp (c) sp^2 (d) sp^3 (e) sp^3d^2

d 29. Which of the following molecules has 120° bond angles?

(a) BeI_2 (b) CH_4 (c) H_2S (d) BCl_3 (e) NF_3

c 30. Which one of the following molecules has only 120° bond angles?

(a) PF_3 (b) SF_6 (c) BI_3 (d) PF_5 (e) NF_3

b 31. Which one of the following molecules has a central atom that is *sp²* hybridized?

(a) HF (b) BCl₃ (c) PCl₃ (d) H₂O (e) NH₃

b 32. Which of the following is a **false** statement about BF₃?

(a) BF₃ has trigonal planar molecular geometry.
(b) BF₃ has trigonal pyramidal electronic geometry.
(c) All three bond angles in BF₃ are 120°.
(d) The B atom does not satisfy the octet rule.
(e) Although the electronegativity difference between B and F is large (2.0 units), BF₃ is a covalent compound.

b 33. Which response includes all the following statements about BCl₃ molecules that are true, and no false statements?

I. The B-Cl bonds are quite polar.
II. The bond dipoles exactly cancel.
III. The Cl-B-Cl bond angles are 109.5°.
IV. BCl₃ molecules are polar.
V. BCl₃ molecules are pyramidal.

(a) I, III, and IV   (b) I and II   (c) II, IV, and V
(d) I, III, and V   (e) II, III, IV, and V

Tetrahedral Electronic Geometry (AB₄)

c 34. What kind of hybrid orbitals are utilized by the carbon atom in CF₄ molecules?

(a) *sp* (b) *sp²* (c) *sp³* (d) *sp³d* (e) *sp³d²*

c 35. What is the hybridization at the central atom in SiH₄?

(a) *sp* (b) *sp²* (c) *sp³* (d) *sp³d* (e) *sp³d²*

c 36. The silicon atom in the SiH₄ molecule is _____ hybridized and the H-Si-H bond angles are _____.

(a) *sp²*, 120°   (b) *sp²*, 109.5°   (c) *sp³*, 109.5°
(d) *sp³*, 120°   (e) *sp²*, 90°

b 37. Consider CH₄ and CF₄. Electronegativities: C = 2.5, H = 2.1, F = 4.0. Which statement is **false**?

(a) Both are *sp³* hybridized at carbon.
(b) The bond angles in CF₄ are smaller than those in CH₄.
(c) The C-F bonds are more polar than the C-H bonds.
(d) Both molecules are nonpolar.
(e) The bond dipoles in CF₄ are directed toward the fluorine, but those in CH₄ are directed toward the carbon atom.

e 38. Which statement concerning the ion, SO_4^{2-}, is true?

  (a) The ion's charge results in there being one lone pair of electrons on the S atom.
  (b) There are five regions of high electron density on the S atom.
  (c) The bond angles are 90°.
  (d) The electronic geometry is octahedral.
  (e) The hybridization of the S atom is sp^3.

c 39. Consider the bonding in the SiH_4 molecule. Which response includes all the following statements that are true, and no false statements?

  I. SiH_4 is a polar molecule.
  II. The H-Si-H bond angles are 120°.
  III. The SiH_4 molecule is tetrahedral.
  IV. The hybridization of Si is sp^2.

  (a) I and III   (b) II and IV   (c) III
  (d) I, II, and IV   (e) I and II

b 40. Which response contains all the characteristics listed that should apply to silicon tetrafluoride, SiF_4, and no other characteristics?

  I.   tetrahedral   II. 120° bond angles
  III. sp hybridized at C   IV. polar molecule
  V.   one unshared pair of electrons on Si

  (a) II and III   (b) I   (c) I and V
  (d) II, IV, and V   (e) none of these

Tetrahedral Electronic Geometry (AB_3U)

c 41. Molecules such as NF_3 are best described by assuming what kind of hybridization at the central atom?

  (a) sp (b) sp^2 (c) sp^3 (d) sp^3d (e) sp^3d^2

c 42. What is the hybridization of the nitrogen atom in NH_3?

  (a) sp (b) sp^2 (c) sp^3 (d) sp^3d (e) sp^3d^2

c 43. What is the hybridization of the sulfur atom in SO_3^{2-}?

  (a) sp (b) sp^2 (c) sp^3 (d) sp^3d (e) sp^3d^2

c 44. The **electronic** geometry of the central atom in PCl_3 is _____.

  (a) pyramidal   (b) trigonal planar   (c) tetrahedral
  (d) octahedral   (e) trigonal bipyramidal

a 45. The **molecular** geometry of PCl_3 is _____.

  (a) pyramidal   (b) trigonal planar   (c) tetrahedral
  (d) octahedral   (e) trigonal bipyramidal

Harcourt, Inc

Chapter 8   123

d 46. Which statement is **false**?

(a) Lone pairs of electrons repel each other more strongly than bonding pairs repel each other.
(b) Lone pairs of electrons repel each other more strongly than they repel bonding pairs.
(c) Lone pairs of electrons are considered a part of the electronic geometry about a central atom.
(d) Lone pairs of electrons have no effect on the molecular geometry about an atom.
(e) Lone pairs of electrons are not considered a part of the molecular geometry about a central atom.

a 47. Many simple molecules contain two **lone pairs** of electrons (unshared pairs) that occupy hybrid orbitals on the central element. Which of the following kinds of electron-electron repulsions are **smallest** in such molecules?

(a) bonding pair-bonding pair
(b) bonding pair-lone pair
(c) lone pair-lone pair
(d) depends on hybrid orbitals
(e) repulsions between all types of pairs of electrons are the same

d 48. Many simple molecules contain **lone pairs** of electrons (also referred to as **unshared pairs**) which occupy hybrid orbitals of the central element in a molecule. If an atom of the central element utilizes sp^3 hybrid orbitals in a compound, which one of the following types of repulsions would be **greater**?

(a) bonding pair-bonding pair
(b) bonding pair-lone pair
(c) lone pair-bonding pair
(d) lone pair-lone pair
(e) repulsions between all types of pairs of electrons are the same

a 49. Four of the following statements about the ammonia molecule, NH_3, are correct. One is not. Which one?

(a) The ammonia molecule is tetrahedral.
(b) Since nitrogen is more electronegative than hydrogen, the bond dipoles are directed toward the nitrogen atoms.
(c) The bond dipoles reenforce the effect of the unshared pair of electrons on the nitrogen atom.
(d) The bond angles in the ammonia molecule are less than 109°.
(e) The nitrogen atom can be described as utilizing sp^3 hybrid orbitals in the ammonia molecule.

e 50. Four of the following statements about the ammonia molecule, NH_3, and the nitrogen trifluoride molecule, NF_3, are correct. One is not. Which one?

(a) The NH_3 molecule is more polar than the NF_3 molecule.
(b) The bond dipoles in NF_3 are directed toward the more electronegative fluorine atoms.
(c) The bond dipoles in NF_3 oppose the effect of the unshared pair of electrons.
(d) The nitrogen atom can be described as utilizing sp^3 hybrid orbitals in the nitrogen trifluoride molecule.
(e) Fluorine atoms are larger than hydrogen atoms, and therefore the bond angles in NF_3 are greater than in NH_3.

Harcourt, Inc

c 51. Which statement for NH$_3$ and NF$_3$ is false? Electronegativities: N = 3.0, H = 2.1, F = 4.0.

    (a) Both are sp^3 hybridized at nitrogen.
    (b) The bond angles in NF$_3$ are smaller than those in NH$_3$.
    (c) The NF$_3$ molecule is more polar than the NH$_3$ molecule.
    (d) Both molecules have one unshared pair of electrons in the outer shell of nitrogen.
    (e) The bond dipoles of NF$_3$ are directed toward fluorine, whereas those in NH$_3$ are directed toward nitrogen.

d 52. Which response contains all the characteristics listed that should apply to phosphorus trichloride, PCl$_3$, and no other characteristics?

    I. trigonal planar   II. polar bonds
    III. sp^2 hybridized at P   IV. polar molecule
    V. one unshared pair of electrons on P

    (a) I, IV, and V   (b) II, III, and IV   (c) I, II, and IV
    (d) II, IV, and V   (e) none of these

Tetrahedral Electronic Geometry (AB$_2$U$_2$)

b 53. The **molecular** geometry of the hydrogen sulfide molecule is _____.

    (a) linear   (b) angular   (c) trigonal planar
    (d) pyramidal   (e) trigonal bipyramidal

a 54. Which response contains all of the following molecules that can be described as bent or angular molecules and none that have other shapes?

$$BeI_2, SO_2, H_2S, PF_3, CO_2$$

    (a) SO$_2$, H$_2$S   (b) BeI$_2$, SO$_2$ (c) BeI$_2$, PF$_3$, CO$_2$
    (d) PF$_3$, CO$_2$ (e) another combination

c 55. Which of the following molecules has the smallest bond angle(s)?

    (a) CH$_4$ (b) NH$_3$ (c) H$_2$O   (d) CF$_4$ (e) CO$_2$

c 56. What is the hybridization of the oxygen atom in H$_3$O$^+$?

    (a) sp (b) sp^2 (c) sp^3 (d) sp^3d (e) sp^3d^2

b 57. Why are hybrid orbitals not proposed in describing the bonding in H$_2$S, H$_2$Se and H$_2$Te?

    (a) They are all linear molecules.
    (b) Their bond angles are close to the 90° angle between two unhybridized p orbitals.
    (c) All three of these molecules are nonpolar.
    (d) They have d orbitals in their valence shells.
    (e) They each have two lone pairs of electrons on the central atom.

d 58. Which statement about the water molecule is **incorrect**?

(a) The decrease in bond angle from the tetrahedral angle (109.5°) is greater for H_2O than for NH_3.
(b) The bond angle for H_2O is smaller than for NH_3 because of the lone pair–lone pair repulsion in H_2O.
(c) The water molecule is very polar.
(d) The effect of the very polar bonds in water is opposed by the effect of the 2 lone pairs of electrons.
(e) The O atom is sp^3 hybidized.

Trigonal Bipyramidal Electronic Geometry

e 59. An element that is sp^3d hybridized and which has no lone pairs of electrons around it in a molecule is at the center of a(an) _____ described by imaginary lines connecting the identical surrounding atoms.

(a) triangular plane (b) tetrahedron (c) octahedron
(d) pyramid (e) trigonal bipyramid

d 60. What is the hybridization at arsenic in AsF_5 molecules?

(a) sp (b) sp^2 (c) sp^3 (d) sp^3d (e) sp^3d^2

d 61. What is the hybridization at the central atom (P) of PF_3Cl_2?

(a) sp (b) sp^2 (c) sp^3 (d) sp^3d (e) sp^3d^2

a 62. Which one of the following molecules has sp^3d hybridization at the central atom?

(a) PF_5 (b) NH_3 (c) SF_6 (d) HF (e) CF_4

d 63. How many unshared electrons are there in the highest energy level of phosphorus in the covalent compound PF_5? (HINT: draw Lewis dot formula.)

(a) eight (b) two (c) three (d) none (e) five

a 64. The PF_5 molecule exists whereas NF_5 does not. Which one of the following is the **best** explanation for this fact?

(a) Phosphorus atoms can undergo hybridization involving *d* orbitals, whereas nitrogen atoms cannot.
(b) The first five ionization energies of nitrogen are too high but those of phosphorus are not.
(c) The electron affinity of nitrogen is too high.
(d) NF_5 would be too polar to be stable.
(e) Simple P^{5+} ions exist but simple N^{5+} ions are too small to exist.

d 65. Draw the dot formula for SF_4. What is the hybridization at S?

(a) sp (b) sp^2 (c) sp^3 (d) sp^3d (e) sp^3d^2

d 66. What is the hybridization of the central I atom in I_3^-?

(a) sp (b) sp^2 (c) sp^3 (d) sp^3d (e) sp^3d^2

e 67. Which of the following species has trigonal bipyramidal electronic geometry **and** trigonal bipyramidal molecular geometry?

(a) SF_4 (b) BrF_3 (c) I_3^- (d) XeF_2 (e) $AsCl_5$

Octahedral Electronic Geometry

b 68. Which one of the following molecules is octahedral?

(a) $BeCl_2$ (b) SeF_6 (c) BF_3 (d) PF_5 (e) CF_4

e 69. What is the hybridization at the central element in SeF_6?

(a) sp (b) sp^2 (c) sp^3 (d) sp^3d (e) sp^3d^2

e 70. What is the hybridization at the sulfur atom in SF_6?

(a) sp (b) sp^2 (c) sp^3 (d) sp^3d (e) sp^3d^2

e 71. The F-S-F bond angles in SF_6 are _____.

(a) 109.5°   (b) 120°   (c) 90° and 120°
(d) 45° and 90°   (e) 90° and 180°

e 72. Draw the dot formula for the noble gas compound xenon tetrafluoride, XeF_4. What is the hybridization of xenon?

(a) sp (b) sp^2 (c) sp^3 (d) sp^3d (e) sp^3d^2

c 73. Which of the following species has square planar molecular geometry?

(a) CH_4 (b) SF_4 (c) XeF_4 (d) NH_4^+ (e) SO_4^{2-}

d 74. Which of the following species has square pyramidal molecular geometry?

(a) PF_5 (b) NH_4^+ (c) $SbCl_5$ (d) IF_5 (e) $SiCl_4$

e 75. Which of the following species does **not** have octahedral electronic geometry?

(a) SeF_6 (b) BrF_5 (c) XeF_4 (d) PF_6^- (e) SF_4

d 76. The central atom of which of the following species does **not** have sp^3d^2 hybridization?

(a) XeF_4 (b) SeF_6 (c) PF_6^- (d) IF_7 (e) ClF_5

Compounds Containing Double or Triple Bonds

d 77. Chose the following false statement.

(a) A sigma bond is a bond resulting from head-on overlap of atomic orbitals.
(b) A pi bond is a bond resulting from side-on overlap of atomic orbitals.
(c) A double bond consists of one sigma bond and one pi bond.
(d) A triple bond may consist of one sigma bond and two pi bonds or of two sigma bonds and one pi bond.
(e) A carbon atom involved in a double bond may not be sp^3 hybridized.

c 78. How many sigma(s) bonds and how many pi (π) bonds does the ethene molecule contain?

$$C_2H_4, \quad \begin{array}{c} H \quad\quad H \\ \ddot{C}::\ddot{C} \\ H \quad\quad H \end{array}$$

(a) 4 σ and 2 π (b) 5 σ and 2 π (c) 5 σ and 1 π
(d) 5 σ and 0 π (e) 8 σ and 2 π

e 79. How many sigma(s) bonds and how many pi (π) bonds does the acetylene molecule contain?

(a) 5 σ and 1 π (b) 2 σ and 3 π (c) 3 σ and 1 π
(d) 2 σ and 2 π (e) 3 σ and 2 π

b 80. What is the hybridization of a carbon atom involved in a double bond?

(a) sp (b) sp^2 (c) sp^3 (d) sp^3d (e) sp^3d^2

a 81. What is the hybridization of a carbon atom involved in a triple bond?

(a) sp (b) sp^2 (c) sp^3 (d) sp^3d (e) sp^3d^2

d 82. Which of the following responses contains **all** of the true statements for ethyne?

I. The ethyne molecule has a double bond between its 2 carbon atoms.
II. The ethyne molecule has a triple bond between its 2 carbon atoms.
III. One pi bond is formed between the 2 carbon atoms by head-on overlap of the sp hybrid orbitals.
IV. One sigma bond is formed between the 2 carbon atoms by head-on overlap of the sp^2 hybrid orbitals.
V. The ethyne molecule is linear.

(a) I and III (b) II and IV (c) I and IV
(d) II and V   (e) II and III

Summary of Electronic and Molecular Geometries

e 83. Choose the species that is **incorrectly** matched with the **electronic** geometry about the central atom.

	Molecule	Electronic Geometry
(a)	CF_4	tetrahedral
(b)	$BeBr_2$	linear
(c)	H_2O	tetrahedral
(d)	NH_3	tetrahedral
(e)	PF_3	pyramidal

e 84. Choose the species that is **incorrectly** matched with the **electronic** geometry about the central atom.

	Molecule	Electronic Geometry
(a)	PCl_5	trigonal bipyramidal
(b)	SF_6	octahedral
(c)	XeF_2	trigonal bipyramidal
(d)	XeF_4	octahedral
(e)	$GeCl_4$	trigonal bipyramidal

c 85. Choose the species that is **incorrectly** matched with the **electronic** geometry about the central atom.

	Ion	Electronic Geometry
(a)	NO_2^-	trigonal planar
(b)	ClO_4^-	tetrahedral
(c)	SO_3^{2-}	pyramidal
(d)	ClO_3^-	tetrahedral
(e)	BrO_4^-	tetrahedral

e 86. Which molecule is **incorrectly** matched with the **molecular** geometry?

	Molecule	Molecular Geometry
(a)	CCl_4	tetrahedral
(b)	PH_3	pyramidal
(c)	$BeCl_2$	linear
(d)	BBr_3	trigonal planar
(e)	SO_2	linear

b 87. Which molecule is **incorrectly** matched with the **molecular** geometry?

	Molecule	Molecular Geometry
(a)	TeF_6	octahedral
(b)	NH_3	tetrahedral
(c)	AsF_5	trigonal bipyramidal
(d)	SiF_4	tetrahedral
(e)	H_2O	angular

a 88. Which molecule or polyatomic ion is **incorrectly** matched with the **molecular (or ionic)** geometry?

	Ion	Molecular Geometry
(a)	$SiCl_6^{2-}$	trigonal bipyramidal
(b)	PH_4^+	tetrahedral
(c)	ClO_2^-	angular
(d)	NH_4^+	tetrahedral
(e)	SO_4^{2-}	tetrahedral

d 89. Which molecule is **incorrectly** matched with the **molecular** geometry?

	Molecule	Molecular Geometry
(a)	SeF_6	octahedral
(b)	CCl_4	tetrahedral
(c)	SO_3	pyramidal
(d)	SF_4	tetrahedral
(e)	SbH_3	pyramidal

e 90. Which molecule is **incorrectly** matched with the **molecular** geometry?

	Molecule	Molecular Geometry
(a)	SF_6	octahedral
(b)	AsH_3	pyramidal
(c)	BCl_3	trigonal planar
(d)	AsF_5	trigonal bipyramidal
(e)	H_2S	linear

d 91. Which one of the following pairs of molecules and molecular geometries is **incorrectly** matched?

	Compound	Molecular Geometry
(a)	NF_3	pyramidal
(b)	H_2O	angular
(c)	BF_3	trigonal planar
(d)	AsF_5	pentagonal planar
(e)	SeF_6	octahedral

d 92. Which one of the following pairs of molecules and molecular geometries is **incorrectly** matched?

	Compound	Molecular Geometry
(a)	HBr	linear
(b)	CBr_4	tetrahedral
(c)	AsH_3	pyramidal
(d)	$BeBr_2$	angular
(e)	CH_4	tetrahedral

a 93. Which response includes all of the molecules listed that have the same molecular geometry as electronic geometry, and only those molecules?

BCl_3, AsF_5, NF_3, H_2S

(a) BCl_3, AsF_5
(b) AsF_5, NF_3, H_2S
(c) NF_3, H_2S
(d) BCl_3, H_2S
(e) BCl_3, NF_3

e 94. Which one of the following molecules has a central atom that is **not** sp^3 hybridized?

(a) H_2O (b) NH_3 (c) CH_4 (d) $CHCl_3$ (e) SF_4

e 95. Which, if any, of the compounds listed are **not** sp^3 hybridized at the central atom?

I. CF_4 II. H_2O III. SiF_4 IV. $CHCl_3$ V. NH_3

(a) III and IV
(b) I, II, and III
(c) II, IV, and V
(d) III and V
(e) all are sp^3 hybridized

a 96. Which species is **incorrectly** matched with the **hybridization** at the central atom?

	Species	Hybridization at Central Atom
(a)	SeF_6	sp^3d
(b)	CO_2	sp
(c)	NH_4^+	sp^3
(d)	BCl_3	sp^2
(e)	$BeBr_2$	sp

c 97. Which species is **incorrectly** matched with the **hybridization** at the central atom?

	Species	Hybridization at Central Atom
(a)	SO_2	sp^2
(b)	CF_4	sp^3
(c)	PF_5	sp^3d^2
(d)	SeO_4^{2-}	sp^3
(e)	HCN	sp

c 98. Which species is **incorrectly** matched with the **hybridization** at the central atom?

	Compound	Hybridization at Central Atom
(a)	BF_3	sp^2
(b)	$SnCl_4$	sp^3
(c)	AsF_5	sp^3d^2
(d)	H_2O	sp^3
(e)	$BeCl_2$	sp

b 99. Which species is **incorrectly** matched with **bond angles**?

	Molecule	Bond Angles
(a)	SiCl$_4$	109.5°
(b)	BeI$_2$	slightly less than 109°
(c)	SF$_6$	90° (and 180°)
(d)	AsF$_5$	90°, 120° (and 180°)
(e)	BF$_3$	120°

e 100. Which species is **incorrectly** matched with **bond angles**?

	Molecule	Bond Angles
(a)	HCN	180°
(b)	ClO$_3^-$	slightly less than 109°
(c)	NH$_3$	107°
(d)	SeO$_4^-$	109.5°
(e)	SnCl$_4$	90°, 120° (and 180°)

a 101. Which species is **incorrectly** matched with **bond angles**?

	Molecule	Bond Angles
(a)	NCl$_3$	120°
(b)	BeBr$_2$	180°
(c)	CCl$_4$	109.5°
(d)	PCl$_5$	120° and 90° (and 180°)
(e)	SeF$_6$	90° (and 180°)

Polarity of Molecules

d 102. Which one of the following molecules is polar?

(a) BCl$_3$ (b) CdI$_2$ (c) CCl$_4$ (d) NCl$_3$ (e) Br$_2$

c 103. Which one of the following molecules is polar?

(a) SiH$_4$ (b) SF$_6$ (c) H$_2$S (d) PCl$_5$ (e) SiF$_4$

b 104. Which one of the following molecules is polar?

(a) AsF$_5$ (b) PCl$_3$ (c) CO$_2$ (d) CF$_4$ (e) SiH$_4$

d 105. Which one of the following molecules is polar?

(a) BCl$_3$ (b) BeI$_2$ (c) CCl$_4$ (d) CHCl$_3$ (e) Br$_2$

e 106. Which one of the following molecules is polar?

(a) SiH$_4$ (b) SF$_6$ (c) H$_2$ (d) PCl$_5$ (e) SF$_4$

c 107. Which response includes all the polar molecules listed and no nonpolar molecules?

BCl_3, AsF_5, NF_3, H_2S

(a) BCl_3, AsF_5 (b) AsF_5, NF_3, H_2S (c) NF_3, H_2S
(d) BCl_3, H_2S (e) BCl_3, NF_3

d 108. Which response contains all of the **polar** molecules below and no nonpolar molecules?

I. BF_3 II. PF_3 III. H_2O IV. PF_5 V. TeF_6 (Te is in Group VIA)

(a) I, II, and IV   (b) II, III, and V   (c) III
(d) II and III   (e) another one or another combination

d 109. Which response contains all the covalent molecules below that are polar, and no others?

I. CCl_4 II. HBr   III. N_2 IV. C_2H_6 V. H_2S

(a) II, III, and IV   (b) I and II   (c) III, IV, and V
(d) II and V   (e) I, III, and V

d 110. Which compound consists of **nonpolar molecules**? (It may help to draw dot formulas.)

(a) H_2S (b) PH_3 (c) $AsCl_3$ (d) SiH_4 (e) SF_4

d 111. Which one of the following is a **nonpolar** molecule with **polar** covalent bonds?

(a) NH_3 (b) H_2Te (c) $SOCl_2$ (S is central atom)
(d) $BeBr_2$ (e) HF

c 112. Which one of the following is a **nonpolar** molecule with **polar** bonds?

(a) H_2O (b) NH_3 (c) PF_5 (d) $CHCl_3$ (e) none of these

e*113. Which one of the following is a **polar** molecule with **nonpolar** bonds?

(a) H_2O (b) NH_3 (c) PF_5 (d) $CHCl_3$ (e) none of these

b*114. Which one of the following is the **most polar molecule**?
(Electronegativities: H = 2.1, Be = 1.5, B = 2.0, F = 4.0, S = 2.5, Cl = 3.0, Te = 2.1, P = 2.1)

(a) BF_3 (b) H_2S (c) $BeCl_2$ (d) H_2Te (e) PF_5

9 Molecular Orbitals in Chemical Bonding

Molecular Orbitals

d 1. Which one of the following statements is **false**?

(a) Valence bond theory and molecular orbital theory can be described as two different views of the same thing.
(b) When one considers the molecular orbitals resulting from the overlap of any two specific atomic orbitals, the bonding orbitals are always lower in energy than the antibonding orbitals.
(c) Molecular orbitals are generally described as being more delocalized than hybridized atomic orbitals.
(d) One of the shortcomings of molecular orbital theory is its inability to account for a triple bond in the N_2 molecule.
(e) One of the shortcomings of valence bond theory is its inability to account for the paramagnetism of O_2.

a 2. Which statement is **false**? A **sigma** molecular orbital

(a) may result from side-on overlap of p atomic orbitals.
(b) may result from head-on overlap of p atomic orbitals.
(c) may result from overlap of an s atomic orbital with a p atomic orbital.
(d) appears spherical when viewed along its major axis (axis of overlap).
(e) may be either bonding or antibonding.

a 3. Which statement is **true**? A **pi** molecular orbital

(a) may result from side-on overlap of p atomic orbitals.
(b) may result from head-on overlap of p atomic orbitals.
(c) may result from overlap of an s atomic orbital with a p atomic orbital.
(d) appears spherical when viewed along its major axis (internuclear axis).
(e) can only be an antibonding orbital.

c 4. Which statement is **false**?

(a) The energy of a bonding orbital is always lower than the energies of the combining atomic orbitals.
(b) The energy of an antibonding orbital is always higher than the energies of the combining atomic orbitals.
(c) The number of molecular orbitals formed is equal to twice the number of atomic orbitals that have combined because both a bonding and an antibonding orbital are formed.
(d) An antibonding orbital is always less stable than the original combining atomic orbitals.
(e) A bonding molecular orbital has a high electron density between the two nuclei while an antibonding molecular orbital has a node between the nuclei.

d 5. How many MOs are formed by the combination of the p atomic orbitals of 2 atoms?

(a) 1   (b) 2   (c) 3   (d) 6   (e) 12

134

b 6. The out-of-phase head-on overlap of a pair of 2p orbitals forms a ____ molecular orbital.

(a) σ_{2p} (b) $\sigma_{2p}^{\star}$ (c) π_{2p} (d) $\pi_{2p}^{\star}$ (e) σ_{2s}

c 7. The in-phase side-on overlap of a pair of 2p orbitals forms a ____ molecular orbital.

(a) σ_{2p} (b) $\sigma_{2p}^{\star}$ (c) π_{2p} (d) $\pi_{2p}^{\star}$ (e) σ_{2s}

Molecular Orbital Energy Level Diagrams

d 8. The molecular orbital energy level diagram has the σ_{2p} orbital at a lower energy level that the two π_{2p} orbitals for which of the following molecules?

(a) C_2 (b) N_2 (c) B_2 (d) O_2 (e) Li_2

c 9. The molecular orbital energy level diagram has the σ_{2p} orbital at a higher energy level that the two π_{2p} orbitals for which of the following species?

(a) O_2 (b) O_2^- (c) N_2 (d) F_2 (e) O_2^+

e 10. Which statement is **false**?

(a) Electrons occupy MOs by following the Aufbau Principle.
(b) Electrons occupy MOs by following the Pauli Exclusion Principle.
(c) Electrons occupy MOs by following Hund's Rule.
(d) In applying MO theory we usually account for all electrons, not just the valence electrons.
(e) No two molecular orbitals for any molecule ever have the same energy.

d 11. Which of the following statements concerning the molecular orbital energy level diagrams for first and second period homonuclear diatomic molecules is **false**?

(a) The two π_{2p} orbitals have the same energy.
(b) The two $\pi_{2p}^{\star}$ orbitals have the same energy.
(c) The diagram for H_2, He_2, Li_2, Be_2, B_2, C_2, and N_2 molecules has the two π_{2p} orbitals at a lower energy than the σ_{2p} orbital.
(d) The diagram for O_2, F_2, and Ne_2 molecules has the $\sigma_{2p}^{\star}$ orbital at a lower energy than the two $\pi_{2p}^{\star}$ orbitals.
(e) The bonding orbitals always have lower energy than the antibonding orbitals formed from the same set of atomic orbitals.

Bond Order and Bond Stability

c 12. What is the bond order for a molecule with 10 electrons in bonding molecular orbitals and 5 electrons in antibonding molecular orbitals?

(a) 1 (b) 2 (c) 2.5 (d) 3 (e) 5

Harcourt, Inc

b 13. What is the bond order of a species with 7 bonding electrons and 4 antibonding electrons?

(a) 1   (b) 1.5   (c) 2   (d) 2.5   (e) 3

d 14. Which statement concerning bond order is **false**?

(a) The bond order usually corresponds to the number of bonds described by the valence bond theory.
(b) A diatomic molecule with a fractional bond order contains an odd number of electrons.
(c) A bond order equal to zero means that a molecule is no more stable than the separate atoms.
(d) The greater the bond order for a bond between 2 atoms, the longer the length of the bond between them.
(e) The greater the bond order for a bond between 2 atoms, the greater the bond energy for that bond.

c 15. Which statement is **false**?

(a) Bond energy is a measure of bond strength.
(b) A diatomic molecule with a fractional bond order must be paramagnetic.
(c) A diatomic molecule with an integral bond order must be diamagnetic.
(d) A molecule with a bond order greater than zero would be more stable than the separate atoms.
(e) The larger the bond order of a diatomic species, the more stable it is expected to be.

Homonuclear Diatomic Molecules

c *16. The molecular orbital electron configuration given below is for what **neutral homonuclear** diatomic molecule?

$$\sigma_{1s}^2 \; \sigma_{1s}^{\star 2} \; \sigma_{2s}^2 \; \sigma_{2s}^{\star 2} \; \pi_{2p_y}^2 \; \pi_{2p_z}^2$$

(a) N_2 (b) N_2^{2+} (c) C_2 (d) BN   (e) B_2^{2-}

c 17. Which of the molecular orbital electron configurations given below is **correct** for O_2?

(a) $\sigma_{1s}^2 \; \sigma_{1s}^{\star 2} \; \sigma_{2s}^2 \; \sigma_{2s}^{\star 2} \; \sigma_{2p}^2 \; \pi_{2p_y}^2 \; \pi_{2p_z}^2 \; \pi_{2p}^{\star 2}$
(b) $\sigma_{1s}^2 \; \sigma_{1s}^{\star 2} \; \sigma_{2s}^2 \; \sigma_{2s}^{\star 2} \; \pi_{2p_y}^2 \; \pi_{2p_z}^2 \; \sigma_{2p}^2 \; \pi_{2p_y}^{\star 1} \; \pi_{2p_z}^{\star 1}$
(c) $\sigma_{1s}^2 \; \sigma_{1s}^{\star 2} \; \sigma_{2s}^2 \; \sigma_{2s}^{\star 2} \; \sigma_{2p}^2 \; \pi_{2p_y}^2 \; \pi_{2p_z}^2 \; \pi_{2p_y}^{\star 1} \; \pi_{2p_z}^{\star 1}$
(d) $\sigma_{1s}^2 \; \sigma_{1s}^{\star 2} \; \sigma_{2s}^2 \; \sigma_{2s}^{\star 2} \; \pi_{2p_y}^2 \; \pi_{2p_z}^2 \; \sigma_{2p}^2 \; \pi_{2p}^{\star 2}$
(e) $\sigma_{1s}^2 \; \sigma_{1s}^{\star 2} \; \sigma_{2s}^2 \; \sigma_{2s}^{\star 2} \; \pi_{2p_y}^2 \; \pi_{2p_z}^2 \; \pi_{2p_y}^{\star 2} \; \pi_{2p_z}^{\star 2}$

a 18. Draw the molecular orbital diagram for B_2. The number of electrons in the π_{2p_z} molecular orbital is _____.

(a) 1   (b) 2   (c) 3   (d) 4   (e) zero

b 19. Draw the molecular orbital diagram for C_2. The number of electrons in the π_{2p_z} molecular orbital is _____.

(a) 1   (b) 2   (c) 3   (d) 4   (e) zero

e 20. Draw the molecular orbital diagram for C_2. The number of electrons in the σ_{2p} molecular orbital is _____.

(a) 1   (b) 2   (c) 3   (d) 4   (e) zero

e 21. Draw the molecular orbital diagram for N_2. The number of electrons in the $\pi^*_{2p_z}$ molecular orbital is _____.

(a) 1   (b) 2   (c) 3   (d) 4   (e) zero

b 22. Draw the molecular orbital diagram for N_2. The number of electrons in the σ_{2p} molecular orbital is _____.

(a) 1   (b) 2   (c) 3   (d) 4   (e) zero

a 23. Draw the molecular orbital diagram for O_2. The number of electrons in the $\pi^*_{2p_z}$ molecular orbital is _____.

(a) 1   (b) 2   (c) 3   (d) 4   (e) zero

e 24. Draw the molecular orbital diagram for O_2. The number of electrons in the σ^*_{2p} molecular orbital is _____.

(a) 1   (b) 2   (c) 3   (d) 4   (e) zero

b 25. Draw the molecular orbital diagram for F_2. The number of electrons in the $\pi^*_{2p_z}$ molecular orbital is _____.

(a) 1   (b) 2   (c) 3   (d) 4   (e) zero

e 26. Draw the molecular orbital diagram for F_2. The number of electrons in the σ^*_{2p} molecular orbital is _____.

(a) 1   (b) 2   (c) 3   (d) 4   (e) zero

b 27. The total numbers of electrons in the π_{2p} molecular orbitals of B_2 is _____.

(a) 1   (b) 2   (c) 3   (d) 4   (e) zero

d 28. The total numbers of electrons in the π^*_{2p} molecular orbitals of F_2 is _____.

(a) 1   (b) 2   (c) 3   (d) 4   (e) zero

a 29. What is the bond order for B_2?

(a) 1   (b) 2   (c) 3   (d) 1.5   (e) 2.5

b 30. What is the bond order for C_2?

(a) 1   (b) 2   (c) 3   (d) 1.5   (e) zero

c 31. What is the bond order for N_2?

(a) 1   (b) 2   (c) 3   (d) 1.5   (e) 2.5

b 32. What is the bond order in O_2?

(a) 1   (b) 2   (c) 3   (d) 2.5   (e) 1.5

a 33. What is the bond order for F_2?

(a) 1   (b) 2   (c) 3   (d) 4.5   (e) 2.5

b 34. The number of unpaired electrons in the O_2 molecule is _____.

(a) 1   (b) 2   (c) 3   (d) 4   (e) zero

e 35. The number of unpaired electrons in the C_2 molecule is _____.

(a) 1   (b) 2   (c) 3   (d) 4   (e) zero

b 36. The number of unpaired electrons in the B_2 molecule is _____.

(a) 1   (b) 2   (c) 3   (d) 4   (e) zero

a 37. Which response lists only the molecules given below that have unpaired electrons?

Li_2, Be_2, B_2, C_2, N_2

(a) B_2 (b) Be_2 (c) N_2 (d) Li_2 and B_2 (e) Be_2 and N_2

b 38. Which response lists only the molecules given below that are paramagnetic?

B_2, C_2, N_2, O_2, F_2

(a) B_2 and N_2 (b) B_2 and O_2 (c) N_2 and O_2
(d) N_2 and F_2 (e) C_2, O_2, and F_2

e 39. Which of the molecular orbital electron configurations given below is **correct** for C_2^{2+}?

(a) $\sigma_{1s}^2 \; \sigma_{1s}^{\star 2} \; \sigma_{2s}^2 \; \sigma_{2s}^{\star 2} \; \sigma_{2p}^2$ (b) $\sigma_{1s}^2 \; \sigma_{1s}^{\star 2} \; \sigma_{2s}^2 \; \sigma_{2s}^{\star 2} \; \pi_{2p_y}^2$

(c) $\sigma_{1s}^2 \; \sigma_{1s}^{\star 2} \; \sigma_{2s}^2 \; \sigma_{2s}^{\star 2}$ (d) $\sigma_{1s}^2 \; \sigma_{1s}^{\star 2} \; \sigma_{2s}^2 \; \sigma_{2s}^{\star 2} \; \pi_{2p_y}^2 \; \pi_{2p_z}^1$

(e) $\sigma_{1s}^2 \; \sigma_{1s}^{\star 2} \; \sigma_{2s}^2 \; \sigma_{2s}^{\star 2} \; \pi_{2p_y}^1 \; \pi_{2p_z}^1$

b 40. Which of the molecular orbital electron configurations given below is **correct** for F_2^-?

(a) $\sigma_{1s}^2\ \sigma_{1s}^{\star 2}\ \sigma_{2s}^2\ \sigma_{2s}^{\star 2}\ \pi_{2p_y}^2\ \pi_{2p_z}^2\ \sigma_{2p}^2\ \sigma_{2p}^{\star 2}\ \pi_{2p_y}^{\star 2}\ \pi_{2p_z}^{\star 1}$

(b) $\sigma_{1s}^2\ \sigma_{1s}^{\star 2}\ \sigma_{2s}^2\ \sigma_{2s}^{\star 2}\ \sigma_{2p}^2\ \pi_{2p_y}^2\ \pi_{2p_z}^2\ \pi_{2p_y}^{\star 2}\ \pi_{2p_z}^{\star 2}\ \sigma_{2p}^{\star 1}$

(c) $\sigma_{1s}^2\ \sigma_{1s}^{\star 2}\ \sigma_{2s}^2\ \sigma_{2s}^{\star 2}\ \pi_{2p_y}^2\ \pi_{2p_z}^2\ \sigma_{2p}^2\ \pi_{2p_y}^{\star 2}\ \pi_{2p_z}^{\star 2}\ \sigma_{2p}^{\star 1}$

(d) $\sigma_{1s}^2\ \sigma_{1s}^{\star 2}\ \sigma_{2s}^2\ \sigma_{2s}^{\star 2}\ \sigma_{2p}^2\ \pi_{2p_y}^2\ \pi_{2p_z}^2\ \pi_{2p_y}^{\star 2}\ \pi_{2p_z}^{\star 1}$

(e) $\sigma_{1s}^2\ \sigma_{1s}^{\star 2}\ \sigma_{2s}^2\ \sigma_{2s}^{\star 2}\ \pi_{2p_y}^2\ \sigma_{2p}^2\ \pi_{2p_z}^2\ \sigma_{2p}^{\star 2}\ \pi_{2p_y}^{\star 2}\ \pi_{2p_z}^{\star 1}$

d 41. What diatomic molecule and/or ion(s) would have the molecular orbital electron configuration given below?

$$\sigma_{1s}^2\ \sigma_{1s}^{\star 2}\ \sigma_{2s}^2\ \sigma_{2s}^{\star 2}\ \pi_{2p_y}^2\ \pi_{2p_z}^1$$

(a) O_2^+ and N_2 (b) C_2^- and B_2^+ (c) N_2^+ and C_2
(d) C_2^+ and B_2^- (e) C_2^- and N_2^+

a 42. What diatomic molecule and/or ion(s) would have the molecular orbital electron configuration given below?

$$\sigma_{1s}^2\ \sigma_{1s}^{\star 2}\ \sigma_{2s}^2\ \sigma_{2s}^{\star 2}\ \sigma_{2p}^2\ \pi_{2p_y}^2\ \pi_{2p_z}^2\ \pi_{2p_y}^{\star 2}\ \pi_{2p_z}^{\star 1}$$

(a) O_2^- and F_2^+ (b) O_2 and F_2 (c) F_2^- and Ne_2^+
(d) F_2 and O_2^- (e) O_2^+ and F_2^-

c 43. What diatomic molecule and/or ion(s) would have the molecular orbital electron configuration given below?

$$\sigma_{1s}^2\ \sigma_{1s}^{\star 2}\ \sigma_{2s}^2\ \sigma_{2s}^{\star 2}\ \sigma_{2p}^2\ \pi_{2p_y}^2\ \pi_{2p_z}^2\ \pi_{2p_y}^{\star 2}\ \pi_{2p_z}^{\star 2}$$

(a) F_2 and O_2^{2+} (b) O_2 and F_2^{2+} (c) F_2 and O_2^{2-}
(d) F_2 and Ne_2^{2-} (e) O_2 and F_2

d 44. What diatomic molecule and/or ion(s) would have the molecular orbital electron configuration given below?

$$\sigma_{1s}^2\ \sigma_{1s}^{\star 2}\ \sigma_{2s}^2\ \sigma_{2s}^{\star 2}\ \pi_{2p_y}^2\ \pi_{2p_z}^2$$

(a) C_2 and B_2^- (b) C_2 and N_2^+ (c) B_2^- and N_2^+
(d) B_2^{2-} and N_2^{2+} (e) B_2 and Be_2^{2-}

a 45. Which response lists all the following diatomic molecules and ions that are paramagnetic?

$Li_2,\ Be_2^-,\ B_2,\ C_2,\ N_2,\ N_2^+,\ N_2^{2+}$

(a) Be_2^-, B_2, and N_2^+ (b) Be_2^-, N_2, and N_2^{2+} (c) Be_2^-, B_2, and N_2
(d) Li_2, B_2, and N_2 (e) Li_2, C_2, and N_2^{2+}

e 46. Which response lists all the following diatomic molecules and ions that are paramagnetic?

$$Be_2, B_2, B_2^{2+}, C_2^{2+}, C_2^{2-}, O_2^-, O_2^{2-}$$

(a) B_2^{2+}, C_2^{2+}, and O_2^{2-}
(b) Be_2 and B_2
(c) B_2, C_2^{2+}, and C_2^{2-}
(d) B_2, C_2^{2-}, and O_2^{2-}
(e) B_2, C_2^{2+}, and O_2^-

c 47. Which response lists all the following diatomic molecules and ions that have at least one unpaired electron?

$$Be_2, B_2, B_2^+, C_2, N_2, N_2^+$$

(a) Be_2 and B_2
(b) B_2^+ and C_2
(c) B_2, B_2^+, and N_2^+
(d) B_2^+ and N_2
(e) B_2 and N_2

e 48. What is the bond order for N_2^+?

(a) 1   (b) 2   (c) 3   (d) 1.5   (e) 2.5

d 49. What is the bond order for O_2^-?

(a) 1   (b) 2   (c) 3   (d) 1.5   (e) 2.5

e 50. What is the bond order for B_2^{2+}?

(a) 0.5   (b) 1   (c) 1.5   (d) 2   (e) zero

a 51. What is the bond order for O_2^{2-}?

(a) 1   (b) 1.5   (c) 2   (d) 2.5   (e) zero

d 52. What is the bond order for O_2^{2+}?

(a) 1   (b) 2   (c) 2.5   (d) 3   (e) zero

b 53. What is the bond order for B_2^-?

(a) 1   (b) 1.5   (c) 2   (d) 2.5   (e) 3

c 54. Which of the following diatomic species has the highest bond order?

(a) B_2^+ (b) C_2^- (c) N_2 (d) O_2 (e) F_2

d 55. Which of the following diatomic species has the highest bond order?

(a) B_2^- (b) O_2^{2-} (c) O_2^- (d) O_2^{2+} (e) B_2^{2+}

a 56. Which of the following diatomic species has the lowest bond order?

(a) B_2^+ (b) C_2^- (c) N_2 (d) O_2 (e) F_2

d 57. What is the bond order for each of the following species and which one would be predicted to be **most** stable?

H_2^+, H_2, H_2^-

	bond order			most stable
	H_2^+	H_2	H_2^-	
(a)	0	1	0	H_2
(b)	0.5	1	1.5	H_2^-
(c)	0	1	2	H_2^-
(d)	0.5	1	0.5	H_2
(e)	1.5	1	0.5	H_2^+

b 58. What is the bond order for each of the following species and which one would be predicted to have the **shortest** bond length?

N_2, N_2^-, N_2^+

	bond order			shortest bond length
	N_2	N_2^-	N_2^+	
(a)	3	3.5	2.5	N_2^+
(b)	3	2.5	2.5	N_2
(c)	3	4	2	N_2^-
(d)	2	3	1	N_2^-
(e)	3	4	2	N_2^-

c 59. What is the bond order for each of the following species and which is(are) **paramagnetic**?

B_2, B_2^+, B_2^-

	bond order			paramagnetic
	B_2	B_2^+	B_2^-	
(a)	1	0.5	1.5	none
(b)	1	0.5	0.5	B_2^+
(c)	1	0.5	1.5	all three
(d)	1	1.5	1.5	B_2^-
(e)	2	3	1	B_2

e 60. What is the bond order for each of the following species and which one would be predicted to have the **largest** bond energy?

F_2, F_2^+, F_2^-

	bond order			largest bond energy
	F_2	F_2^+	F_2^-	
(a)	1	1.5	0.5	F_2^-
(b)	1	0.5	1.5	F_2^-
(c)	1	0.5	0.5	F_2
(d)	2	3	1	F_2^+
(e)	1	1.5	0.5	F_2^+

e 61. Which of the following statements concerning homonuclear diatomic molecules of the third and subsequent periods is **false**?

 (a) The heavier halogens, Cl_2, Br_2, and I_2, which contain only sigma (single) bonds, are the only well-characterized examples at room temperature.
 (b) Other homonuclear diatomic species (other than the halogens) are neither common nor very stable.
 (c) The heavier elements are unstable because they cannot form strong pi bonds with each other.
 (d) For these larger atoms the sigma bond length is too great to allow the atomic *p* orbitals on different atoms to overlap side-on-side effectively.
 (e) All of the above statements explain why P_2 is much more stable than N_2.

Heteronuclear Diatomic Molecules

d 62. Which statement regarding stable **heteronuclear** diatomic molecules is false?

 (a) The bonding molecular orbitals have more of the character of the more electronegative element than of the less electronegative element.
 (b) The antibonding molecular orbitals have more of the character of the more electropositive element than of the more electronegative element.
 (c) All have bond orders greater than zero.
 (d) Their molecular orbital diagrams are more symmetrical than those of homonuclear diatomic molecules.
 (e) The greater the difference in energy between two overlapping atomic orbitals, the more polar is the bond resulting from the electrons occupying that bonding molecular orbital.

e 63. Which of the following statements about nitrogen oxide, NO, is **false**?

 (a) Nitrogen oxide is isoelectronic with the N_2^- ion.
 (b) Nitrogen oxide has a bond order of 2.5.
 (c) The bonding MOs of NO have more oxygen-like atomic orbital character.
 (d) The antibonding MOs of NO have more nitrogen-like atomic orbital character.
 (e) Nitrogen oxide is a very unstable molecule.

a 64. Draw the molecular orbital diagram for NO. The total number of electrons in the π_{2p}^* molecular orbitals is _____.

 (a) 1 (b) 2 (c) 3 (d) 4 (e) zero

e 65. What is the bond order for NO+?

 (a) 1 (b) 1.5 (c) 2 (d) 2.5 (e) 3

c 66. What is the bond order for NO−?

 (a) 1 (b) 1.5 (c) 2 (d) 2.5 (e) 3

a *67. What is the bond order for HF−?

 (a) 0.5 (b) 1 (c) 1.5 (d) 2 (e) 2.5

Harcourt, Inc

Delocalization and the Shapes of Molecular Orbitals

e 68. Which statement concerning the carbonate ion, CO_3^{2-}, is **false**?

    (a) Valence bond theory describes the ion in terms of 3 contributing resonance structures.
    (b) All of the carbon-oxygen bonds in the ion have the same bond length and bond energy.
    (c) The carbon-oxygen bonds in the ion are intermediate in length and energy between those of C–O and C=O bonds.
    (d) The bonding in the isoelectronic nitrate ion, NO_3^-, can be described in a manner similar to the carbonate ion.
    (e) MO terminology describes the formation of a delocalized bonding pi molecular orbital system extending above the plane of the sigma system of the carbonate ion and an antibonding pi molecular orbital system extending below the plane of the sigma system.

a 69. Which statement concerning the benzene molecule, C_6H_6, is **false**?

    (a) Valence bond theory describes the molecule in terms of 3 resonance structures.
    (b) All six of the carbon–carbon bonds have the same length.
    (c) The carbon–carbon bond lengths are intermediate between those for single and double bonds.
    (d) The entire benzene molecule is planar.
    (e) The valence bond description involves sp^2 hybridization at each carbon atom.

10 Reactions in Aqueous Solutions I: Acids, Bases, and Salts

Properties of Aqueous Solutions of Acids and Bases

a 1. Which of the following is **not** a common property of most protonic acids?

    (a) They have a bitter taste.
    (b) They change the colors of many indicators (highly colored dyes).
    (c) Nonoxidizing acids react with metals more active than hydrogen to liberate H_2.
    (d) They react with metal hydroxides and metal oxides to form salts and water.
    (e) They react with salts of more volatile acids to form the more volatile acid and a new salt.

c 2. The following describe properties of substances. Which one is **not** a property of acids?

    (a) They have a sour taste.
    (b) They react with metal oxides to form salts and water.
    (c) They react with other acids to form salts and water.
    (d) Their aqueous solutions conduct an electric current.
    (e) They react with active metals to liberate H_2.

d 3. Which of the following is **not** a common property of aqueous solutions of most bases?

    (a) They have a bitter taste.
    (b) They have a slippery feeling.
    (c) They change the colors of many indicators.
    (d) They react with protonic acids to form salts and stronger bases.
    (e) Their aqueous solutions conduct an electric current.

b 4. Which of these common substances is **not** acidic?

    (a) vinegar   (b) Drano®   (c) "digestive juice"
    (d) human blood   (e) car battery solution

The Arrhenius Theory

c 5. Which statement is **not** consistent with the Arrhenius theory of acid-base reactions?

    (a) An acid is a substance that contains hydrogen and produces H+ in aqueous solution.
    (b) A base is a substance that contains the OH group and produces OH- in aqueous solution.
    (c) Ammonia is classified as a base.
    (d) Neutralization is the combination of H+ ions with OH- ions to form H_2O.
    (e) The Arrhenius theory satisfactorily explained the reactions of protonic acids with metal hydroxides.

c 6. According to the Arrhenius theory, which of the following is an acid?

    (a) NH_3 (b) CH_4 (c) HCl   (d) H_2 (e) NaOH a 7. According to the Arrhenius theory, which of the following is a base?

(a) KOH (b) H_2O_2 (c) NaCl (d) HClO (e) CH_3COOH

The Hydronium Ion (Hydrated Hydrogen Ion)

c 8. Which statement concerning the hydrated hydrogen ion is **incorrect**?

(a) Although we often represent hydrogen ions as bare protons (H^+), we know they are hydrated in aqueous solutions.
(b) It is now known that hydrogen ions exist as $H^+(H_2O)_n$ where n is a small integer.
(c) The value of n for $H^+(H_2O)_n$ can be calculated for almost all solutions.
(d) The hydrated hydrogen ion is the species that gives aqueous solutions of acids their characteristic acidic properties.
(e) The hydrogen ion is often represented in equations by $H^+(aq)$, H_3O^+, or even H^+.

The Brønsted-Lowry Theory

b 9. According to the Brønsted-Lowry theory, a base is defined as _____.

(a) an electron pair acceptor   (b) a proton acceptor
(c) an electron pair donor   (d) a proton donor
(e) any species that can produce hydroxide ions in aqueous solution

a 10. A Brønsted-Lowry acid is defined as a _____.

(a) species that donates a proton   (b) species that accepts a proton
(c) species that accepts a share in an electron pair
(d) species that makes available a share in an electron pair
(e) species that produces hydroxide ions in aqueous solution

d 11. Which of the following can be a Brønsted-Lowry acid?

(a) $AlCl_3$ (b) CH_4 (c) NaOH (d) H_2O (e) H_2

e 12. Which one of the following could **not** be a Brønsted-Lowry acid?

(a) H_2O (b) HN_3 (c) H_3O^+ (d) NH_4^+ (e) BF_3

c 13. Which one of the following species could **not** react as a Brønsted-Lowry acid?

(a) HCl (b) H_2O (c) CaO (d) NH_3 (e) CH_3COOH

a 14. Which one of the following species could **not** react as a Brønsted-Lowry base?

(a) NH_4^+ (b) NH_2^- (c) CN^- (d) F^- (e) CH_3COO^-

e 15. Which **cannot** be both a Brønsted-Lowry acid and a Brønsted-Lowry base?

(a) H_2O (b) HCO_3^- (c) $H_2PO_4^-$ (d) HPO_4^{2-} (e) PO_4^{3-}

c 16. In the equation
$$HF + H_2O \rightleftharpoons H_3O^+ + F^-$$

(a) H_2O is a base, and HF is its conjugate acid.
(b) H_2O is an acid, and HF is the conjugate base.
(c) HF is an acid, and F^- is its conjugate base.
(d) HF is a base, and H_3O^+ is its conjugate acid.
(e) HF is a base, and F^- is its conjugate acid.

c 17. Which one of the following statements is **true**?
$$CH_3NH_2 + H_2O \rightleftharpoons CH_3NH_3^+ + OH^-$$
methylamine   water   methylammonium ion   hydroxide ion

(a) CH_3NH_2 is the conjugate base of H_2O.
(b) $CH_3NH_3^+$ is the conjugate base of CH_3NH_2.
(c) H_2O is the conjugate acid of OH^-.
(d) OH^- is the conjugate acid of H_2O.
(e) There are no conjugate acid-base pairs.

b 18. Which of the following does **not** represent a conjugate acid-base pair?

(a) H_2O/OH^- (b) H_3O^+/OH^- (c) HCl/Cl^-
(d) HNO_3/NO_3^- (e) NH_4^+/NH_3

d 19. Which one of the following pairs of acids and conjugate bases is incorrectly labeled or **incorrectly** matched?

	Acid	Conjugate Base
(a)	HF	F^-
(b)	HClO	ClO^-
(c)	H_2O	OH^-
(d)	NH_4^+	NH_2^-
(e)	H_3O^+	H_2O

d 20. Which one of the following pairs of acids and conjugate bases is incorrectly labeled or **incorrectly** matched?

	Acid	Conjugate Base
(a)	$HClO_2$	ClO_2^-
(b)	NH_4^+	NH_3
(c)	H_2S	HS^-
(d)	CO_3^{2-}	HCO_3^-
(e)	HS^-	S^{2-}

a 21. The following acids are listed in order of decreasing acid strength in water.
$$HI > HNO_2 > CH_3COOH > HClO > HCN$$
According to the Brønsted-Lowry theory, which one of the following anions is the weakest base?

(a) I^- (b) NO_2^- (c) CH_3COO^- (d) ClO^- (e) CN^-

The Autoionization of Water

d 22. Which statement concerning the autoionization (self-ionization) of water is **false**?

$$2H_2O(\ell) \rightleftharpoons H_3O^+(aq) + OH^-(aq)$$

(a) This reaction is an acid-base reaction according to the Brønsted-Lowry theory.
(b) Water is amphiprotic.
(c) A H_2O molecule may react as an acid by donating a proton.
(d) In this reaction H_3O^+ and OH^- are a conjugate acid-base pair.
(e) A H_2O molecule may react as a base by accepting a proton.

Amphoterism

d 23. Which of the following compounds will dissolve in (by reaction with) hydrochloric acid and will also dissolve in (by reaction with) an excess of 6 M NaOH solution?

(a) KOH   (b) $Mg(OH)_2$ (c) $Ba(OH)_2$
(d) $Sn(OH)_2$ (e) $Sr(OH)_2$

c 24. Which one of the following is an amphoteric metal hydroxide?

(a) KOH   (b) $Ba(OH)_2$ (c) $Be(OH)_2$
(d) LiOH   (e) $Mg(OH)_2$

e 25. Which one of the following hydroxides is amphoteric?

(a) $Ca(OH)_2$ (b) $Ba(OH)_2$ (c) $Fe(OH)_3$
(d) NaOH   (e) $Zn(OH)_2$

a *26. Which of the following is **not** an amphoteric metal hydroxide?

(a) $La(OH)_3$ (b) $Be(OH)_2$ (c) $Cr(OH)_3$
(d) $Al(OH)_3$ (e) $Zn(OH)_2$

d 27. Which one of the following hydroxides is **not** amphoteric?

(a) $Cr(OH)_3$ (b) $Sn(OH)_2$ (c) $Zn(OH)_2$
(d) $Ca(OH)_2$ (e) $Al(OH)_3$

d 28. Which response lists all the following compounds that are amphoteric and no others?

I. RbOH   II. $Fe(OH)_3$ III. $Al(OH)_3$ IV. $Be(OH)_2$ V. $Mn(OH)_2$

(a) I, II, and V   (b) II, III, IV, and V   (c) II and V
(d) III and IV   (e) IV and V

d 29. Which one of the following cations forms an amphoteric hydroxide?

(a) Na^+ (b) Fe^{3+} (c) Ba^{2+} (d) Zn^{2+} (e) Ag^+

d 30. Write the **net ionic** equation for the reaction of aluminum hydroxide with an excess of concentrated sodium hydroxide solution. The complex ion that is formed in this reaction has _____ hydroxide ions bonded to each aluminum, and the charge on the complex ion is _____.

(a) one, 2+   (b) two, 1+   (c) three, 0   (d) four, 1–   (e) four, 4+

e 31. Write the **net ionic** equation for the reaction of beryllium hydroxide with an excess of concentrated sodium hydroxide solution. The complex ion that is formed in this reaction has _____ hydroxide ions bonded to each beryllium, and the charge on the complex ion is _____.

(a) one, 2+   (b) two, 1+   (c) three, 0   (d) four, 1–   (e) four, 2–

a 32. Which of the following oxides is amphoteric?

(a) BeO   (b) MgO   (c) CaO   (d) SrO   (e) BaO

Strengths of Acids

b 33. Which one of the following is a weak acid?

(a) H_2SO_4 (b) HF   (c) HCl   (d) HBr   (e) HI

b 34. Which one of the following is a weak acid?

(a) HNO_3 (b) H_3PO_4 (c) $HClO_3$ (d) $HClO_4$ (e) HI

a 35. Which one of the following is a strong acid?

(a) HI   (b) HF   (c) HCN   (d) HNO_2 (e) CH_3COOH

d 36. Which of the following is the strongest acid?

(a) HF   (b) HCl   (c) HBr   (d) HI   (e) CH_3COOH

d 37. Which of the following is the strongest acid?

(a) HClO   (b) $HClO_2$ (c) $HClO_3$ (d) $HClO_4$ (e) HF

b 38. Which of the following is the weakest acid?

(a) HCl   (b) HClO   (c) $HClO_2$ (d) $HClO_3$ (e) $HClO_4$

c 39. Which response includes all the acids listed below that are strong acids, and no weak acids?

H_2SO_4, HI, HF, H_3PO_4, HNO_3

(a) HI, HF, HNO_3
(b) H_2SO_4, HNO_3
(c) H_2SO_4, HI, HNO_3
(d) H_2SO_4, H_3PO_4, HNO_3
(e) another one or another combination

d 40. Which response includes all the following compounds that are weak acids, and no strong acids?

H_2SO_4, HNO_2, H_2SO_3, HF, HI

(a) H_2SO_4, HI
(b) H_2SO_4, H_2SO_3
(c) HNO_2, HF
(d) HNO_2, H_2SO_3, HF
(e) HNO_2, H_2SO_3

c 41. Which response includes all of the following that are strong acids, and no others?

I. $HClO_4$ II. HNO_2 III. H_3AsO_4 IV. HCl   V. H_3PO_4

(a) I, III, IV, and V
(b) I and V
(c) I and IV
(d) II and III
(e) IV and V

e 42. Which response includes all of the acids listed below that are strong acids, and no others?

I. HI   II. HF   III. H_2SO_3 IV. HNO_2 V. H_3PO_4 VI. HNO_3

(a) I, II, and V
(b) II, IV, and V
(c) I, III, IV, and VI
(d) II and IV
(e) I and VI

e 43. Which response includes all the weak acids listed below, and no strong acids?

I. HF   II. HI   III. HNO_3 IV. HBrO   V. $HClO_4$

(a) II and III
(b) I, II, and IV
(c) IV and V
(d) II, III, and IV
(e) I and IV

d 44. Arrange the following in order of increasing acid strength.

H_2O, H_2Se, H_2Te, H_2S

Increasing Strength $\longrightarrow$

(a) $H_2O < H_2Te < H_2Se < H_2S$
(b) $H_2Se < H_2Te < H_2O < H_2S$
(c) $H_2S < H_2Se < H_2Te < H_2O$
(d) $H_2O < H_2S < H_2Se < H_2Te$
(e) $H_2Se < H_2Te < H_2S < H_2O$

a 45. Which one of the following pairs of acids is **incorrectly** listed?

	Stronger Acid	Weaker Acid
(a)	HClO	$HClO_2$
(b)	HNO_3	HNO_2
(c)	H_2SO_4	H_2SO_3
(d)	H_2SO_3	H_2SeO_3
(e)	$HClO_4$	HIO_4

Harcourt, Inc

d 46. Four of the following pairs of acids list the weaker acid at the left and the stronger acid at the right. Which pair is reversed?

	Weaker Acid	Stronger Acid
(a)	HIO_3	$HClO_3$
(b)	HNO_2	HNO_3
(c)	H_3PO_4	HNO_3
(d)	H_3PO_4	H_3AsO_4
(e)	$HBrO_3$	$HClO_3$

e 47. Which indication of relative acid strength is **incorrect**?

(a) $HNO_3 > HNO_2$ (b) $HI > HF$ (c) $H_2PO_4^- > HPO_4^{2-}$
(d) $HClO_3 > HBrO_3$ (e) $HClO > HCl$

c 48. Which of these anions is the strongest base, according to the Brønsted-Lowry theory?

(a) Cl^- (b) NO_3^- (c) F^- (d) I^- (e) ClO_3^-

c 49. According to the Brønsted-Lowry theory, which of these anions is the strongest base?

(a) NO_3^- (b) Cl^- (c) CN^- (d) ClO_4^- (e) HSO_4^-

a 50. According to the Brønsted-Lowry theory, which of these anions is the weakest base?

(a) Br^- (b) ClO^- (c) NO_2^- (d) F^- (e) CH_3COO^-

c *51. According to the Brønsted-Lowry theory, which of these acids would have the weakest conjugate base?

(a) $HBrO_3$ (b) HNO_2 (c) $HClO_4$ (d) $HClO_2$ (e) HF

Acid–Base Reactions in Aqueous Solutions

d 52. Which one of the following is a soluble, strong base?

(a) $Cu(OH)_2$ (b) $Fe(OH)_3$ (c) $Al(OH)_3$ (d) $Sr(OH)_2$ (e) $Mg(OH)_2$

a 53. Which one of the following is a soluble, strong base?

(a) $CsOH$ (b) $Cu(OH)_2$ (c) $Fe(OH)_3$ (d) $Mn(OH)_2$ (e) $Al(OH)_3$

c 54. Which one of the following compounds is **not** a soluble, strong base?

(a) $Ba(OH)_2$ (b) $Ca(OH)_2$ (c) $Cu(OH)_2$ (d) $RbOH$ (e) KOH

d 55. Which one of the following is **not** a soluble, strong base?

(a) $NaOH$ (b) KOH (c) $RbOH$ (d) $Fe(OH)_3$ (e) $Ba(OH)_2$

d 56. Which one of the following is **not** a soluble, strong base?

(a) KOH   (b) CsOH   (c) NH$_3$ (d) Ca(OH)$_2$ (e) Sr(OH)$_2$

d 57. Which response includes all of the following that are insoluble bases, and no others?

I. LiOH   II. KOH   III. Ni(OH)$_2$ IV. Mg(OH)$_2$ V. Ca(OH)$_2$

(a) IV   (b) IV and V   (c) I and III   (d) III and IV   (e) I, IV, and V

a 58. Which response includes all of the following that are insoluble bases, and no others?

I. Co(OH)$_2$ II. Ba(OH)$_2$ III. Al(OH)$_3$ IV. LiOH   V. Zn(OH)$_2$

(a) I, III, and V   (b) I, II, and IV   (c) II, III, and IV
(d) IV and V   (e) I and III

e 59. Which response includes all of the following that are insoluble bases, and no others?

LiOH, Cu(OH)$_2$, Be(OH)$_2$, Ba(OH)$_2$, KOH

(a) LiOH, Ba(OH)$_2$ (b) Cu(OH)$_2$, KOH, Ba(OH)$_2$
(c) LiOH, Be(OH)$_2$ (d) Cu(OH)$_2$, Ba(OH)$_2$
(e) another one or another combination

e 60. Which one of the following compounds is **incorrectly** classified?

	Compound	Classification
(a)	(C$_2$H$_5$)$_2$NH	weak base
(b)	HCN	weak acid
(c)	Sr(OH)$_2$	soluble, strong base
(d)	Cr(OH)$_3$	insoluble base
(e)	Fe(OH)$_2$	soluble base

b 61. Which one of the following compounds is **not** a salt?

(a) Fe(NO$_3$)$_3$ (b) Cr(OH)$_3$ (c) KMnO$_4$ (d) Na$_2$SO$_3$ (e) Ca(ClO$_4$)$_2$

b 62. Which one of the following compounds is **not** a salt?

(a) K$_2$SO$_4$ (b) HClO$_4$ (c) CaCrO$_4$ (d) CoCl$_2$ (e) NaMnO$_4$

a 63. Which one of the following salts is **insoluble** in water?

(a) MgCO$_3$ (b) (NH$_4$)$_2$SO$_4$ (c) KCH$_3$COO
(d) Cu(NO$_3$)$_2$ (e) Fe$_2$(SO$_4$)$_3$

d 64. Which one of the following salts is **insoluble** in water?

(a) AgNO$_3$ (b) KI   (c) NH$_4$Cl   (d) PbS   (e) NaBr

a 65. Write the balanced **formula unit** equation for the reaction of hydrobromic acid with calcium hydroxide. What is the sum of the coefficients? (Do not forget coefficients of one.)

(a) 6   (b) 7   (c) 3   (d) 4   (e) 5

a 66. Write the balanced **formula unit** equation for the complete neutralization of dilute sulfuric acid with potassium hydroxide in aqueous solution. What is the sum of the coefficients? (Do not forget coefficients of one.)

(a) 6   (b) 8   (c) 9   (d) 4   (e) 5

a 67. Write the balanced **formula unit** equation for the reaction of iron(II) hydroxide with hydrochloric acid. What is the sum of the coefficients? (Do not forget coefficients of one.)

(a) 6   (b) 7   (c) 8   (d) 4   (e) 5

a *68. Write the balanced **net ionic** equation for the complete neutralization of chromium(III) hydroxide with sulfuric acid. Use H^+ rather than H_3O^+. What is the sum of the coefficients? (Do not forget coefficients of one.)

(a) 8   (b) 11   (c) 16   (d) 18   (e) 22

c 69. Write the **net ionic** equation for the complete neutralization of HBr by $Ca(OH)_2$. Use H^+ rather than H_3O^+. What is the sum of the coefficients? (Do not forget coefficients of one.)

(a) 6   (b) 8   (c) 3   (d) 4   (e) 5

b 70. Consider the complete neutralization reactions between the following acid-base pairs in dilute aqueous solutions. In which of these have strong acids listed as reactants?

I. $CH_3COOH + NaOH \rightarrow$
II. $HNO_3 + Mg(OH)_2 \rightarrow$
III. $H_3PO_4 + Ca(OH)_2 \rightarrow$
IV. $HCl + KOH \rightarrow$
V. $H_2CO_3 + Sr(OH)_2 \rightarrow$

(a) I and V
(b) II and IV
(c) II, III, and IV
(d) II, III, and V
(e) III, IV, and V

e 71. Consider the complete neutralization reactions between the following acid-base pairs in dilute aqueous solutions. In which of these are soluble, strong bases listed as reactants?

I. $CH_3COOH + NaOH \rightarrow$
II. $HNO_3 + Mg(OH)_2 \rightarrow$
III. $H_3PO_4 + Ca(OH)_2 \rightarrow$
IV. $HCl + KOH \rightarrow$
V. $H_2CO_3 + Sr(OH)_2 \rightarrow$

(a) I, IV, and V
(b) II and V
(c) I and V
(d) II, III, and IV
(e) I, III, IV, and V

d 72. Consider the complete neutralization reactions between the following acid-base pairs in dilute aqueous solutions. Which of the reactions produce insoluble salts?

I. $CH_3COOH + NaOH \rightarrow$
II. $HNO_3 + Mg(OH)_2 \rightarrow$
III. $H_3PO_4 + Ca(OH)_2 \rightarrow$
IV. $HCl + KOH \rightarrow$
V. $H_2CO_3 + Sr(OH)_2 \rightarrow$

(a) I, II, and III (b) II and IV (c) I, IV, and V
(d) III and V (e) III

b 73. For the complete neutralization reactions between the following pairs of acid and bases, which pair is a soluble, strong base and a weak acid that react to form a soluble salt?

(a) $Ca(OH)_2$, HNO_3 (b) KOH, H_3PO_4 (c) NH_3, HF
(d) $Ba(OH)_2$, H_2CO_3 (e) $Fe(OH)_2$, H_2S

d 74. For the complete neutralization reactions between the following pairs of acid and bases, which pair is a soluble, strong base and a weak acid that react to form an insoluble salt?

(a) $Ca(OH)_2$, HNO_3 (b) KOH, H_3PO_4 (c) NH_3, HF
(d) $Ba(OH)_2$, H_2CO_3 (e) $Fe(OH)_2$, H_2S

e 75. For the complete neutralization reactions between the following pairs of acid and bases, which pair is an insoluble base and a weak acid that react to form an insoluble salt?

(a) $Ca(OH)_2$, HNO_3 (b) KOH, H_3PO_4 (c) NH_3, HF
(d) $Ba(OH)_2$, H_2CO_3 (e) $Pb(OH)_2$, H_2S

a 76. For which one of the pairs of acids and bases does the following represent the **net ionic** equation?

$$H_3O^+ + OH^- \rightarrow 2H_2O$$

(a) NaOH, HBr (b) CH_3COOH, $Ca(OH)_2$ (c) HNO_3, $Mg(OH)_2$
(d) CH_3COOH, $Cu(OH)_2$ (e) CH_3COOH, $Cr(OH)_3$

b 77. For which one of the pairs of acid and bases does
$$CH_3COOH + OH^- \rightarrow CH_3COO^- + H_2O$$
represent the correct **net ionic** equation?

(a) NaOH, HBr (b) CH_3COOH, $Ca(OH)_2$ (c) HNO_3, $Mg(OH)_2$
(d) CH_3COOH, $Cu(OH)_2$ (e) CH_3COOH, $Cr(OH)_3$

c 78. Which pair of acids would **each** react with barium hydroxide and have the **net ionic** equation:

$$H^+ + OH^- \rightarrow H_2O$$

(a) HCl and HNO_2 (b) HBr and HF (c) HCl and HNO_3
(d) HNO_3 and H_2CO_3 (e) $HClO_4$ and H_2SO_4

d 79. Consider the complete neutralization reactions between the following acid-base pairs in dilute aqueous solutions. For which of the reactions is the net ionic equation the following?

$$H^+ + OH^- \rightarrow H_2O$$

I. $CH_3COOH + NaOH \rightarrow$
II. $HNO_3 + Mg(OH)_2 \rightarrow$
III. $H_3PO_4 + Ca(OH)_2 \rightarrow$
IV. $HCl + KOH \rightarrow$
V. $H_2CO_3 + Sr(OH)_2 \rightarrow$

(a) I and III   (b) I   (c) II and III   (d) IV   (e) I, IV, and V

a 80. Consider the following pairs of acids and bases:

I. HBr, KOH   II. HF, TlOH   III. HNO_2, NaOH   IV. $HClO_4$, LiOH

If net ionic equations are written for reactions between each pair of acids and bases, for which two pairs will the net ionic equations be identical?

(a) I and IV   (b) II and III   (c) II and IV   (d) I and II   (e) III and IV

b 81. Which response indicates all of the reactants listed below that react according to identical **net ionic** equations?

I. $HClO_4$ with KOH
II. H_2CO_3 with $Ca(OH)_2$
III. HBr with $Ca(OH)_2$
IV. CH_3COOH with $Ba(OH)_2$
V. HNO_3 with $Ba(OH)_2$

(a) I and III   (b) I, III, and V   (c) II, III, and V
(d) I, II, and IV   (e) another one or another combination

Acidic Salts and Basic Salts

d 82. Which response gives all of the members of the following list that are **acidic** salts and none that are not acidic salts?

KH_2PO_4, K_2HPO_4, K_3PO_4, $NaHCO_3$, LiOH, LiH

(a) KH_2PO_4, K_2HPO_4, LiH
(b) KH_2PO_4, $NaHCO_3$, LiOH
(c) K_3PO_4, LiH
(d) KH_2PO_4, K_2HPO_4, $NaHCO_3$
(e) KH_2PO_4, K_2HPO_4, $NaHCO_3$, LiH

b 83. Which response gives all of the members of the following list that are **basic** salts and none that are not basic salts?

$Al(OH)Cl_2$, $Al(OH)_2Cl$, $AlCl_3$, $Ca(OH)Cl$, $CaCl_2 \cdot 2H_2O$

(a) $Al(OH)Cl_2$, $Al(OH)_2Cl$, $AlCl_3$
(b) $Al(OH)Cl_2$, $Al(OH)_2Cl$, $Ca(OH)Cl$
(c) $Al(OH)_2Cl$, $CaCl_2 \cdot 2H_2O$
(d) $Al(OH)Cl_2$, $Ca(OH)Cl$
(e) $Al(OH)Cl_2$, $Al(OH)_2Cl$, $Ca(OH)Cl$, $CaCl_2 \cdot 2H_2O$

The Lewis Theory

c 84. According to the Lewis theory, a base _____.

(a) is a proton acceptor
(b) is a proton donor
(c) makes available a share in a pair of electrons
(d) is any compound that contains electron pairs
(e) accepts a share in a pair of electrons

d 85. Neutralization, according to the Lewis theory, involves _____.

(a) proton transfer
(b) the formation of a gas
(c) the formation of an ionic solid
(d) the formation of a coordinate covalent bond
(e) the combination of a hydrogen ion with a hydroxide ion to form water

a 86. According to the Lewis theory, which one of the following compounds would be expected to react as a Lewis acid, but not as a Lewis base?

(a) BCl_3 (b) NH_3 (c) PH_3 (d) H_2O (e) all of these

b 87. According to the Lewis theory, which one of the following species would be expected to react as a Lewis base, but not as a Lewis acid?

(a) BF_3 (b) F^- (c) $AlBr_3$ (d) $SnCl_4$ (e) H^+

d 88. In the reaction $BF_3 + NH_3 \rightarrow F_3B:NH_3$, the BF_3 functions as a(an) _____.

(a) Brønsted-Lowry acid (b) Brønsted-Lowry base (c) Arrhenius base
(d) Lewis acid   (e) Lewis base

e 89. In the reaction $AlCl_3 + Cl^- \rightarrow AlCl_4^-$, the Cl^- functions as a(an) _____.

(a) Brønsted-Lowry acid (b) Brønsted-Lowry base (c) Arrhenius base
(d) Lewis acid   (e) Lewis base

d 90. In the reaction $SnCl_4 + 2Cl^- \rightarrow SnCl_6^{2-}$, the $SnCl_4$ functions as a(an) _____.

(a) Brønsted-Lowry acid (b) Brønsted-Lowry base (c) Arrhenius base
(d) Lewis acid   (e) Lewis base

a 91. Which one of the following reactions can be classified as a Lewis acid-base reaction but not as a Brønsted-Lowry acid-base reaction?

(a) $Al(OH)_3(s) + OH^-(aq) \rightarrow Al(OH)_4^-(aq)$
(b) $NH_3(g) + H_2O(\ell) \rightarrow NH_4^+(aq) + OH^-(aq)$
(c) $CH_3COO^-(aq) + H_2O(\ell) \rightarrow CH_3COOH(aq) + OH^-(aq)$
(d) $H^+(g) + H^-(g) \rightarrow H_2(g)$
(e) $Ca(OH)_2(s) + 2HCl(g) \rightarrow CaCl_2(s) + 2H_2O(\ell)$

e 92. Gaseous methylamine, CH_3NH_2, can be classified as _____.

(a) a Lewis base only
(b) Arrhenius and Brønsted-Lowry bases
(c) a Brønsted-Lowry base only
(d) Arrhenius and Lewis bases
(e) Brønsted-Lowry and Lewis bases

d 93. In this reaction CH_3NH_2 can be classified as _____.

$$CH_3NH_2 + H_2O \rightleftharpoons CH_3NH_3^+ + OH^-$$
methylamine   water   methylammonium ion   hydroxide ion

(a) only an Arrhenius base
(b) only a Lewis base
(c) only a Brønsted-Lowry base
(d) a Brønsted-Lowry base and a Lewis base
(e) Arrhenius, Brønsted-Lowry, and Lewis bases

e 94. Based on the reactions we have studied, ammonia can be considered as _____.

(a) an Arrhenius base (only)
(b) a Lewis base (only)
(c) a Brønsted-Lowry base (only)
(d) both an Arrhenius base and a Lewis base
(e) both a Brønsted-Lowry base and a Lewis base

a 95. In aqueous solution, which of the following would be an acid according to the definitions of the Arrhenius, the Brønsted-Lowry, and the Lewis theories? Stated differently, which of the following behaves as an acid when dissolved in water according to all three of these theories?

(a) HCl   (b) $AlCl_3$ (c) BF_3 (d) $NaNH_2$ (e) CN^-

The Preparation of Acids

b 96. Which pair of reactants can be used in the preparation of hydrogen bromide, HBr, a volatile acid?

(a) H_2SO_4 and NaBr
(b) H_3PO_4 and NaBr
(c) $HBrO_4$ and NaOH
(d) $HBrO_3$ and NaOH
(e) $NaBrO_3$ and H_2O

d 97. Which ternary acid is incorrectly matched with its anhydride?

	Acid	Anhydride
(a)	H_2CO_3	CO_2
(b)	HNO_2	N_2O_3
(c)	H_3PO_4	P_4O_{10}
(d)	H_2SO_3	SO_3
(e)	$HClO_4$	Cl_2O_7

b 98. Write the balanced **formula unit** equation for the reaction of dinitrogen pentoxide, N_2O_5, with water to form an acid. What is the sum of the coefficients in the balanced equation? (Use the smallest whole numbers possible and don't forget coefficients of one.)

(a) 3   (b) 4   (c) 6   (d) 8   (e) 10

d *99. Write the balanced **formula unit** equation for the reaction of tetraphosphorus decoxide, P_4O_{10}, with water to form an acid. What is the sum of the coefficients in the balanced equation? (Use the smallest whole numbers possible and don't forget coefficients of one.)

(a) 8   (b) 9   (c) 10   (d) 11   (e) 12

b 100. The reaction of PCl_5 with water produces two acids, __a__ and __b__.

	a	b
(a)	H_3PO_3	HCl
(b)	H_3PO_4	HCl
(c)	H_3PO_3	HClO
(d)	H_3PO_4	HClO
(e)	H_3PO_3	$HClO_2$

c 101. The products of the reaction of PCl_3 with water are _____ and _____.

(a) PCl_5, HCl   (b) HPO_2, HCl   (c) H_3PO_3, HCl
(d) PH_3, HClO   (e) H_3PO_4, HCl

b*102. The products of the reaction of $AsCl_3$ with water are _____ and _____.

(a) $AsCl_5$, HCl   (b) H_3AsO_3, HCl   (c) AsH_3, HClO
(d) AsH_3, $HClO_4$ (e) H_3AsO_4, HCl

11 Reactions in Aqueous Solutions II: Calculations

Calculations Involving Molarity

a 1. What volume of 12.6 M HCl must be added to enough water to prepare 5.00 liters of 3.00 M HCl?

(a) 1.19 L (b) 21.0 L (c) 0.840 L (d) 7.56 L (e) 2.14 L

b 2. What volume of 12.0 M HNO₃ is required to prepare 900. mL of 2.0 M HNO₃ solution?

(a) 100. mL (b) 150. mL (c) 200. mL (d) 250. mL (e) 300. mL

b 3. How many grams of Ca(OH)₂ are contained in 1500. mL of 0.0250 M Ca(OH)₂ solution?

(a) 3.17 g (b) 2.78 g (c) 1.85 g (d) 2.34 g (e) 3.17 g

a 4. What mass of Ca(OH)₂ is present in 500. mL of 0.00500 molar Ca(OH)₂?

(a) 0.185 g (b) 0.282 g (c) 0.370 g (d) 0.416 g (e) 0.615 g

b 5. How many grams of AgNO₃ are needed to make 500. mL of 0.200 M solution?

(a) 170. g (b) 17.0 g (c) 85.0 g (d) 34.0 g (e) 85000 g

d 6. What mass of NaOH must be dissolved in water to prepare 1.50 L of 0.500 M solution?

(a) 20.0 g (b) 13.3 g (c) 43.5 g (d) 30.0 g (e) 120. g

e 7. What is the molarity of 1600. mL of a solution that contains 3.25 g of H₃PO₄?

(a) 2.03 x 10⁻³ M (b) 6.90 x 10⁻³ M (c) 6.22 x 10⁻² M
(d) 6.09 x 10⁻³ M (e) 2.07 x 10⁻² M

c 8. Calculate the molarity of a solution that contains 4.5 g of (COOH)₂ in 3000 mL of solution.

(a) 0.033 M (b) 0.045 M (c) 0.017 M (d) 0.090 M (e) 0.12 M

a 9. What is the molarity of 600. mL of solution containing 6.72 grams of sulfuric acid?

(a) 0.114 M (b) 0.288 M (c) 0.342 M (d) 0.180 M (e) 0.360 M

d 10. If 75.0 mL of 0.250 M HNO₃ and 75.0 mL of 0.250 M KOH are mixed, what is the molarity of the salt in the resulting solution?

(a) 0.250 M (b) 0.500 M (c) 0.333 M (d) 0.125 M (e) 0.167 M

c 11. What is the molarity of the NaBr produced when 50.0 mL of 4.80 M NaOH is mixed with 150.0 mL of 1.60 M HBr?

(a) 2.40 M (b) 3.20 M (c) 1.20 M (d) 1.33 M (e) 480 M

d 12. What is the molarity of $Ca(NO_3)_2$ in a solution resulting from mixing 150.0 mL of 0.0200 M HNO_3 with 150.0 mL of 0.0100 M $Ca(OH)_2$?

(a) 0.00670 M (b) 0.00330 M (c) 0.00250 M
(d) 0.00500 M (e) 0.0100 M

c 13. If 400. mL of 0.20 M HCl solution is added to 800. mL of 0.050 M $Ba(OH)_2$ solution, the resulting solution will be _____ in $BaCl_2$.

(a) 0.067 M (b) 0.26 M (c) 0.033 M (d) 0.47 M (e) 0.67 M

c 14. If 200. mL of 0.800 molar $HClO_4$ is added to 400. mL of 0.200 molar $Ba(OH)_2$ solution, the resulting solution will be _____ in $Ba(ClO_4)_2$.

(a) 0.100 M (b) 0.0750 M (c) 0.133 M (d) 0.0670 M (e) 0.0250 M

b 15. What is the molarity of the sodium sulfate in the solution resulting from the addition of 100. mL of 0.0500 M H_2SO_4 to 200. mL of 0.0500 M NaOH?

(a) 0.00833 M (b) 0.0167 M (c) 0.0333 M
(d) 0.0250 M (e) 0.0500 M

a 16. If 200. mL of 2.00 M H_3PO_4 solution is added to 600. mL of 2.00 M NaOH solution, the resulting solution will be _____ molar in Na_3PO_4.

(a) 0.500 (b) 0.667 (c) 0.800 (d) 1.00 (e) 0.100

c 17. If 150.0 mL of 0.100 M HBr and 50.0 mL of 0.100 M KOH solutions are mixed, what are the molarities of the ions in the resulting solution?

(a) 0.0750 M in H^+, 0.0750 M in Br^-, 0.0250 M in K^+
(b) 0.0500 M in H^+, 0.0750 M in Br^-, 0.0500 M in K^+
(c) 0.0500 M in H^+, 0.0750 M in Br^-, 0.0250 M in K^+
(d) 0.0667 M in H^+, 0.100 M in Br^-, 0.0333 M in K^+
(e) 0.0500 M in H^+, 0.100 M in Br^-, 0.100 M in K^+

e 18. What is the molarity of the salt produced in the reaction of 50.0 mL of 0.0400 M HBr with 100.0 mL of 0.0300 M $Ca(OH)_2$?

(a) 0.0150 M (b) 0.075M (c) 0.0200 M
(d) 0.0133 M (e) 0.00667 M

d 19. If 50.0 mL of 0.0400 M HBr is added to 100.0 mL of 0.0300 M $Ba(OH)_2$, what is the molarity of the acid or base that remains unreacted?

(a) 0.0150 M (b) 0.0250M (c) 0.0200 M
(d) 0.0133 M (e) 0.00667 M

a 20. What volume of 0.122 M HCl would react with 26.2 mL of 0.176 M NaOH?

(a) 37.8 mL (b) 18.2 mL (c) 5.56 mL (d) 26.3 mL (e) 75.6 mL

a 21. What volume of 0.0750 M Ba(OH)$_2$ will completely react with 125 mL of 0.0350 M HCl?

(a) 29.2 mL (b) 42.0 mL (c) 230. mL (d) 536 mL (e) 53.6 mL

d 22. What volume of 0.50 M KOH would be required to neutralize completely 500. mL of 0.25 M H$_3$PO$_4$ solution?

(a) 2.5 x 10^2 mL (b) 1.4 x 10^3 mL (c) 83 mL
(d) 7.5 x 10^2 mL (e) 5.2 x 10^2 mL

Titrations

a 23. What is the molarity of a potassium hydroxide solution if 38.65 mL of the KOH solution is required to titrate 25.84 mL of 0.1982 M hydrochloric acid solution?

(a) 0.1325 M (b) 0.2963 M (c) 0.8817 M (d) 0.2648 M (e) 0.01324 M

c 24. What is the molarity of a sulfurous acid solution if 23.7 mL of this H$_2$SO$_3$ solution requires 16.8 mL of 0.296 M NaOH for titration to the equivalence point?

(a) 0.210 M (b) 0.358 M (c) 0.105 M (d) 0.421 M (e) 0.0525 M

b 25. What is the molarity of a barium hydroxide solution if 18.62 mL of this Ba(OH)$_2$ solution requires 35.84 mL of 0.2419 M HCl for titration to the equivalence point?

(a) 0.4656 M (b) 0.2328 M (c) 0.1164 M (d) 0.3492 M (e) 0.6984 M

d 26. What is the molarity of a phosphoric acid solution if 15.66 mL of this H$_3$PO$_4$ solution requires 43.81 mL of 0.2283 M NaOH for titration to the equivalence point?

(a) 1.925 M (b) 0.1597 M (c) 0.3193 M (d) 0.2129 M (e) 0.6387 M

e 27. If 20 mL of 0.010 M H$_3$PO$_4$ solution is completely neutralized by 60.0 mL of Ca(OH)$_2$ solution, what is the molarity of the Ca(OH)$_2$ solution?

(a) 0.010 M (b) 0.200 M (c) 0.030 M (d) 0.040 M (e) 0.0050 M

d 28. Which of the following is **not** a property of an ideal primary standard?

(a) It must not react with or absorb components of the atmosphere.
(b) It must react according to one invariable reaction.
(c) It must have a high percentage purity.
(d) It must have a low formula weight.
(e) It should be nontoxic.

b 29. During a titration the point at which stoichiometrically equivalent amounts of acid and base have reacted is called the _____ point.

(a) indicator (b) equivalence (c) end
(d) standardization (e) primary standard

The Mole Method and Molarity

b 30. Calculate the molarity of an H_2SO_4 solution if 40.0 mL of the H_2SO_4 solution neutralizes 0.212 g of Na_2CO_3.

(a) 0.0250 M   (b) 0.0500 M   (c) 0.0750 M   (d) 0.100 M   (e) 0.0900 M

e 31. Calculate the molarity of an HCl solution if 38.65 mL of this HCl solution is required to neutralize 0.2152 g of Na_2CO_3.

(a) 0.05251 M   (b) 0.02625 M   (c) 0.2104 M
(d) 0.05427 M   (e) 0.1050 M

d 32. Potassium hydrogen phthalate is often used to standardize solutions of strong bases. Highly pure samples of this solid weak acid are readily available. It is stable and easy to use in the laboratory. Calculate the molarity of a sodium hydroxide solution if 23.73 mL are required to titrate 0.5816 g of potassium hydrogen phthalate ($C_8H_5O_4K$).

$$C_8H_5O_4K + NaOH \rightarrow C_8H_4O_4KNa + H_2O$$

(a) 0.1042 M   (b) 0.2208 M   (c) 0.3116 M   (d) 0.1200 M   (e) 0.06825 M

b 33. Calculate the molarity of a NaOH solution if 35.42 mL of this NaOH solution is required to titrate 0.7566 g of KHP, $KC_6H_4(COO)(COOH)$.

(a) 0.05278 M   (b) 0.1046 M   (c) 0.2092 M
(d) 0.3128 M   (e) 0.0105 M

c 34. A 0.742-gram sample of KHP, $KC_6H_4(COO)(COOH)$, reacts with 35.0 mL of $Ba(OH)_2$ solution. What is the molarity of the $Ba(OH)_2$ solution?

(a) 0.127 M   (b) 0.0636 M   (c) 0.0520 M   (d) 0.208 M   (e) 0.104 M

a 35. A 0.5240-g sample of impure solid Na_2CO_3 is neutralized by 43.60 mL of 0.1077 M H_2SO_4. What mass of Na_2CO_3 was contained in the sample? There are no acidic or basic impurities present in the sample.

(a) 0.4977 g   (b) 0.5210 g   (c) 0.5106 g   (d) 0.5048 g   (e) 0.4624 g

c 36. A 0.7860-g sample of impure solid Na_2CO_3 is neutralized by 23.48 mL of 0.1082 M HCl. What percentage purity of Na_2CO_3 was contained in the sample? There are no acidic or basic impurities present in the sample.

(a) 34.24%   (b) 12.16%   (c) 17.13%   (d) 32.32%   (e) 68.52%

b 37. An impure sample of $(COOH)_2$ which weighed 1.432 g was dissolved in water and titrated with standard KOH solution. The titration required 41.66 mL of 0.2008 M KOH solution. Calculate the % $(COOH)_2$ in the sample. There are no acidic or basic impurities present in the sample.

(a) 12.62%   (b) 26.30%   (c) 30.65%   (d) 15.32%   (e) 21.25%

Equivalent Weights and Normality

d 38. How many equivalents weights are there in 150. g of CH_3COOH?

(a) 3.41 eq   (b) 1.25 eq   (c) 0.83 eq   (d) 2.50 eq   (e) 0.62 eq

c 39. How many equivalent weights are there in 20.0 g of $Mg(OH)_2$ assuming the $Mg(OH)_2$ is to be completely neutralized?

(a) 0.343 eq   (b) 2.91 eq   (c) 0.686 eq   (d) 1.46 eq   (e) 1.10 eq

a 40. Calculate the normality of a solution that contains 4.5 g of $(COOH)_2$ in 3000. mL of solution. (Assume the $(COOH)_2$ is to be completely neutralized.)

(a) 0.033 N (b) 0.045 N (c) 0.066 N (d) 0.090 N (e) 0.12 N

c 41. What is the normality of 1600. mL of a solution that contains 3.25 g of H_3PO_4? (Assume that the acid is to be neutralized completely.)

(a) $2.03 \times 10^{-3}\ N$ (b) $6.90 \times 10^{-3}\ N$ (c) $6.22 \times 10^{-2}\ N$
(d) $6.09 \times 10^{-3}\ N$ (e) $2.07 \times 10^{-2}\ N$

a 42. What is the normality of 150. mL of a solution that contains 5.00 grams of H_3PO_4? (Assume the acid is to be completely neutralized.)

(a) 1.02 N (b) 0.870 N (c) 0.340 N (d) 0.113 N (e) 0.667 N

c 43. A solution of H_2SO_4 contains 9.11 mg of H_2SO_4 per mL. What is the normality of the solution? (Assume the H_2SO_4 is to be completely neutralized by a base.)

(a) 0.0482 N (b) 0.684 N (c) 0.186 N (d) 0.372 N (e) 0.0930 N

c 44. What is the normality of a 0.0500 M H_2SO_4 solution? (Assume complete neutralization.)

(a) 0.0125 N (b) 0.0250 N (c) 0.100 N
(d) 0.0175 N (e) 0.0500 N

c 45. How many equivalent weights of phosphoric acid are contained in 500. mL of 2.00 M phosphoric acid? (Assume the acid is to be completely neutralized by a base.)

(a) 1.00 eq   (b) 2.00 eq   (c) 3.00 eq   (d) 4.00 eq   (e) 6.00 eq

a 46. Calculate the volume of 2.00 M H_2SO_4 required to prepare 200. mL of 0.100 N H_2SO_4. (Assume the acid is to be completely neutralized.)

(a) 5.00 mL   (b) 7.50 mL   (c) 10.0 mL   (d) 12.5 mL   (e) 15.0 mL

b 47. What volume of 0.1012 N HCl solution is required to neutralize 25.86 mL of 0.1134 N $Ba(OH)_2$?

(a) 23.08 mL   (b) 28.98 mL   (c) 46.16 mL   (d) 57.96 mL   (e) 14.49 mL

e 48. If 38.46 mL of KOH solution is required to titrate 22.68 mL of 0.1088 N H$_3$PO$_4$ solution, what is the normality of the KOH solution? (Assume the acid is completely neutralized by the base.)

(a) 0.1925 N   (b) 0.1604 N   (c) 0.1283 N   (d) 0.09624 N   (e) 0.06416 N

a 49. What is the normality of 35.0 mL of a solution of Ba(OH)$_2$ that exactly neutralizes 10.0 mL of 0.150 molar H$_2$SO$_4$?

(a) 0.0857 N   (b) 0.105 N   (c) 0.0525 N   (d) 0.214 N   (e) 0.107 N

a 50. If 20.0 mL of 0.010 M H$_3$PO$_4$ solution is just neutralized by 60.0 mL of Ca(OH)$_2$ solution, what is the normality of the Ca(OH)$_2$ solution?

(a) 0.010 N   (b) 0.200 N   (c) 0.030 N   (d) 0.040 N   (e) 0.600 N

a 51. What mass of NaOH would be required to neutralize the acid in 75.0 mL of 0.880 N HCl?

(a) 2.64 g   (b) 5.28 g   (c) 2.6 kg   (d) 3.83 g   (e) 6.60 g

d 52. What mass of NaOH would neutralize all of the acid in 90.0 mL of 0.430 N H$_2$SO$_4$?

(a) 3.10 g   (b) 0.77 g   (c) 2.17 g   (d) 1.55 g   (e) 5.26 g

a 53. If 30.25 mL of HCl solution reacts with 0.2064 g of Na$_2$CO$_3$, what is the normality of the HCl solution?

(a) 0.1287 N   (b) 0.0644 N   (c) 0.4291 N
(d) 0.0858 N   (e) 0.2575 N

a 54. What is the normality of an H$_2$SO$_4$ solution if 40.0 mL of the solution reacts with 0.414 g of Na$_2$CO$_3$?

(a) 0.195 N   (b) 0.250 N   (c) 0.150 N   (d) 0.300 N   (e) 0.350 N

b 55. Calculate the normality of a KOH solution if 20.58 mL of the KOH solution reacts with 0.6128 g of KHP, KC$_6$H$_4$(COO)(COOH).

(a) 0.2916 N   (b) 0.1458 N   (c) 0.2056 N
(d) 0.2542 N   (e) 0.7291 N

e 56. If 20.8 mL of Ba(OH)$_2$ solution reacts with 0.306 g of KHP, KC$_6$H$_4$(COO)(COOH), what is the normality of the Ba(OH)$_2$ solution?

(a) 0.0361 N   (b) 0.144 N   (c) 0.300 N   (d) 0.175 N   (e) 0.0721 N

b 57. What mass of BaSO$_4$ will be produced by the reaction of excess BaCl$_2$ with 50.0 mL of 0.400 N H$_2$SO$_4$ solution?

$$H_2SO_4 + BaCl_2 \rightarrow BaSO_4(s) + 2HCl$$

(a) 9.32 g   (b) 2.33 g   (c) 6.72 g   (d) 4.66 g   (e) 10.4 g

Balancing Redox Reactions

e 58. What is the oxidation number of bromine in KBrO$_3$?

    (a) +1   (b) +7   (c) +3   (d) –1   (e) +5

e 59. What is the oxidation number of phosphorus in KH$_2$PO$_4$?

    (a) +1   (b) +2   (c) +3   (d) +4   (e) +5

d 60. What is the oxidation number of Ce in Ce(SO$_4$)$_2$·4H$_2$O?

    (a) +1   (b) +2   (c) +6   (d) +4   (e) +8

e 61. What is the oxidation number of phosphorus in Mg$_3$(PO$_4$)$_2$?

    (a) +6   (b) +7   (c) +3   (d) +4   (e) +5

d 62. What is the oxidation number of S in Na$_2$SO$_3$·12H$_2$O?

    (a) +6   (b) +2   (c) +3   (d) +4   (e) +5

e 63. What is the oxidation number of P in the H$_2$PO$_4^-$ ion?

    (a) –3   (b) +2   (c) +3   (d) +4   (e) +5

d 64. What is the oxidation number of tin in the HSnO$_3^-$ ion?

    (a) +1   (b) +2   (c) +3   (d) +4   (e) +5

c 65. In which of the following is the oxidation number of the underlined element given **incorrectly**?

	oxidation number
(a) K$_2$<u>Cr</u>$_2$O$_7$	+6
(b) Na<u>Al</u>(OH)$_4$	+3
(c) H<u>I</u>O$_4$	+5
(d) (NH$_4$)$_2$<u>S</u>O$_4$	+6
(e) NaH<u>S</u>O$_3$	+4

e 66. What is the coefficient of NaCl in the balanced formula unit equation?

    NaClO$_3$ + H$_2$O + I$_2$ → HIO$_3$ + NaCl

    (a) 1   (b) 2   (c) 3   (d) 4   (e) 5

e 67. Balance the following equation. How many HCl are there on the left side of the balanced equation?

    K$_2$Cr$_2$O$_7$ + Na$_2$SO$_3$ + HCl → KCl + Na$_2$SO$_4$ + CrCl$_3$ + H$_2$O

    (a) 1   (b) 2   (c) 3   (d) 4   (e) 8

c 68. Balance the following equation for a reaction that occurs in basic medium in a nickel-cadmium alkaline battery. What is the coefficient of H_2O in the balanced equation?

$$H_2O + Cd + Ni_2O_3 \rightarrow Cd(OH)_2 + Ni(OH)_2$$

(a) 1 (b) 2 (c) 3 (d) 4 (e) none of these

c 69. Balance this equation for a reaction in basic solution. What is the coefficient of H_2O?

$$KOH + Cl_2 \rightarrow KClO_3 + KCl + H_2O$$

(a) 8 (b) 2 (c) 3 (d) 4 (e) 6

a 70. Balance the following formula unit equation. What is the sum of all of the coefficients? Don't forget coefficients of one.

$$H_2SO_4 + HI \rightarrow I_2 + SO_2 + H_2O$$

(a) 7 (b) 9 (c) 11 (d) 13 (e) 5

c 71. Balance the following formula unit equation. What is the sum of all of the coefficients?

$$KMnO_4 + H_2O_2 + H_2SO_4 \rightarrow O_2 + MnSO_4 + K_2SO_4 + H_2O$$

(a) 22 (b) 24 (c) 26 (d) 28 (e) 30

d 72. Balance the following formula unit equation for a reaction. What is the sum of all of the coefficients?

$$K_2Cr_2O_7 + HCl \rightarrow KCl + CrCl_3 + Cl_2 + H_2O$$

(a) 20 (b) 22 (c) 26 (d) 29 (e) 32

a 73. Balance the following formula unit equation. What is the sum of all of the coefficients?

$$HNO_3 + ZnS \rightarrow S + NO + Zn(NO_3)_2 + H_2O$$

(a) 23 (b) 25 (c) 27 (d) 29 (e) 31

d 74. Balance the following formula unit equation for a reaction in basic solution. What is the sum of all of the coefficients? Don't forget coefficients of one.

$$KMnO_4 + H_2O + Cu \rightarrow Cu(OH)_2 + MnO_2 + KOH$$

(a) 8 (b) 12 (c) 14 (d) 16 (e) 22

a 75. Balance the following equation. How many H^+ are there in the balanced equation?

$$(COOH)_2 + MnO_4^- + H^+ \rightarrow CO_2 + Mn^{2+} + H_2O$$

(a) 6 (b) 2 (c) 3 (d) 4 (e) 8

a 76. Balance the following net ionic equation. What is the coefficient of H_2O?

$$MnO_4^- + H^+ + Br^- \rightarrow Mn^{2+} + Br_2 + H_2O$$

(a) 8 (b) 9 (c) 3 (d) 4 (e) 6

Chapter 11   165

c 77. Balance the following net ionic equation. What is the coefficient of S?
$$H_2S + NO_3^- + H^+ \rightarrow S + NO + H_2O$$
(a) 1   (b) 2   (c) 3   (d) 4   (e) none of these

d 78. Balancing the following net ionic equation, how many H+ are there?
$$Sb + H^+ + NO_3^- \rightarrow Sb_4O_6 + NO + H_2O$$
(a) 1   (b) 6   (c) 3   (d) 4   (e) 5

d 79. How many H_2O are there in the balanced equation?
$$CuS + NO_3^- + H^+ \rightarrow Cu^{2+} + S + NO + H_2O$$
(a) one   (b) two   (c) three   (d) four   (e) five

e 80. Balance the following net ionic equation. What is the coefficient of OH-?
$$Cl_2 + OH^- \rightarrow Cl^- + ClO_3^- + H_2O$$
(a) 1   (b) 2   (c) 4   (d) 5   (e) 6

d 81. Balance the following net ionic equation. What is the sum of the coefficients?
$$H_2S + H^+ + Cr_2O_7^{2-} \rightarrow Cr^{3+} + S + H_2O$$
(a) 18   (b) 20   (c) 22   (d) 24   (e) 26

d 82. Balance the following net ionic equation. What is the sum of the coefficients?
$$(COOH)_2 + MnO_4^- + H^+ \rightarrow CO_2 + Mn^{2+} + H_2O$$
(a) 27   (b) 29   (c) 31   (d) 33   (e) 35

b 83. Balance the following equation. What is the sum of all coefficients?
$$Cu + H^+ + SO_4^{2-} \rightarrow Cu^{2+} + H_2O + SO_2$$
(a) 9   (b) 10   (c) 11   (d) 12   (e) 13

c 84. Balance the following net ionic equation. What is the sum of the coefficients?
$$MnO_4^- + I^- + H_2O \rightarrow I_2 + MnO_2 + OH^-$$
(a) 21   (b) 23   (c) 25   (d) 27   (e) 29

c 85. Balance the following net ionic equation. Use H+ rather than H_3O^+. What is the coefficient of NO_3^-?
$$Cr_2O_7^{2-} + HNO_2 \rightarrow Cr^{3+} + NO_3^- \text{ (acidic solution)}$$
(a) 1   (b) 2   (c) 3   (d) 4   (e) 5

e 86. Balance the following net ionic equation. Use H⁺ rather than H_3O^+. What is the coefficient of H_2S?

$$H_2S + MnO_4^- \rightarrow SO_4^{2-} + Mn^{2+} \text{ (acidic solution)}$$

(a) 1   (b) 2   (c) 3   (d) 4   (e) 5

c 87. Balance the following net ionic equation. What is the coefficient of OH⁻?

$$MnO_4^- + NH_3 \rightarrow NO_3^- + MnO_2 \text{ (basic solution)}$$

(a) 2   (b) 3   (c) 5   (d) 7   (e) 9

d 88. Balance the following net ionic equation. Use H⁺ rather than H_3O^+. What is the sum of the coefficients?

$$Zn + NO_3^- \rightarrow Zn^{2+} + N_2 \text{ (acidic solution)}$$

(a) 25   (b) 27   (c) 29   (d) 31   (e) 33

d 89. Balance the following net ionic equation. Use H⁺ rather than H_3O^+. What is the sum of the coefficients?

$$H_2O_2 + Fe^{2+} \rightarrow Fe^{3+} \text{ (in acidic solution)}$$

(a) 13   (b) 15   (c) 7   (d) 9   (e) 11

d 90. Balance the following net ionic equation. Use H⁺ rather than H_3O^+. What is the sum of the coefficients?

$$MnO_4^- + SO_3^{2-} \rightarrow Mn^{2+} + SO_4^{2-} \text{ (acidic solution)}$$

(a) 27   (b) 29   (c) 21   (d) 23   (e) 25

b 91. Balance the following net ionic equation. Use H⁺ rather than H_3O^+. What is the sum of the coefficients?

$$I^- + NO_2^- \rightarrow NO + I_2 \text{ (acidic solution)}$$

(a) 11   (b) 13   (c) 15   (d) 17   (e) 19

a 92. Balance the following net ionic equation. Use H⁺ rather than H_3O^+. What is the sum of the coefficients?

$$MnO_4^- + Fe^{2+} \rightarrow Mn^{2+} + Fe^{3+} \text{ (acid solution)}$$

(a) 24   (b) 26   (c) 28   (d) 30   (e) 32

d 93. Balance the following net ionic equation. What is the sum of the coefficients?

$$PbO_2 + SeO_3^{2-} \rightarrow PbO + SeO_4^{2-} \text{ (in basic solution)}$$

(a) eight   (b) ten   (c) twelve   (d) four   (e) six

b 94. Balance the following net ionic equation. What is the sum of the coefficients?

$MnO_4^- + Se^{2-} \rightarrow MnO_2 + Se$ (basic solution)

(a) 20   (b) 22   (c) 24   (d) 26   (e) 28

b 95. Balance the following net ionic equation. What is the coefficient of H_2O?

$MnO_4^- + Bi \rightarrow MnO_2 + Bi(OH)_3$ (basic solution)

(a) 1   (b) 2   (c) 3   (d) 4   (e) 5

e 96. In acidic solution MnO_4^- oxidizes H_3AsO_3, a weak acid, to H_3AsO_4, a weak acid, and is reduced to Mn^{2+}. Write the balanced **net ionic** equation for this reaction. How many H^+ are there in the balanced equation?

(a) 1   (b) 2   (c) 3   (d) 4   (e) 6

e 97. Write the balanced **net ionic** equation for the reaction of hydrogen sulfide with bromate ions, BrO_3^-, in acidic solution to form sulfur and bromide ions. How many H^+ are there in the balanced equation?

(a) 1   (b) 2   (c) 3   (d) 6   (e) zero

b 98. In basic solution CrO_4^{2-} oxidizes $HSnO_2^-$ to $HSnO_3^-$ and is reduced to CrO_2^-. Write the balanced **net ionic** equation for this reaction. How many OH^- are there in the balanced equation?

(a) 1   (b) 2   (c) 3   (d) 4   (e) 5

a 99. In basic solution MnO_4^- oxidizes NO_2^- to NO_3^- and is reduced to MnO_2. How many H_2O are there in the balanced **net ionic** equation?

(a) 1   (b) 2   (c) 3   (d) 4   (e) 5

e 100. Iron(II) ions reduce dichromate ions in acidic solution to chromium(III) ions and form ferric ions. Write the balanced **net ionic** equation. Use H^+ rather than H_3O^+. What is the sum of all the coefficients?

(a) 24   (b) 26   (c) 28   (d) 32   (e) 36

b 101. Chlorous acid, $HClO_2$, oxidizes sulfur dioxide, SO_2, to sulfate ions, SO_4^{2-}, and is reduced to chloride ions, Cl^-, in acidic solution. Use H^+ rather than H_3O^+. Write the balanced **net ionic** equation for the reaction. What is the sum of the coefficients?

(a) 11   (b) 13   (c) 15   (d) 17   (e) 19

c 102. In basic solution CrO_4^{2-} oxidizes $HSnO_2^-$ to $HSnO_3^-$ and is reduced to CrO_2^-. Write the balanced **net ionic** equation for this reaction. What is the sum of the coefficients?

(a) 9   (b) 15   (c) 13   (d) 6   (e) 10

Stoichiometry of Redox Reactions

b 103. What volume of 0.0100 M KMnO$_4$ solution is required to oxidize 42.5 mL of 0.0100 M FeSO$_4$ sulfuric acid solution?

$$MnO_4^- + 8H^+ + 5Fe^{2+} \rightarrow Mn^{2+} + 5Fe^{3+} + 4H_2O$$

(a) 16.0 mL (b) 8.5 mL (c) 21.3 mL (d) 80.0 mL (e) 31.9 mL

a 104. What volume of 0.100 M (COOH)$_2$ solution would be required to react with 0.164 grams of KMnO$_4$?

$$5(COOH)_2 + 2MnO_4^- + 6H^+ \rightarrow 10CO_2 + 2Mn^{2+} + 8H_2O$$

(a) 25.9 mL (b) 51.8 mL (c) 77.8 mL (d) 104 mL (e) 90.8 mL

e 105. What is the molarity of a solution of FeSO$_4$ if 25.06 mL of it reacts with 38.19 mL of 0.1214 M KMnO$_4$?

$$MnO_4^- + 8H^+ + 5Fe^{2+} \rightarrow Mn^{2+} + 5Fe^{3+} + 4H_2O$$

(a) 0.1854 M (b) 0.3992 M (c) 0.07985 M
(d) 0.4267 M (e) 0.9250 M

c 106. What mass of molybdenum(III) oxide, Mo$_2$O$_3$, will be completely oxidized to potassium molybdate, K$_2$MoO$_4$, by 35.0 mL of 0.200 M KMnO$_4$?

$$3MnO_4^- + 5Mo^{3+} + 8H_2O \rightarrow 3Mn^{2+} + 5MoO_4^{2-} + 16H^+$$

(a) 8.40 g (b) 4.20 g (c) 1.40 g (d) 2.80 g (e) 6.40 g

c 107. An impure 0.500-gram sample of FeSO$_4$ reacts with 20.0 mL of 0.0200 M KMnO$_4$. Assuming that the impurities do not react with KMnO$_4$, what is the percentage of FeSO$_4$ in the sample?

$$MnO_4^- + 8H^+ + 5Fe^{2+} \rightarrow Mn^{2+} + 5Fe^{3+} + 4H_2O$$

(a) 52.3% (b) 56.6% (c) 60.8% (d) 64.5% (e) 69.2%

a 108. What volume of 0.1066 M Na$_2$S$_2$O$_3$ solution would be required to react with 24.32 mL of 0.1008 M I$_2$ solution?

$$Na_2S_2O_3 + I_2 \rightarrow Na_2S_4O_6 + NaI \text{ (unbalanced)}$$

(a) 45.99 mL (b) 5.75 mL (c) 11.50 mL
(d) 23.00 mL (e) 34.50 mL

a 109. What volume of 0.1084 M (COOH)$_2$ solution would be required to react with 0.1268 grams of KMnO$_4$?

$$(COOH)_2 + MnO_4^- + H^+ \rightarrow CO_2 + Mn^{2+} + H_2O \text{ (unbalanced)}$$

(a) 18.50 mL (b) 37.02 mL (c) 55.51 mL
(d) 24.02 mL (e) 66.61 mL

c 110. What is the molarity of a $K_2Cr_2O_7$ solution if 20.0 mL of the solution requires 60.4 mL of 0.200 M KI solution for complete reaction?

$$K_2Cr_2O_7 + KI + H^+ \rightarrow Cr^{3+} + I_2 + H_2O \text{ (unbalanced)}$$

(a) 1.55 M (b) 3.63 M (c) 0.101 M (d) 0.603 M (e) 0.280 M

b 111. A 30.0 mL sample of a solution of $NaClO_3$ that contains 1.06 g of $NaClO_3$ is to react with a solution of I_2. What is the molarity of the $NaClO_3$ solution?

$$NaClO_3 + H_2O + I_2 \rightarrow HIO_3 + NaCl \text{ (unbalanced)}$$

(a) 5.53 x 10^{-2} M   (b) 0.332 M   (c) 2.00 M
(d) 0.249 M   (e) 2.67 x 10^{-2} M

e 112. A 25.0-mL sample of 0.0833 M $NaClO_3$ reacted with 30.0 mL of an aqueous solution of I_2. How many grams of I_2 were contained in the I_2 solution?

$$NaClO_3 + H_2O + I_2 \rightarrow HIO_3 + NaCl \text{ (unbalanced)}$$

(a) 0.264 g (b) 0.397 g (c) 0.236 g (d) 0.159 g (e) 0.317 g

a*113. What volume of 0.1125 M $K_2Cr_2O_7$ would be required to oxidize 48.16 mL of 0.1006 M Na_2SO_3 in acidic solution? The products include Cr^{3+} and SO_4^{2-} ions.

(a) 14.36 mL   (b) 28.75 mL   (c) 43.12 mL
(d) 56.12 mL   (e) 32.15 mL

b*114. In basic solution MnO_4^- oxidizes NO_2^- to NO_3^- and is reduced to MnO_2. Calculate the volume of 0.1152 M $KMnO_4$ solution that would be required to oxidize 30.48 mL of 0.1024 M $NaNO_2$.

(a) 9.03 mL   (b) 18.06 mL   (c) 27.09 mL
(d) 45.16 mL   (e) 29.42 mL

b*115. In acidic solution IO_3^- oxidizes Ti^{3+} to Ti^{4+} and is reduced to I^-. How many mL of 0.02015 M KIO_3 would be required to oxidize 18.04 mL of 0.05012 M $Ti_2(SO_4)_3$ in acidic solution?

(a) 7.48 mL   (b) 14.96 mL   (c) 22.44 mL
(d) 29.88 mL   (e) none of these

b*116. In basic solution MnO_4^- oxidizes NO_2^- to NO_3^- and is reduced to MnO_2. Calculate the volume of 0.1025 M $KMnO_4$ solution that would be required to oxidize 30.55 mL of 0.1075 M $NaNO_2$.

(a) 10.68 mL   (b) 21.36 mL   (c) 32.04 mL
(d) 16.06 mL   (e) 42.75 mL

b*117. In acidic solution IO$_3^-$ oxidizes Ti^{3+} to Ti^{4+} and is reduced to I$^-$. How many mL of 0.0200 M KIO$_3$ would be required to oxidize 0.932 moles of Ti$_2$(SO$_4$)$_3$ in acidic solution?

(a) 7.50 mL   (b) 15.5 mL   (c) 21.5 mL   (d) 87.5 mL   (e) 75.0 mL

a*118. What is the molarity of a K$_2$Cr$_2$O$_7$ solution if 35.0 mL of it are required for complete reaction with 25.0 mL of 0.161 M Fe(NH$_4$)$_2$(SO$_4$)$_2$ solution? The products of the reaction include Fe^{3+} and Cr^{3+}.

(a) 0.0192 M (b) 0.0383 M (c) 0.155 M
(d) 0.345 M (e) 0.690 M

c*119. What is the molarity of a Na$_2$Cr$_2$O$_7$ solution if 22.5 mL of it is required to oxidize 33.0 mL of 0.150 M Fe(NH$_4$)$_2$(SO$_4$)? The reaction products include Cr^{3+} and Fe^{3+}.

(a) 0.184 M (b) 0.220 M (c) 0.0367 M
(d) 0.0440 M (e) 0.0660 M

a*120. A 50.0-mL sample of K$_2$Cr$_2$O$_7$ solution oxidizes 1.500 g of Na$_2$SO$_3$ to Na$_2$SO$_4$. Cr$_2$(SO$_4$)$_3$ is also produced. What is the molarity of the K$_2$Cr$_2$O$_7$ solution?

(a) 0.0794 M (b) 0.476 M (c) 2.96 M
(d) 0.0952 M (e) 0.0158 M

b*121. A solution of Fe(NH$_4$)$_2$(SO$_4$)$_2$·6H$_2$O containing 2.10 grams of the compound is titrated with acidic Na$_2$Cr$_2$O$_7$ solution. If 41.6 mL of the Na$_2$Cr$_2$O$_7$ solution is required, what is the molarity of the Na$_2$Cr$_2$O$_7$ solution? The reaction products include Cr^{3+} and Fe^{3+}.

(a) 0.129 M (b) 0.0215 M (c) 0.0644 M
(d) 0.0387 M (e) 0.0744 M

12 Gases and the Kinetic-Molecular Theory

Comparison of Solids, Liquids, and Gases

e 1. Which of the following statements is **false**?

(a) Condensed states have much higher densities than gases.
(b) Molecules are very far apart in gases and closer together in liquids and solids.
(c) Gases completely fill any container they occupy and are easily compressed.
(d) Vapor refers to a gas formed by evaporation of a liquid or sublimation of a solid.
(e) Solid water (ice), unlike most substances, is denser than its liquid form (water).

c 2. Which of the following substances would have the greatest molar volume?

(a) $H_2O(s)$ (b) $H_2O(\ell)$ (c) $H_2O(g)$ (d) $CCl_4(s)$ (e) $CCl_4(\ell)$

Composition of the Atmosphere and Some Common Properties of Gases

e 3. Which of the following statements is **false**?

(a) The earth's atmosphere is a mixture of gases and particles of liquids and solids.
(b) The three major gaseous components are N_2, O_2, and Ar.
(c) Atmospheric moisture varies.
(d) All gases mix completely unless they react with each other.
(e) The concentration of the fourth largest component of the atmosphere, CO_2, is the same for all locations on the planet, but is increasing.

a 4. Which statement is **false**?

(a) The density of a gas is constant as long as its temperature remains constant.
(b) Gas expands without limit.
(c) Gases diffuse into each other and mix almost immediately when put into the same container.
(d) The molecular mass of a gaseous compound is an invariable quantity.
(e) Pressure must be exerted on a sample of a gas in order to confine it.

Pressure

c 5. Which statement concerning pressure and its measurement is **false**?

(a) Pressure is defined as force per unit area.
(b) For a simple mercury barometer, the pressure exerted by the vertical column of mercury inside the closed tube equals the atmospheric pressure.
(c) The torr is a common unit for pressure and is defined as 1 torr = 1 cm Hg.
(d) A mercury manometer consists of a glass U-tube partially filled with mercury with one arm open to the atmosphere and the other connected to a container of gas.
(e) When the atmospheric pressure is greater than the pressure of the gas in the manometer, P_{gas} (in torr) = P_{atm} (in torr) – h (difference in height of mercury columns of U-tube, in mm)

c*6. Which statement is **false**?

(a) Atmospheric pressure varies with atmospheric conditions and distance above sea level.
(b) Pilots use portable barometers to determine their altitudes.
(c) A tire gauge measures relative pressure; i.e., it measures how much lower, than atmospheric pressure, the pressure in the tire is.
(d) At sea level, at a latitude of 45° and at 0°C, the average atmospheric pressure supports a 760 mm high column of mercury in a simple barometer; this average sea-level pressure of 760 mm Hg is also called one atmosphere of pressure.
(e) The SI unit of pressure is the pascal (Pa), defined as the pressure exerted by a force of one newton acting on an area of one square meter.

Boyle's Law

b 7. The statement "At constant temperature, the volume occupied by a definite mass of a gas is inversely proportional to the applied pressure." is a statement of _____ Law.

(a) Charles's (b) Boyle's (c) Graham's
(d) Avogadro's (e) Gay-Lussac's

b 8. A sample of a gas occupies 1600. milliliters at 20.0°C and 600. torr. What volume will it occupy at the same temperature and 800. torr?

(a) 1.00 x 10³ mL (b) 1.20 x 10³ mL (c) 1.45 x 10³ mL
(d) 2.02 x 10³ mL (e) 2.13 x 10³ mL

e 9. A sample of oxygen occupies 47.2 liters under a pressure of 1240. torr at 25°C. What volume would it occupy at 25°C if the pressure were decreased to 730. torr?

(a) 27.8 L   (b) 29.3 L   (c) 32.3 L   (d) 47.8 L   (e) 80.2 L

b 10. A gas occupies 1.75 liters and exerts a pressure of 730. torr at a certain temperature. If the temperature remains constant, what volume must it occupy to exert a pressure of 820. torr?

(a) 0.641 L   (b) 1.56 L   (c) 3.44 L   (d) 0.291 L   (e) 1.08 L

e 11. The volume of a sample of a gas is 405 mL at 10.0 atm and 467 K. What volume will it occupy at 4.29 atm and the same temperature?

(a) 17.4 L   (b) 189 mL   (c) 174 mL   (d) 1047 mL   (e) 944 mL

e 12. The volume of a sample of a gas is 415 mL at 10.0 atm and 337 K. What volume will it occupy at 5.32 atm and 337 K?

(a) 14.4 mL   (b) 156 mL   (c) 144 mL   (d) 865 mL   (e) 780 mL

c*13. A 5.00-L sample of a gas exerts a pressure of 1040. torr at 50.0°C. In what volume would the same sample exert a pressure of 1.00 atm at 50.0°C?

(a) 0.146 L   (b) 3.65 L   (c) 6.84 L   (d) 8.72 L   (e) 5.20 L

e*14. A sample of oxygen occupies 47.2 liters under a pressure of 1240. torr at 25°C. By what percent does the volume of this sample change if the pressure were decreased to 730. torr at 25°C?

    (a) 37.8%   (b) 41.1%   (c) 43.4%   (d) 46.1%   (e) 69.9%

d 15. A sample of gas occupies a volume of 1.60 L at a pressure of 720. torr. What would be the pressure of this gas if it were compressed to 1.20 L at the same temperature?

    (a) 540. torr   (b) 360. torr   (c) 180. torr   (d) 960. torr   (e) 1440 torr

c 16. A sample of carbon monoxide, CO, occupies a volume of 350. mL and exerts a pressure of 1020. torr at 25°C. If the volume expands to 500. mL with no temperature change, what pressure will the gas exert?

    (a) 617 torr   (b) 827 torr   (c) 714 torr   (d) 938 torr   (e) 1457 torr

a 17. A 6.25-L sample of gas exerts a pressure of 1.46 atm at 25°C. What would be the pressure of this gas sample at 25°C if it were compressed to a volume of 5.05 L?

    (a) 1.81 atm   (b) 1.18 atm   (c) 21.6 atm   (d) 1.70 atm   (e) 0.280 atm

d 18. The volume of a gas sample is 600. mL at 25°C and a pressure of 2.04 atm. What would be the pressure of this sample if it were allowed to expand to 850. mL at 25°C?

    (a) 2.89 atm   (b) 3.10 atm   (c) 4.40 atm   (d) 1.44 atm   (e) 0.346 atm

Charles's Law; the Absolute Temperature Scale

a 19. The statement "At constant pressure, the volume occupied by a definite mass of a gas is directly proportional to the absolute temperature." is a statement of _____ Law.

    (a) Charles's   (b) Boyle's   (c) Graham's
    (d) Avogadro's   (e) Gay-Lussac's

c*20. Absolute zero is the temperature at which

    (a) a graph of V versus 1/P intersects the 1/P-axis.
    (b) gaseous helium liquefies.
    (c) the straight line graph of V versus T intersects the T-axis.
    (d) a graph of P versus 1/V intersects the 1/V-axis.
    (e) none of the above

d*21. The Rankine temperature scale is an absolute temperature scale based on the Fahrenheit degree rather than the Celsius degree. What number (nearest whole number) goes into the blank in the formula below that allows one to convert directly from degrees Fahrenheit to degrees Rankine?

$$°R = °F + (\underline{\quad})°$$

    (a) 273   (b) –273   (c) 470   (d) 460   (e) – 460

a 22. Choose the response that includes all of the following statements that are correct, and no incorrect statements. **Assume constant pressure in each case**. If a sample of gas is

    I.   heated from 100°C to 200°C the volume will double.
    II.   heated from 0°C to 273°C the volume will double.
    III.   cooled from 400°C to 200°C the volume will decrease by a factor of two.
    IV.   cooled from 1000°C to 200°C the volume will decrease by a factor of five.
    V.   heated from 200°C to 2000°C the volume will increase by a factor of ten.

(a) II   (b) I and II   (c) III and IV
(d) II and III   (e) none of these

b 23. A gas sample occupies 4.20 L under a pressure of 800. torr at 350. K. At what temperature will it occupy 3.60 L at the same pressure?

(a) 7 K   (b) 300. K   (c) 408 K   (d) 300.°C   (e) 770. K

d 24. A sample of nitrogen occupies 5.50 liters under a pressure of 900. torr at 25.0°C. At what temperature will it occupy 10.0 liters at the same pressure?

(a) 32°C   (b) –109°C   (c) 154°C   (d) 269°C   (e) 370.°C

b 25. A given mass of xenon gas has a volume of 200. mL at 25.0°C and 760. torr. To what temperature must the xenon be heated to occupy 450. mL at 760. torr?

(a) 376°C   (b) 398°C   (c) 416°C   (d) 512°C   (e) 492°C

c 26. A sample of a gas occupies 460. mL at 70.0°C and 1.00 atmosphere. At what temperature would the gas occupy 650. mL at the same pressure?

(a) 154°C   (b) –30.3°C   (c) 212°C   (d) 101°C   (e) 188°C

d 27. A 300. mL sample of neon is confined under a pressure of 0.420 atm at 15.0°C. At what temperature would it occupy 1.00 L at the same pressure?

(a) –187°C   (b) 86.4°C   (c) 380.°C   (d) 687°C   (e) 960.°C

b 28. The volume of a sample of a gas is 405 mL at 10.0 atm and 467 K. What volume will the sample occupy at 10.0 atm and 1025 K?

(a) 644 mL   (b) 889 mL   (c) 405 mL   (d) 434 mL   (e) 185 mL

e 29. If 5.50 L of argon gas at 273°C is heated at constant pressure to a temperature of 546°C, what would be its new volume?

(a) 11.0 L   (b) 16.5 L   (c) 2.75 L   (d) 3.67 L   (e) 8.25 L

c 30. The temperature of a fixed amount of hydrogen gas is reduced from 20.0°C to –20.0°C at constant pressure. If the original volume of gas is 420. mL, what is the final volume?

(a) 466 mL   (b) 257 mL   (c) 363 mL   (d) 406 mL   (e) 392 mL

e 31. Suppose you inhaled 1.00 liter of air at –30.0°C (–22°F), and held this air long enough to heat it to body temperature, 37.0°C. What would be the new volume of the air in the lungs assuming constant pressure?

(a) 0.972 L   (b) 0.811 L   (c) 1.23 L   (d) 1.02 L   (e) 1.28 L

Standard Temperature and Pressure

b 32. By international agreement the standard temperature and pressure (STP) for gases is

(a) 25°C and one atmosphere.
(b) 273.15 K and 760. torr.
(c) 298.15 K and 760. torr.
(d) 0°C and 700. torr.
(e) 293 K and one atmosphere.

e 33. A sample of neon, Ne, occupies 1700. mL at STP. At constant pressure, what temperature is necessary to increase the volume to 2100. mL?

(a) –14°C   (b) –52°C   (c) 32°C   (d) 52°C   (e) 64°C

b 34. A sample of gas occupies 650. mL at STP. If the pressure remains constant, what volume will it occupy at a temperature of 65.0°C?

(a) 776 mL   (b) 805 mL   (c) 487 mL   (d) 525 mL   (e) 910. mL

d 35. A sample of helium occupies 5.00 L at STP. What volume will the sample occupy at 0.0°C and 890. torr?

(a) 3.56 L   (b) 3.72 L   (c) 4.06 L   (d) 4.27 L   (e) 5.86 L

b 36. A sample of a gas occupies 20.0 liters at STP. Under what pressure would it occupy 20.0 liters at 50.0°C?

(a) 0.183 atm   (b) 1.18 atm   (c) 0.732 atm   (d) 3.71 atm   (e) 5.46 atm

c 37. A gas sample has a volume of 6.30 L at STP. What would be its pressure if it occupied 5.70 L at 0.0°C?

(a) 688 torr   (b) 7.13 torr   (c) 840. torr   (d) 1.51 atm   (e) 0.475 torr

e 38. A gas sample occupies 1.00 L at 120.°C and 1.00 atm. What volume will it occupy at STP?

(a) 1.14 L   (b) 1.44 L   (c) 0.846 L   (d) 0.782 L   (e) 0.695 L

e 39. A gaseous sample has a volume of 3.88 L at 115°C and 760. torr. What would be its volume at STP?

(a) 5.51 L   (b) 2.98 L   (c) 3.63 L   (d) 3.37 L   (e) 2.73 L

a 40. A sample of He gas has a volume of 373 mL at 0.0°C and 825 torr. What would be its volume at STP?

(a) 405 mL   (b) 508 mL   (c) 344 mL   (d) 235 mL   (e) 1680 mL

The Combined Gas Law Equation

c*41. Which possible response includes all the following statements that are **correct**, and no incorrect statements?

    I. For a sample of gas held at constant pressure, a plot of volume versus temperature is a straight line.
    II. For a sample of gas held at constant pressure, a plot of volume versus temperature is one branch of a hyperbola.
    III. For a sample of gas held at constant temperature, a plot of volume versus pressure is a straight line.
    IV. For a sample of gas held at constant temperature, a plot of volume versus pressure is one branch of a hyperbola.
    V. At constant pressure the volume occupied by a definite mass of a gas varies inversely with the absolute temperature.
    VI. At constant pressure the volume occupied by a definite mass of a gas varies directly with the absolute temperature.
    VII. At constant temperature the volume occupied by a definite mass of a gas varies inversely with the applied pressure.
    VIII. At constant temperature the volume occupied by a definite mass of a gas varies directly with the applied pressure.

(a) I, III, V, and VII   (b) II, IV, VI, and VIII   (c) I, IV, VI, and VII
(d) II, III, V, and VIII   (e) none of these

e 42. A sample of nitrogen occupies a volume of 546 mL at STP. What volume would the nitrogen occupy at 177°C under a pressure of 380. torr?

(a) 91 mL   (b) 273 mL   (c) 900. mL
(d) 1.09×10^3 mL   (e) 1.80×10^3 mL

d 43. A sample of nitrogen occupies a volume of 300. mL under a pressure of 380. torr at −177°C. What volume would the gas occupy at STP?

(a) 91 mL   (b) 1.8×10^2 mL   (c) 3.6×10^2 mL
(d) 4.3×10^2 mL   (e) 8.1×10^2 mL

a 44. A sample of a gas occupies 3.24 liters at 25.0°C and 734 torr. What volume does it occupy at STP?

(a) 2.87 L   (b) 2.96 L   (c) 3.07 L   (d) 3.42 L   (e) 3.66 L

e 45. The volume of a sample of nitrogen is 6.00 liters at 35°C and 740. torr. What volume will it occupy at STP?

(a) 6.59 L   (b) 5.46 L   (c) 6.95 L   (d) 5.46 L   (e) 5.18 L

a 46. A sample of nitrogen occupies 200. mL at 27°C when the pressure is 0.500 atm. What volume would it occupy at standard conditions?

(a) 91 mL   (b) 110. mL   (c) 127 mL   (d) 135 mL   (e) 440. mL

e 47. A sample of neon occupies 4.50 liters at 30.0°C and 2.50 atmospheres. What volume does the sample occupy at standard conditions?

(a) 6.42 L   (b) 5.36 L   (c) 12.5 L   (d) 2.00 L   (e) 10.1 L

c 48. A sample of helium occupies 1.40 L at standard conditions. What pressure will it exert in a 0.500 L vessel at 100.°C?

(a) 624 torr   (b) 1.56 x 10³ torr   (c) 2.91 x 10³ torr
(d) 371 torr   (e) 4.05 x 10³ torr

c 49. A sample of a gas occupies 200. mL at STP. Under what pressure would this sample occupy 300. mL if the temperature were changed to 718°C?

(a) 0.841 atm   (b) 0.966 atm   (c) 2.42 atm
(d) 1.76 atm   (e) 1.42 atm

d 50. A sample of gas occupies 30.0 liters at STP. What will be its temperature in °C if the volume is reduced to 22.5 liters under 3.50 atmospheres pressure?

(a) 212°C   (b) 284°C   (c) 386°C   (d) 444°C   (e) 516°C

a 51. A sample of a gas occupies 1.80 L at –10.°C and 450. torr. What volume will it occupy at 30.°C and 800. torr?

(a) 1.17 L   (b) 0.879 L   (c) 3.69 L   (d) 2.04 L   (e) 2.78 L

c 52. A sample of gas has a volume of 180. mL at –30.°C and 1.05 atm. What will be its volume at 35°C and 0.525 atm?

(a) 52 mL   (b) 280. mL   (c) 456 mL   (d) 42 mL   (e) 114 mL

c 53. A sample of a gas occupies 60.0 mL at –10.0°C and 720. torr. What pressure will the gas exert in a 101 mL sealed bulb at 25°C?

(a) 1373 torr   (b) 171 torr   (c) 485 torr
(d) 377 torr   (e) 1069 torr

b 54. A gas sample occupies 2.50 L at 125°C and 2.06 atm. What will be its volume at 25°C and 1.08 atm?

(a) 6.37 L   (b) 3.57 L   (c) 0.981 L   (d) 1.75 L   (e) 0.954 L

c 55. A sample of gas occupies 20.0 liters at 32°C when the pressure is 0.750 atm. What will be its temperature in °C if the volume increased to 25.0 liters at a pressure of 0.680 atm?

(a) –3.9°C   (b) 309°C   (c) 73°C   (d) 346°C   (e) 36°C

c 56. A sample of neon occupies 4.50 liters at 30.0°C and 2.50 atmospheres. What volume will it occupy at 1000. torr and 300.°C?

(a) 4.52 L   (b) 4.48 L   (c) 16.2 L   (d) 8.96 L   (e) 1.25 L

Avogadro's Law and the Standard Molar Volume

b 57. A 1 L sample of gas at 25°C and 1 atm pressure is subjected to an increase in pressure and a decrease in temperature. The **density** of the gas _____.

(a) decreases
(b) increases
(c) remains the same
(d) becomes zero
(e) either increases or decreases, depending on the sizes of the pressure and temperature changes

a 58. What volume will 12.40 grams of CO_2 occupy at STP, if it behaves ideally?

(a) 6.31 L   (b) 8.46 L   (c) 4.42 L   (d) 11.7 L   (e) 9.68 L

d 59. Calculate the volume occupied by 0.588 g of NH_3 at STP, if it behaves ideally.

(a) 0.502 L   (b) 0.368 L   (c) 0.621 L   (d) 0.775 L   (e) 0.804 L

b 60. What is the volume occupied by 10.0 grams of NH_3 at STP, if it behaves ideally?

(a) 10.6 L   (b) 13.2 L   (c) 14.3 L   (d) 14.9 L   (e) 15.8 L

e 61. What is the molecular weight of a gas if 0.104 gram of the gas occupies 48.7 mL at STP?

(a) 5.06 g/mol   (b) 28.2 g/mol   (c) 34.5 g/mol
(d) 40.0 g/mol   (e) 47.8 g/mol

a 62. What is the molecular weight of a gas if 5.00 grams of it occupies 2200. mL at STP?

(a) 50.9 g/mol   (b) 42.6 g/mol   (c) 59.6 g/mol
(d) 63.4 g/mol   (e) 68.0 g/mol

d 63. A 4.28-gram sample of a gas occupies 800. mL at STP. What is its molecular weight?

(a) 60.0 g/mol   (b) 153 g/mol   (c) 76.5 g/mol
(d) 120. g/mol   (e) 140. g/mol

a 64. At STP, 4.00 L of a pure gaseous substance has a mass of 10.0 grams. The molecular weight of this substance is _____.

(a) 56.0 amu   (b) 67.4 amu   (c) 73.9 amu   (d) 89.6 amu   (e) 120. amu

c 65. The simplest formula for a compound is NO_2. If 46.0 grams of the gas occupies 11.2 liters at standard conditions, the mass of one mole of the compound is _____.

(a) 46.0 g/mol   (b) 23.0 g/mol   (c) 92.0 g/mol
(d) 30.0 g/mol   (e) 62.0 g/mol

d 66. A sample of nitrous oxide, N_2O, occupies 1.65 L at STP. What is the mass of the sample?

(a) 18.9 g   (b) 22.1 g   (c) 28.6 g   (d) 32.4 g   (e) 46.0 g

c 67. What mass of methane, CH_4, is contained in a 500.-mL sample of CH_4 at STP?

(a) 0.212 g   (b) 0.268 g   (c) 0.357 g   (d) 0.390 g   (e) 0.407 g

c 68. A sample of NO_2 occupies 16,500 mL at STP. What is the mass of the sample?

(a) 21.0 g   (b) 28.6 g   (c) 33.9 g   (d) 58.6 g   (e) 62.4 g

d 69. If 4.00 g of a gas occupies 1.12 liters at STP, what is the mass of 2.00 moles of the gas?

(a) 20.0 g   (b) 40.0 g   (c) 80.0 g   (d) 160. g   (e) 120. g

e 70. If 1.00 L of a gas is 4.40 times as heavy as 1.00 L of O_2 at the same temperature and pressure, then the molecular weight of the gaseous substance is _____.

(a) 50.0 amu   (b) 67.0 amu   (c) 70.4 amu   (d) 88.0 amu   (e) 141 amu

d 71. Given the mass of one mole of gaseous NO is 30.0 grams/mole, calculate the density of NO in grams/liter at STP.

(a) 1.12 g/L   (b) 0.987 g/L   (c) 2.17 g/L   (d) 1.34 g/L   (e) 1.60 g/L

b 72. Calculate the density of O_2, in g/L, at STP.

(a) 0.810 g/L   (b) 1.43 g/L   (c) 1.62 g/L   (d) 0.714 g/L   (e) 1.14 g/L

b 73. The simplest formula for a compound is NO_2. If 46.0 grams of the gas occupies 11.2 liters at standard conditions, the molecular formula of the gas is _____.

(a) NO_2 (b) N_2O_4 (c) N_3O_6 (d) NO   (e) N_2O

The Ideal Gas Equation

a 74. Evaluate R in units of mL torr/mol K. (Hint: Consider one mole of an ideal gas at STP.)

$$R = \underline{} \frac{mL \cdot torr}{mol \cdot K}$$

(a) 6.24×10^4 (b) 82.1   (c) 0.108
(d) 4.02×10^2 (e) 7.67×10^2

d 75. Which set of **units** for the following values of R is incorrect?

(a) 0.08206 L·atm/mol·K   (b) 8.314 kPa·dm^3/mol·K
(c) 62.4 L·torr/mol·K   (d) 22.4 L/mol·K
(e) 82.1 mL·atm/mol·K

e 76. Ten (10.0) moles of a gas are contained in a 10.0 L container at 273 K. Calculate the pressure of the gas.

(a) 2.24 atm   (b) 4.48 atm   (c) 11.2 atm   (d) 15.7 atm   (e) 22.4 atm

b 77. What is the pressure of 64.0 g of oxygen gas in a 1.50-L container at –37°C?

(a) 4.12 atm (b) 25.8 atm (c) 51.6 atm (d) 19.6 atm (e) 8.2 atm

b 78. Assume that 454 g (1.00 lb) of Dry Ice (solid CO_2) is placed into an evacuated 50.0 L closed tank. What is the pressure in the tank in atmospheres at 45.0°C after all the solid has vaporized?

(a) 14.6 atm (b) 5.38 atm (c) 3.14 atm (d) 8.41 atm (e) 10.6 atm

c 79. What is the volume that 57.0 grams of F_2 would occupy at 227°C and 1.50 atm?

(a) 1.06 L (b) 18.6 L (c) 41.0 L (d) 82.1 L (e) 106 L

e 80. What is the volume occupied by 3.50 grams of Cl_2 gas at 45°C and 745 torr?

(a) 1.17 L (b) 0.961 L (c) 1.05 L (d) 1.46 L (e) none of these

c 81. Calculate the volume that 256 grams of SO_2 occupies at 227°C under a pressure of 1520 torr.

(a) 1.94 L (b) 22.4 L (c) 81.9 L (d) 146.7 L (e) 194 L

d 82. Calculate the volume occupied by 37.0 grams of SO_2 at 215°C and 1340 torr.

(a) 7.62 L (b) 9.09 L (c) 11.6 L (d) 13.1 L (e) 15.0 L

c 83. What is the volume occupied by 10.0 grams of NH_3 at 500.°C and 1140 torr?

(a) 14.9 L (b) 16.1 L (c) 24.9 L (d) 32.7 L (e) 46.2 L

c 84. How many moles of an ideal gas are contained in 8.21 L at 73°C and 380. torr?

(a) 0.250 (b) 1.5 x 10^{23} (c) 0.144
(d) 7.5 x 10^{23} (e) 4.2 x 10^{-25}

a 85. A sample of a gas occupies a volume of 350. mL at 3000. torr and 25°C. Calculate the number of moles of gas present.

(a) 5.65 x 10^{-2} mol (b) 8.39 x 10^{-3} mol (c) 1.25 x 10^{-2} mol
(d) 1.38 x 10^{-2} mol (e) 2.19 x 10^{-3} mol

e*86. How many oxygen atoms are contained in a sample of O_2 that occupies 1500. mL at 50.0°C and 1.00 atm pressure?

(a) 1.75 x 10^{23} (b) 3.87 x 10^{23} (c) 3.40 x 10^{22}
(d) 8.91 x 10^{22} (e) 6.81 x 10^{22}

b 87. The volume of a sample of nitrogen is 6.00 liters at 35°C and 740. torr. How many grams of nitrogen, N_2, are present in the sample?

(a) 8.28 g (b) 6.48 g (c) 6.82 g (d) 3.24 g (e) 4.14 g

b 88. Calculate the mass of a 10.-liter sample of C_2H_6 at 459°C under a pressure of 760. torr.

(a) 2.5 g   (b) 5.0 g   (c) 7.5 g   (d) 10 g   (e) 8.2 g

b 89. What mass of CH_4 is contained in a 500.-mL sample of CH_4 at −40.0°C and 300. torr?

(a) 0.0870 g   (b) 0.165 g   (c) 0.322 g   (d) 0.734 g   (e) 1.06 g

b 90. Calculate the density of O_2 gas (in g/L) at 1.22 atm and 25°C.

(a) 1.74 g/L   (b) 1.60 g/L   (c) 0.871 g/L   (d) 19.0 g/L   (e) 0.798 g/L

d 91. Calculate the density of F_2 gas (in g/L) at 780. torr and 37°C.

(a) 1.40 g/L   (b) 1.74 g/L   (c) 1.08 g/L   (d) 1.53 g/L   (e) 0.766 g/L

Determinations of Molecular Weights and Molecular Formulas of Gaseous Substances

e 92. At 150.°C and 1.00 atm, 500. mL of a vapor has a mass of 0.4183 g. What is the molecular weight of the compound?

(a) 55.0 g/mol   (b) 58.0 g/mol   (c) 62.0 g/mol
(d) 65.0 g/mol   (e) 29.1 g/mol

c 93. A 0.397-g sample of a gas occupies a volume of 275 mL at 10.°C and 365 torr. What is the molecular weight of the gas?

(a) 85.9 g/mol   (b) 72.8 g/mol   (c) 69.8 g/mol
(d) 61.3 g/mol   (e) 59.8 g/mol

d 94. A gaseous compound is 30.4% nitrogen and 69.6% oxygen by mass. A 5.25-gram sample of the gas occupies a volume of 1.00 liter and exerts a pressure of 1.26 atmospheres at −4.0°C. What is its molecular formula?

(a) NO   (b) NO_2 (c) N_3O_6 (d) N_2O_4 (e) N_2O_5

d 95. A gaseous compound is 30.4% nitrogen and 69.6% oxygen. A 6.06-gram sample of gas occupies a volume of 1.00 liter and exerts a pressure of 1.26 atmospheres at −40.0°C. What is its molecular formula?

(a) NO   (b) NO_2 (c) N_3O_6 (d) N_2O_4 (e) N_2O_5

b 96. Analysis of a volatile liquid shows that it contains 62.04% carbon, 10.41% hydrogen, and 27.54% oxygen by mass. At 150.°C and 1.00 atm, 500. mL of the vapor has a mass of 0.8365 g. What is the molecular formula of the compound?

(a) C_2H_5OH (b) C_3H_6O (c) $C_6H_{12}O_2$
(d) $C_{12}H_{24}O_4$ (e) C_4H_6O

b 97. Analysis of a volatile liquid showed that it is 54.5% carbon, 9.1% hydrogen, and 36.4% oxygen by mass. A separate 0.345-gram sample of its vapor occupied 120. mL at 100.°C and 1.00 atm. What is the molecular formula for the compound?

(a) C_2H_4O (b) $C_4H_8O_2$ (c) $C_3H_6O_2$
(d) $C_6H_{12}O_4$ (e) $C_5H_{12}O$

c 98. Analysis of a sample of a gaseous compound shows that it contains 81.8% C and 18.2% H by mass. At standard conditions, 112 mL of the compound weighs 0.220 g. What is the molecular formula for the compound?

(a) C_2H_4 (b) C_4H_8 (c) C_3H_8 (d) C_3H_6 (e) C_6H_{16}

b 99. A gaseous compound is 78.26% boron and 21.74% hydrogen by mass. A 0.600-g sample of it occupies 487 mL at STP. What is the molecular formula for the gas?

(a) BH_3 (b) B_2H_6 (c) B_3H_9 (d) B_2H_4 (e) B_3H_6

b 100. A gaseous compound containing only nitrogen and oxygen is 63.64% N and 36.36% O by weight. At standard conditions 196 mL of the gas weighs 0.385 grams. What is the molecular formula of the gas?

(a) NO   (b) N_2O (c) N_2O_4 (d) NO_2 (e) N_4O_2

b 101. A gaseous compound is 23.79% C, 5.99% H and 70.22% Cl by mass. At STP 258 mL of the gas weighs 0.581 g. What is the molecular formula for this compound?

(a) CH_3Cl_2 (b) CH_3Cl (c) CH_2Cl_2 (d) C_2HCl (e) $C_2H_7Cl_2$

a 102. A gaseous compound is 52.14% C, 13.13% H and 34.73% O by mass. At STP 325 mL of the gas weighs 0.668 g. What is the molecular formula for this compound?

(a) C_2H_6O (b) C_3H_5O (c) C_4H_8O (d) $C_4H_{12}O$ (e) $C_4H_{12}O_2$

b 103. A 2.08-gram sample of a gaseous compound contains 1.92 grams of carbon and 0.16 gram of hydrogen. At STP, the same sample occupies a volume of 1793 mL. What is the molecular formula for the compound?

(a) C_4H_8 (b) C_2H_2 (c) C_3H_9 (d) C_4H_4 (e) C_2H_4

d 104. A 0.580-g sample of a compound containing only carbon and hydrogen contains 0.480 g of carbon and 0.100 g of hydrogen. At STP 33.6 mL of the gas has a mass of 0.087 g. What is the molecular formula for the compound?

(a) CH_3 (b) C_2H_6 (c) C_2H_5 (d) C_4H_{10} (e) C_4H_{12}

b 105. A sample of a gas contains 0.305 g of carbon, 0.407 g of oxygen, and 1.805 g of chlorine. A different sample of the same gas having a mass of 8.84 g occupies 2.00 liters at STP. What is the molecular formula for the compound?

(a) COCl   (b) $COCl_2$ (c) $C_2O_2Cl_2$ (d) $C_2O_2Cl_4$ (e) CO_2Cl_2

Dalton's Law of Partial Pressures

c 106. A 10.0-L flask contains 0.400 mole of H_2, 0.300 mole of He and 0.500 mole of Ne at 35.0°C. What is the total pressure in the flask?

(a) 2.53 atm   (b) 4.05 atm   (c) 3.03 atm   (d) 1.01 atm   (e) 0.345 atm

a 107. A 10.0-L flask contains 0.400 mole of H_2, 0.300 mole of He and 0.500 mole of Ne at 35.0°C. What is the partial pressure of the Ne?

(a) 1.26 atm   (b) 3.03 atm   (c) 1.05 atm   (d) 1.69 atm   (e) 0.144 atm

a 108. What is the pressure exerted by a mixture of 14.0 grams of N_2, 71.0 grams of Cl_2, and 16.0 grams of He in a 50.0-liter container at 0°C?

(a) 2.47 atm   (b) 6.80 atm   (c) 14.1 atm   (d) 4.66 atm   (e) 0.687 atm

b 109. What is the pressure exerted by a mixture of 1.0 g H_2 and 5.0 g He when confined to a volume of 5.0 liters at 20.°C?

(a) 12.6 atm   (b) 8.4 atm   (c) 3.61 atm   (d) 10.4 atm   (e) 6.5 atm

a 110. What is the pressure exerted by 26.0 g SO_2, 4.0 g H_2, and 20.2 g Ne in a 10.0-liter container at 50.0°C?

(a) 9.04 atm   (b) 10.1 atm   (c) 14.4 atm   (d) 18.6 atm   (e) 26.3 atm

c 111. A mixture of gases containing 21.0 g of N_2, 106.5 g of Cl_2 and 12.0 g of He at 14°C is in a 50.0-L container. What is the total pressure in the vessel?

(a) 1.8 atm   (b) 2.2 atm   (c) 2.5 atm   (d) 2.7 atm   (e) 3.2 atm

d 112. A mixture of gases containing 21.0 g of N_2, 106.5 g of Cl_2, and 12.0 g of He at 14°C is in a 50.0-L container. What is the partial pressure of N_2?

(a) 0.501 atm   (b) 0.707 atm   (c) 0.448 atm
(d) 0.353 atm   (e) 0.398 atm

b 113. A mixture of gases containing 26.0 g SO_2, 4.0 g H_2, and 20.2 g Ne at 23°C is in a 5.00-L container. What is the partial pressure of H_2?

(a) 1.94 atm   (b) 19.5 atm   (c) 4.86 atm   (d) 9.72 atm   (e) 0.755 atm

c 114. A mixture of gases containing 21.0 g of N_2, 106.5 g of Cl_2, and 12.0 g of He at 21°C is in a 8.00-L container. What is the partial pressure of He?

(a) 2.26 atm   (b) 4.53 atm   (c) 9.05 atm   (d) 5.55 atm   (e) 0.647 atm

b 115. What is the mole fraction of O_2 in a mixture of 2.00 g He, 12.0 g O_2, and 17.0 g N_2?

(a) 0.608   (b) 0.253   (c) 0.410   (d) 0.200   (e) 0.267

d 116. A mixture of gases consists of 2.00 g He, 12.0 g O_2, and 17.0 g N_2. What is the mole fraction of N_2 in this mixture?

(a) 0.267 (b) 0.506 (c) 0.253 (d) 0.410 (e) 0.208

d 117. A mixture of 16.0 g of He, 21.0 g of N_2, and 16.0 g of O_2 at 25°C is in a 75.0-L container. What are the mole fractions of each of these three gases?

	X_{He}	X_{N_2}	X_{O_2}
(a)	0.842	0.158	0.105
(b)	0.889	0.167	0.111
(c)	0.800	0.150	0.100
(d)	0.762	0.143	0.0952
(e)	0.952	0.762	0.143

b 118. A mixture of 16.0 g of O_2, 21.0 g of N_2, and 16.0 g of He at 20.°C is in a 25.0-L container. What are the partial pressures of O_2, N_2, and He in this mixture?

	O_2	N_2	He
(a)	0.481 atm	1.44 atm	3.85 atm
(b)	0.481 atm	0.721 atm	3.85 atm
(c)	0.962 atm	1.44 atm	3.85 atm
(d)	0.481 atm	0.721 atm	0.240 atm
(e)	0.962 atm	0.721 atm	0.240 atm

b 119. A 10.0-liter vessel contains gas A at a pressure of 300. torr. A 3.0-liter vessel contains gas B at a pressure of 400. torr. Gas A is forced into the second vessel. Calculate the resulting pressure in torr. Assume temperature remains constant.

(a) 1000 torr (b) 1400 torr (c) 1800 torr
(d) 2000 torr (e) 2300 torr

c 120. A 151-mL sample of O_2 was collected over water at 21°C when the barometric pressure was 742 torr. What mass of O_2 was contained in the sample? The vapor pressure of H_2O at 21°C is 18.7 torr.

(a) 2.21 g (b) 0.042 g (c) 0.19 g (d) 1.12 g (e) 0.86 g

b 121. A 300.-mL sample of hydrogen, H_2, was collected over water at 21°C on a day when the barometric pressure was 748 torr. What mass of hydrogen is present? The vapor pressure of water is 19 torr at 21°C.

(a) 0.0186 g (b) 0.0241 g (c) 0.0213 g
(d) 0.0269 g (e) 0.0281 g

d 122. A 500.-mL sample of oxygen, O_2, was collected by displacement of water at 24°C under a barometric pressure of 738 torr. What mass of dry oxygen was collected? The vapor pressure of water at 24°C is 22 torr.

(a) 0.245 g (b) 0.310 g (c) 0.319 g (d) 0.619 g (e) 0.637 g

e 123. A 5.42-liter sample of gas was collected over water when the temperature was 24°C and the barometric pressure was 706 torr. The dry sample of gas had a mass of 5.60 g. What is the mass of **three** moles of the dry gas? The vapor pressure of water at 24°C is 22 torr.

(a) 7.0 g   (b) 14 g   (c) 28 g   (d) 81 g   (e) 84 g

d 124. A sample of hydrogen collected by displacement of water occupied 30.0 mL at 24.0°C on a day when the barometric pressure was 736 torr. What volume would the hydrogen occupy if it were dry and at STP? The vapor pressure of water at 24.0°C is 22.4 torr.

(a) 32.4 mL   (b) 21.6 mL   (c) 36.8 mL   (d) 25.9 mL   (e) 27.6 mL

a 125. A 300.-mL sample of hydrogen, H_2, was collected over water at 21°C on a day when the barometric pressure was 748 torr. What volume will the **dry** hydrogen occupy at STP? The vapor pressure of water is 19 torr at 21°C.

(a) 267 mL   (b) 261 mL   (c) 256 mL   (d) 276 mL   (e) 282 mL

a 126. A sample of oxygen collected over water at 32.0°C and 752 torr has a volume of 0.627 liters. What volume would the dry oxygen occupy under standard conditions? The vapor pressure of water at 32.0°C is 35.7 torr.

(a) 0.529 L   (b) 0.644 L   (c) 0.556 L   (d) 0.609 L   (e) 0.487 L

Mass–Volume Relationships in Reactions Involving Gases

c 127. Which one of the following statements about the following reaction is false?

$$CH_4(g) + 2O_2(g) \rightarrow CO_2(g) + 2H_2O(g)$$

(a) Every methane molecule that reacts produces two water molecules.
(b) If 16.0 g of methane react with 32.0 g of oxygen, the maximum amount of CO_2 produced will be 22.0 g.
(c) If 11.2 liters of methane react with an excess of oxygen, the volume of CO_2 produced at STP is (44/16)(11.2) liters.
(d) If 16.0 g of methane react with 64.0 g of oxygen, the combined masses of the products will be 80.0 g.
(e) If 22.4 liters of methane at STP react with 64.0 g of oxygen, 22.4 L (STP) of CO_2 can be produced.

d 128. Assume that ammonia can be prepared by the following reaction in the gas phase at STP. If the reaction conditions are maintained at STP, which one of the following statements is incorrect?

$$N_2 + 3H_2 \rightarrow 2NH_3$$

(a) 22.4 L of N_2 reacts with 3(22.4) L of H_2 to form 2(22.4) L of NH_3.
(b) 273 L of N_2 reacts with 819 L of H_2 to form 546 L of NH_3.
(c) 28.0 g of N_2 reacts with 6.0 g of H_2 to form 44.8 L of NH_3.
(d) A maximum of 273 L of NH_3 can be prepared from 546 L of N_2, given an adequate supply of H_2.
(e) 1 mole of N_2 reacts with 3 moles of H_2 to form 2 moles of NH_3.

a 129. Consider the following **gas phase** reaction. If the reaction conditions are maintained at STP, which response contains all the following statements that are **incorrect** and no others?

$$2CO + O_2 \rightarrow 2CO_2$$

I. 2(22.4 liters) of CO will react with 22.4 liters of O_2 to produce 2(22.4 liters) of CO_2.
II. 412 liters of CO will react with 206 liters of O_2 to produce 412 liters of CO_2.
III. 2 moles of CO will react with one mole of O_2 to produce 2 moles of CO_2.
IV. 28 grams of CO will react with 24 grams of O_2 to produce 52 grams of CO_2.
V. 28 grams of CO will react with 24 grams of O_2 to produce 33.6 liters of CO_2.

(a) IV and V (b) I, II, and V (c) III and V
(d) II, IV, and V (e) I, III, and V

b*130. One half liter of element X (a gas) reacts with one liter of hydrogen gas to form one liter of a gaseous compound. All gas volumes are measured at the same conditions of temperature and pressure. What is the formula of gaseous compound formed in this reaction? (The formula for element X is X_2.)

(a) HX (b) H_2X (c) H_3X (d) H_4X (e) H_5X

b 131. What volume of hydrogen (STP) is produced by dissolving one-half mole of aluminum in sulfuric acid?

$$2Al(s) + 3H_2SO_4(aq) \rightarrow Al_2(SO_4)_3(aq) + 3H_2(g)$$

(a) 14.9 L (b) 16.8 L (c) 22.4 L (d) 29.9 L (e) 67.2 L

d 132. What volume of O_2 would be required to react with excess SO_2 at STP to produce 0.500 mole of SO_3?

$$2SO_2(g) + O_2(g) \rightarrow 2SO_3(g)$$

(a) 44.8 L (b) 22.4 L (c) 33.6 L (d) 5.60 L (e) 11.2 L

c 133. Consider the following gas phase reaction.

$$4HCl(g) + O_2(g) \rightarrow 2Cl_2(g) + 2H_2O(g)$$

What volume of chlorine at STP can be prepared from the reaction of 600. mL of gaseous HCl, measured at STP, with excess O_2, assuming that all the HCl reacts?

(a) 150. mL (b) 267 mL (c) 300. mL (d) 425 mL (e) 600. mL

e 134. If 3.25 liters of ammonia react with an excess of oxygen at 500.°C and 5.00 atm pressure, what volume of steam will be produced at the same temperature and pressure?

$$4NH_3(g) + 5O_2(g) \rightarrow 4NO(g) + 6H_2O(g)$$

(a) 2.04 L (b) 2.17 L (c) 3.33 L (d) 4.14 L (e) 4.88 L

e*135. Calculate the weight of KClO₃ that would be required to produce 29.52 liters of oxygen measured at 127°C and 760. torr.

$$2KClO_3(s) \xrightarrow{heat} 2KCl(s) + 3O_2(g)$$

(a) 7.82 g   (b) 12.2 g   (c) 14.64 g   (d) 24.4 g   (e) 73.5 g

e 136. What volume of NO (STP) can be produced by dissolving 62.1 grams of lead in nitric acid?

$$3Pb + 8HNO_3 \rightarrow 3Pb(NO_3)_2 + 2NO + 4H_2O$$

(a) 1.12 L   (b) 1.68 L   (c) 2.24 L   (d) 3.36 L   (e) 4.48 L

e 137. What volume of dry gaseous HBr at STP can be obtained from the reaction of 81.2 g of PBr₃ with excess water?

$$PBr_3(\ell) + 3H_2O(\ell) \rightarrow H_3PO_3(\ell) + 3HBr(g)$$

(a) 6.72 L   (b) 15.2 L   (c) 17.4 L   (d) 26.4 L   (e) 20.2 L

b 138. If sufficient acid is used to react completely with 3.78 grams of zinc, what volume of hydrogen would be produced at STP?

$$Zn(s) + 2HCl(aq) \rightarrow ZnCl_2(aq) + H_2(g)$$

(a) 1.14 L   (b) 1.29 L   (c) 1.38 L   (d) 1.45 L   (e) 1.53 L

a 139. What volume of H₂ at STP would be required to react completely with 12.4 grams of Fe₃O₄ according to the equation below?

$$Fe_3O_4(s) + 4H_2(g) \xrightarrow{heat} 3Fe(s) + 4H_2O(g)$$

(a) 4.80 L   (b) 16.7 L   (c) 32.0 L   (d) 20.4 L   (e) 1.20 L

a*140. A 0.720-gram sample, known to contain only CaCO₃ and BaCO₃, was heated until the carbonates were decomposed to oxides as indicated. After heating, the mass of the solid residue was 0.516 gram. What volume of CO₂, measured at STP, was produced by the reactions?

$$CaCO_3(s) \xrightarrow{heat} CaO(s) + CO_2(g)$$

$$BaCO_3(s) \xrightarrow{heat} BaO(s) + CO_2(g)$$

(a) 104 mL   (b) 168 mL   (c) 284 mL   (d) 486 mL   (e) 627 mL

d 141. Consider the following gas phase reaction.

$$4HCl(g) + O_2(g) \rightarrow 2Cl_2(g) + 2H_2O(g)$$

What mass of chlorine can be prepared from the reaction of 600. mL of gaseous HCl, measured at STP, with excess O₂, assuming that all the HCl reacts?

(a) 190. g   (b) 0.475 g   (c) 0.686 g   (d) 0.951 g   (e) 1.42 g

b 142. Magnesium reacts with ammonia, NH_3, at high temperatures to produce solid magnesium nitride, Mg_3N_2, and hydrogen. How many grams of magnesium react with 16,400 mL (STP) of ammonia?

$$3Mg(s) + 2NH_3(g) \xrightarrow{heat} Mg_3N_2(s) + 3H_2(g)$$

(a) 11.6 g (b) 26.7 g (c) 34.2 g (d) 22.9 g (e) 19.6 g

e*143. A sample of iron was dissolved in hydrochloric acid. The liberated hydrogen was collected by displacement of water. The hydrogen occupied 262 mL at 26°C. The barometric pressure was 748 torr. What mass of iron was dissolved? The vapor pressure of water is 25 torr at 26°C.

$$Fe(s) + 2HCl(aq) \rightarrow FeCl_2(aq) + H_2(g)$$

(a) 0.284 g (b) 0.414 g (c) 0.496 g (d) 0.549 g (e) 0.565 g

c*144. A 4.52-gram sample containing only $CaCO_3$ and $MgCO_3$ was heated to decompose the carbonates. The liberated CO_2 occupied 1.36 liters at 800.°C under a pressure of 3.30 atmospheres. What mass of $CaCO_3$ was contained in the original sample?

$$CaCO_3(s) \xrightarrow{heat} CaO(s) + CO_2(g)$$

$$MgCO_3(s) \xrightarrow{heat} MgO(s) + CO_2(g)$$

(a) 1.3 g (b) 2.4 g (c) 1.5 g (d) 3.2 g (e) 2.7 g

e*145. A 0.720-gram sample, known to contain only $CaCO_3$ and $BaCO_3$, was heated until the carbonates were decomposed to oxides as indicated. After heating, the mass of the solid residue was 0.516 gram. What mass of $BaCO_3$ was present in the sample?

$$CaCO_3(s) \xrightarrow{heat} CaO(s) + CO_2(g)$$

$$BaCO_3(s) \xrightarrow{heat} BaO(s) + CO_2(g)$$

(a) 0.316 g (b) 0.384 g (c) 0.516 g (d) 0.463 g (e) 0.521 g

d 146. Silver oxide decomposes completely at temperatures in excess of 300°C to produce metallic silver and oxygen gas. A 1.60-gram sample of impure Ag_2O gives 72.1 mL of O_2 measured at STP. What is the percentage of Ag_2O in the original sample?

$$2Ag_2O(s) \xrightarrow{heat} 4Ag(s) + O_2(g)$$

(a) 79.6% (b) 62.0% (c) 86.4% (d) 93.3% (e) 73.3%

b 147. An impure sample of $NaClO_3$ having a mass of 25.0 grams was heated until all the $NaClO_3$ had decomposed. The liberated oxygen occupied 4.64 liters at STP. What percentage of the sample was $NaClO_3$?

$$2NaClO_3(s) \xrightarrow{heat} 2NaCl(s) + 3O_2(g)$$

(a) 88.6% (b) 58.8% (c) 48.6% (d) 77.3% (e) 91.1%

d 148. A 25.0-gram sample of impure $NaClO_3$ produces 6.72 L of oxygen (STP). What is the percentage purity of the $NaClO_3$?

$$2NaClO_3(s) \xrightarrow{heat} 2NaCl(s) + 3O_2(g)$$

(a) 79%   (b) 81%   (c) 83%   (d) 85%   (e) 87%

b 149. An impure sample of $KClO_3$ that had a mass of 30.0 g was heated until all the $KClO_3$ had decomposed. The liberated oxygen occupied 6.72 liters at STP. What percent of the sample was $KClO_3$?

$$2KClO_3(s) \xrightarrow{heat} 2KCl(s) + 3O_2(g)$$

(a) 80.2%   (b) 81.7%   (c) 82.9%   (d) 84.1%   (e) 40.8%

b 150. A 22.0-gram sample of impure sodium chlorate was heated until decomposition was complete. The liberated oxygen occupied a volume of 3.20 L measured at 20.°C and 790. torr. What was the percent by mass of sodium chlorate in the original sample?

$$2NaClO_3(s) \xrightarrow{heat} 2NaCl(s) + 3O_2(g)$$

(a) 22.3%   (b) 44.5%   (c) 54.8%   (d) 72.6%   (e) 92.0%

a*151. If 48.8 grams of calcium hydride is treated with 39.2 grams of H_2O, how many grams of H_2 are produced? The other product of the reaction is calcium hydroxide.

(a) 4.4 g   (b) 3.9 g   (c) 5.5 g   (d) 4.9 g   (e) 2.2 g

d*152. If 36.0 g of C_3H_8 and 112 g of O_2 are placed in a closed container and the mixture is ignited, what is the maximum mass of CO_2 that could be produced? The other product of the reaction is water.

(a) 68.9 g   (b) 74.7 g   (c) 86.8 g   (d) 92.4 g   (e) 108 g

b*153. If 36.0 g of C_3H_8 and 112 g of O_2 are placed in a closed container and the mixture is ignited, the reaction products are CO_2 and H_2O. If 61.6 g of CO_2 are actually produced in the reaction, what is the percent yield of CO_2?

(a) 42.6%   (b) 66.7%   (c) 59.4%   (d) 71.0%   (e) 78.4%

The Kinetic-Molecular Theory

e 154. At the same _____, the molecules of all samples of ideal gases have the same average kinetic energies.

(a) volume   (b) pressure   (c) mass
(d) density   (e) temperature

e 155. The average kinetic energy of ideal gas molecules is directly proportional to the _____.

(a) volume of the sample   (b) pressure of the sample   (c) mass of the molecule
(d) density of the sample   (e) absolute temperature of the sample

c 156. All of the following statements, except one, are important postulates of the kinetic-molecular theory of gases. Which one?

    (a) Gases consist of large numbers of particles in rapid random motion.
    (b) The volume of the molecules of a gas is very small compared to the total volume in which the gas is contained.
    (c) The average kinetic energy of the molecules is inversely proportional to the absolute temperature.
    (d) The time during which a collision between two molecules occurs is negligibly short compared to the time between collisions.
    (e) There are no attractive or repulsive forces between the individual molecules.

a 157. Which one of the following statements is **not consistent** with the kinetic-molecular theory?

    (a) The volume occupied by the molecules (only) of a gas becomes significant only at very low pressures.
    (b) A given sample of a gas is mostly empty space except near the liquefaction point.
    (c) Except near the liquefaction point, the attractive forces between molecules of a gas are very small.
    (d) Collisions between the molecules of a gas are elastic.
    (e) The attractive forces between the molecules of a gas become significant only at very low temperatures.

b 158. Which statement is **false**?

    (a) The molecules of an ideal gas are relatively very far apart on the average.
    (b) The molecules of a gas move in unchanging curved paths until they collide with other molecules or the walls of their containers, according to the kinetic-molecular theory.
    (c) The average kinetic energies of molecules of samples of different ideal gases at the same temperature are the same.
    (d) The average kinetic energies of molecules of samples of different ideal gases at different temperatures are different.
    (e) At a given temperature, according to the kinetic molecular theory, the average velocity of HF molecules is greater than the average velocity of F_2 molecules.

a 159. Which response includes all the following statements that describe correctly the properties of gases, and no others?

    I. At constant temperature, the pressure increases as the volume of a definite mass of a gas increases.
    II. At constant pressure, the volume of a definite mass of a gas increases as the temperature increases.
    III. At reasonable temperatures and pressures, gases consist mostly of empty space.
    IV. Heating a sample of gas forces the particles closer together when the pressure and volume remain constant.
    V. The forces of attraction between individual gaseous particles are relatively weak.

    (a) II, III, and V   (b) I, II, and V   (c) III and V
    (d) II, IV, and V   (e) I, III, and V

d 160. Which one of the following statements does **not** describe ideal gases?

  (a) The average kinetic energies of molecules of different kinds of gases are the same at a given temperature.
  (b) Under ordinary conditions of temperature and pressure, gas molecules are widely separated.
  (c) Gas molecules move in rapid, random, straight-line paths until collision occurs with other molecules or with the walls of the container.
  (d) The volume occupied by the gas molecules themselves under ordinary conditions is large in comparison to the total volume occupied by the gas.
  (e) Gases are "less ideal" near their liquefaction points.

b 161. A gaseous mixture contains equal numbers of HCl and HBr molecules and no other kind of molecules. Which one of the following statements is true? Assume ideal gas behavior for both HCl and HBr.

  (a) The masses of HCl and HBr are equal.
  (b) The partial pressures of HCl and HBr are equal.
  (c) If the temperature is increased while the volume is held constant, the partial pressure of HBr will rise more rapidly than the partial pressure of HCl.
  (d) Since HCl molecules are lighter, they have higher velocities, collide with each other and with the walls of the container more frequently, and therefore contribute more to the total pressure of the mixture than do HBr molecules.
  (e) None of the above statements is true.

Diffusion and Effusion of Gases

b 162. A mixture of 0.75 mol H_2(g) and 0.75 mol N_2(g) is introduced into a 15.0-liter container having a pinhole leak at 30°C. After a period of time which of the following is true?

  (a) The partial pressure of H_2 exceeds that of N_2 in the container.
  (b) The partial pressure of N_2 exceeds that of H_2 in the container.
  (c) The partial pressures of the two gases remain equal.
  (d) The partial pressures of both gases increase above their initial values.
  (e) The partial pressures of the two gases remain unchanged.

e 163. What is the order of increasing rate of effusion for the following gases?
  Ar, CO_2, He, N_2

  (a) N_2<Ar<CO_2<He   (b) Ar<CO_2<He<N_2 (c) Ar<He<CO_2<N_2
  (d) CO_2<N_2<Ar<He   (e) CO_2<Ar<N_2<He

b*164. At a given temperature and pressure, the velocity of an He molecule would be _____ times the velocity of a CH_4 molecule.

  (a) one   (b) two   (c) three   (d) four   (e) five

d*165. Under comparable conditions, how much faster will a sample of neon effuse out through a small opening than a sample of Cl_2?

  (a) 0.533   (b) 1.33   (c) 1.75   (d) 1.87   (e) 3.51

b*166. Calculate the ratio of the rate of effusion of CO_2 to He.

(a) 0.090/1 (b) 0.30/1 (c) 3.3/1 (d) 11/1 (e) 12/1

a*167. What is the molecular weight of a (hypothetical) gas that diffuses 1.414 times faster than nitrogen, N_2?

(a) 14.0 g/mol (b) 4.85 g/mol (c) 23.5 g/mol
(d) 32.6 g/mol (e) 46.6 g/mol

e*168. A gas of unknown composition effuses at the rate of 10 mL/s in an effusion apparatus in which CH_4 gas effuses at the rate of 20 mL/s. What is the approximate molecular weight of the unknown gas?

(a) 4 amu (b) 8 amu (c) 16 amu (d) 32 amu (e) 64 amu

Real Gases: Deviations from Ideality

a 169. The most significant intermolecular forces of attraction between individual nonpolar atoms or molecules in the gas phase are _____ attractions.

(a) dispersion (b) strong electrostatic (c) covalent
(d) ionic (e) strong dipole

c 170. Which one of the responses includes all of the following statements that are correct, and no incorrect statements?

I. At a given temperature and pressure, say 500°C and 1 atmosphere, the attractive forces between water molecules are greater than the attractive forces between ammonia molecules.
II. At 500°C and 1 atmosphere pressure, the attractive forces between ammonia molecules are greater than those between water molecules.
III. By elastic collisions between molecules of a gas, we refer to collisions in which there is no net gain or loss of energy.
IV. If the pressure on a sample of an ideal gas is doubled while the absolute temperature is doubled, the final volume will be unchanged.
V. If the pressure on a sample of an ideal gas is halved while the absolute temperature is halved, the final volume will be unchanged.

(a) I, II, and III (b) III, IV, and V (c) I, III, IV, and V
(d) I, IV, and V (e) II and IV

b 171. The van der Waals constant, a, in the relationship $(P + \frac{n^2a}{V^2})(V - nb) = nRT$ is a factor that corrects for

(a) deviations in the gas constant, R.
(b) the attractive forces between gas molecules.
(c) the tendency of the gas molecules to ionize.
(d) the average velocities of the gas molecules.
(e) the volume occupied by the gas molecules.

e 172. The van der Waals constant, b, in the relationship $(P + \frac{n^2a}{V^2})(V - nb) = nRT$ is a factor that corrects for

(a) deviations in the gas constant, R.
(b) the attractive forces between gas molecules.
(c) the tendency of the gas molecules to ionize.
(d) the average velocities of the gas molecules.
(e) the volume occupied by the gas molecules.

c 173. Which one of the statements below is **false**?

(a) A real gas behaves more nearly as an ideal gas at high temperatures and low pressures.
(b) In the van der Waals equation, the "a" factor corrects for attractive forces, and one would expect a larger value of "a" for HF than for He.
(c) The "b" factor in the van der Waals equation should be larger for He than for Cl_2.
(d) Gases approach their liquefaction points as temperature decreases and as pressure increases.
(e) Both "a" and "b" of the van der Waals equation have values of zero for an ideal gas.

e 174. Which one of the responses includes all of the following statements that are correct, and no incorrect statements?

I. Helium has smaller van der Waals constants than ammonia.
II. The van der Waals constant "a" corrects for the attractive forces between molecules.
III. The van der Waals constant "b" corrects for the volume occupied by the molecules of a gas.
IV. Deviations from ideality are greater at low temperatures and high pressures than at high temperatures and low pressures.

(a) I and II   (b) II and III   (c) III and IV
(d) I, III, and IV (e) I, II, III, and IV

e 175. Which of the following gases is expected to have the largest value for its van der Waals constant "b"?

(a) Ne   (b) O_2 (c) H_2 (d) CO   (e) CO_2

a 176. Which of the following gases is expected to have the smallest value for its van der Waals constant "a"?

(a) Ne   (b) O_2 (c) N_2 (d) H_2O (e) CO_2

a 177. Which of the following gases is expected to have the smallest value for its van der Waals constant "b"?

(a) Ne   (b) O_2 (c) N_2 (d) Cl_2 (e) H_2O

a 178. Which of the following gases would be expected to have a behavior closest to that of an ideal gas under most conditions?

(a) He   (b) O_2 (c) NH_3 (d) Cl_2 (e) H_2O

c*179. Calculate the pressure (in atm) exerted by 1.00 mole of acetylene at 125°C in a 20.0-liter container. The van der Waals constants for acetylene are: a = 20.0 $L^2 \cdot atm/mol^2$, b = 0.100 L/mol.

(a) 0.485   (b) 0.533   (c) 1.59   (d) 1.64   (e) 1.86

b*180. Calculate the pressure (in atm) exerted by 1.00 mole of ammonia at 27°C in a 10.00-liter container. The van der Waals constants for ammonia are: a = 4.00 $L^2 \cdot atm/mol^2$, b = 0.0400 L/mol. (The values for a and b have been rounded off to simplify the arithmetic in this problem.)

(a) 2.38   (b) 2.43   (c) 2.34   (d) 2.46   (e) 2.29

13 Liquids and Solids

Kinetic-Molecular Description of Liquids and Solids

c 1. Which one of the following statements does **not** describe the general properties of liquids accurately?

    (a) Liquids have characteristic volumes that do not change greatly with changes in temperature. (Assuming that the liquid is not vaporized.)
    (b) Liquids have characteristic volumes that do not change greatly with changes in pressure.
    (c) Liquids diffuse only very slowly when compared to solids.
    (d) The liquid state is highly disordered compared to the solid state.
    (e) Liquids have high densities compared to gases.

e 2. Which one of the following statements does **not** describe the general properties of solids accurately?

    (a) Solids have characteristic volumes that do not change greatly with changes in temperature.
    (b) Solids have characteristic volumes that do not change greatly with changes in pressure.
    (c) Solids diffuse only very slowly when compared to liquids and gases.
    (d) Solids are not fluid.
    (e) Most solids have high vapor pressures at room temperature.

Intermolecular Attractions and Phase Changes

a 3. Which of the following responses arranges interactions in order of increasing strength? ("permanent" is used to indicate non-H-bonding permanent dipole-dipole interactions)
        covalent bonds, **dispersion** forces, **hydrogen** bonds, non-H-bonding **permanent dipole**-permanent dipole interactions

    (a) permanent dipole < dispersion < hydrogen < covalent
    (b) hydrogen < permanent dipole < covalent < dispersion
    (c) dispersion < hydrogen < permanent dipole < covalent
    (d) covalent < hydrogen < permanent dipole < dispersion
    (e) hydrogen < dispersion < covalent < permanent dipole

d 4. The boiling points of the halogens increase in the order $F_2 < Cl_2 < Br_2 < I_2$ due to the resulting increasing _____ interactions.

    (a) ion-dipole   (b) hydrogen-bonding   (c) ion-ion
    (d) dispersion forces   (e) permanent dipole-dipole

b 5. For which of the following would permanent dipole-dipole interactions play an important role in determining physical properties in the liquid state?

    (a) BF_3 (b) ClF   (c) $BeCl_2$ (d) F_2 (e) CCl_4

c 6. For which of the following would dispersion forces be the most important factor in determining physical properties in the liquid state?

(a) H_2O (b) NaCl   (c) F_2 (d) HF   (e) NH_4Cl

a 7. For which of the following would hydrogen bonding **not** be an important factor in determining physical properties in the liquid state?

(a) HI   (b) H_2O (c) HF   (d) NH_3 (e) H_2O_2

a 8. Which response includes all of the following substances in which dispersion forces are the **most significant** factors in determining boiling points, and no other substances?

I. Cl_2 II. HF   III. Ne   IV. KNO_2 V. CCl_4

(a) I, III, and V   (b) I, II, and III   (c) II and IV
(d) II and V   (e) III, IV, and V

c 9. In which of the following would dispersion forces be the **only significant** factors in determining boiling point?

I. Ar   II. Li_2SO_4 III. SiF_4 IV. Br_2 V. NH_3

(a) I, II, and III   (b) II, IV, and V   (c) I, III, and IV
(d) I, IV and V   (e) II and V

d 10. Which response includes all of the following substances that can exhibit hydrogen bonding, and no others?

I. H_2 II. CH_4 III. NH_3 IV. SiH_4 V. HF

(a) II and V   (b) I, II, and III   (c) III, IV, and V
(d) III and V   (e) I, III, and IV

a 11. Which response includes all the following substances that exhibit hydrogen bonding, and no other substances?

I. NH_3 II. SiH_4 III. SbH_3 IV. CH_4

(a) I   (b) II   (c) I and III   (d) II and IV   (e) II and III

b 12. Which response includes all of the following compounds that exhibit hydrogen bonding and no other compounds?
CH_4, AsH_3, CH_3NH_2, H_2Te, HF

(a) AsH_3, H_2Te (b) CH_3NH_2, HF   (c) CH_4, AsH_3, H_2Te
(d) AsH_3, CH_3NH_2 (e) HF, H_2Te

d*13. Which of the following substances has the weakest "hydrogen bonding" in the liquid state?

(a) HF   (b) NH_3 (c) CH_3OH (d) HCl   (e) H_2O

Harcourt, Inc

e 14. Which of the following boils at the **highest** temperature?

(a) C_2H_6 (b) C_3H_8 (c) C_4H_{10} (d) C_5H_{12} (e) C_6H_{14}

b 15. Which one of the following boils at the **lowest** temperature?

(a) H_2O (b) H_2S (c) H_2Se (d) H_2Te (e) H_2Po

c 16. Which one of the following boils at the **lowest** temperature?

(a) KNO_3 (b) Ca (c) Kr (d) NH_3 (e) AsH_3

a 17. Which of the following boils at the **lowest** temperature?

(a) CF_4 (b) HF (c) HCl (d) KI (e) SiF_4

d 18. All of the following are gases at room temperature and atmospheric pressure. Which one will liquefy most easily when pressurized at a certain temperature?

(a) H_2 (b) Ar (c) SiH_4 (d) NH_3 (e) F_2

a 19. Which one of the following would have the highest molar heat of vaporization?

(a) CH_3NH_2 (b) CH_4 (c) C_2H_6 (d) SiH_4 (e) H_2S

The Liquid State

c 20. The term used to describe resistance to flow of a liquid is _____.

(a) surface tension   (b) capillary action   (c) viscosity
(d) vapor pressure   (e) vaporization

c 21. On a relative basis, the weaker the intermolecular forces in a substance are,

(a) the larger is its heat of vaporization.
(b) the more it deviates from the ideal gas law.
(c) the greater is its vapor pressure at a particular temperature.
(d) the larger is its molar heat capacity as a liquid.
(e) the higher is its boiling point.

b 22. Which statement is **false**?

(a) In the absence of a phase change, the viscosity of a liquid increases as temperature decreases.
(b) All other factors being equal, if adhesive forces are strong, capillary action is likely to occur less readily than if adhesive forces are weak.
(c) The shape of a meniscus depends on the difference between the strengths of cohesive forces and adhesive forces.
(d) Liquids with strong cohesive forces have high heats of vaporization.
(e) Vaporization of liquids can occur below their normal boiling points at one atmosphere pressure.

c 23. Which one of the following statements does **not** describe the general properties of liquids accurately?

(a) In the liquid state the close spacing of molecules leads to large intermolecular forces that are strongly dependent on the nature of the molecules involved.
(b) Liquids are practically incompressible.
(c) As the temperature of a liquid is increased, the vapor pressure of the liquid decreases.
(d) The normal boiling point of a liquid is the temperature at which the vapor pressure of the liquid becomes equal to exactly 760 torr.
(e) Vapor pressures of liquids at a given temperature differ greatly, and these differences in vapor pressure are due to the nature of the molecules in different liquids.

c 24. As we increase the temperature of a liquid, its properties change. Which of the following would **not** be an expected change in the properties of a typical liquid as we increase its temperature?

(a) decrease in viscosity (b) decrease in density
(c) increase in surface tension (d) increase in vapor pressure
(e) increase in tendency to evaporate

a 25. Which response contains all the **true** statements, and no others?

I. The vapor pressure of ethyl alcohol increases as temperature increases.
II. Liquids of lower molecular weight always exhibit relatively higher vapor pressures than liquids of higher molecular weight.
III. A liquid boils at a higher temperature at sea level than on top of a mountain.
IV. The temperature of boiling water inside a pressure cooker is above 100°C.

(a) I, III, and IV (b) I and II (c) II and IV
(d) III (e) I and IV

Heat Transfer Involving Liquids

a 26. Calculate the amount of heat required to raise the temperature of 165 g of water from 12.0°C to 88.0°C. (Sp. heat of $H_2O(\ell)$ = 4.18 J/g·°C)

(a) 52.4 kJ (b) 1.93 J (c) 60.7 kJ (d) 1.46 kJ (e) 15.6 J

d 27. How much heat is released when 165 g of liquid benzene, $C_6H_6(\ell)$, cools from 78.0°C to 12.0°C? (Sp. heat of $C_6H_6(\ell)$ = 1.74 J/g·°C)

(a) 4.35 J (b) 6.26 kJ (c) 45.5 kJ (d) 18.9 kJ (e) 22.4 kJ

c 28. Calculate the amount of heat (in joules) required to convert 92.5 g of water at 25.0°C to steam at 108.0°C. (Sp. heat of $H_2O(\ell)$ = 4.18 J/g·°C, Sp. heat of $H_2O(g)$ = 2.03 J/g·°C, heat of vap. of $H_2O(\ell)$ = 2.260 kJ/g)

(a) 2.26 x 10^5 J (b) 3.05 x 10^4 J (c) 2.40 x 10^5 J
(d) 2.20 x 10^4 J (e) 6.43 x 10^5 J

d 29. How much heat is released when 40.0 g of steam at 250.0°C cools and condenses to water at 30.0°C? (Sp. heat of $H_2O(\ell)$ = 4.18 J/g·°C, Sp. heat of $H_2O(g)$ = 2.03 J/g·°C, heat of vap. of $H_2O(\ell)$ = 2.260 kJ/g)

(a) 24.0 kJ   (b) 23.0 J   (c) 32.9 kJ   (d) 114 kJ   (e) 122 kJ

a 30. If 100. grams of liquid water at 100.°C and 200. grams of water at 20.0°C are mixed in an insulated container, what will the final temperature of the mixture be? (Sp. heat of $H_2O(\ell)$ = 4.18 J/g·°C)

(a) 46.7°C   (b) 60.0°C   (c) 66.7°C   (d) 73.3°C   (e) 77.8°C

e*31. If 10.0 g of steam at 110.0°C is pumped into an insulated vessel containing 100. g of water at 20.0°C, what will be the equilibrium temperature of the mixture? (Sp. heat of $H_2O(\ell)$ = 4.18 J/g·°C, Sp. heat of $H_2O(g)$ = 2.03 J/g·°C, heat of vap. of $H_2O(\ell)$ = 2.260 kJ/g)

(a) 86.8°C   (b) 54.5°C   (c) 58.4°C   (d) 38.6°C   (e) 76.9°C

c*32. Using the Clausius-Clapeyron equation determine the vapor pressure of water at 50.0°C. The molar heat of vaporization of water is 40.7 kJ/mol.

$$\ln\left(\frac{P_2}{P_1}\right) = \frac{\Delta H_{vap}}{R}\left(\frac{1}{T_1} - \frac{1}{T_2}\right)$$

(a) 700 torr   (b) 450 torr   (c) 100 torr   (d) 80 torr   (e) 55 torr

e 33. Which substance would be expected to have the **highest** heat of vaporization?

(a) F_2 (b) CCl_4 (c) C_2H_6 (d) CH_3OCH_3 (e) CH_2OHCH_2OH

e 34. Which response has the following substances arranged in order of **increasing** boiling point?

Ar, $NaClO_3$, H_2O, H_2Se

(a) $NaClO_3 < H_2O < H_2Se < Ar$
(b) $NaClO_3 < H_2Se < H_2O < Ar$
(c) $Ar < NaClO_3 < H_2Se < H_2O$
(d) $Ar < H_2O < H_2Se < NaClO_3$
(e) $Ar < H_2Se < H_2O < NaClO_3$

Heat Transfer Involving Solids

b 35. The amount of energy associated with holding the individual particles (atoms, ions, or molecules) together in a crystal lattice is most directly related to _____.

(a) the specific heat of the solid
(b) the heat of fusion of the solid
(c) the specific heat of the liquid
(d) the density of the solid
(e) the heat of condensation of the liquid

d 36. What is the heat of fusion of chromium (**in kJ/mol**) if 10.0 grams of solid chromium absorb 2.82 x 10^3 J of heat in melting at 2173 K, its melting point?

(a) 1.34   (b) 6.86   (c) 48.5   (d) 14.7   (e) 34.3

d 37. What would be the final temperature of the system if 30.0 g of lead at 150.°C is dropped into 10.0 g of water at 10.0°C in an insulated container? (Sp. heat of $H_2O(\ell)$ = 4.18 J/g·°C, Sp. heat of Pb(s) = 0.128 J/g·°C)

(a) 12.4°C   (b) 16.8 °C   (c) 19.4°C   (d) 21.7°C   (e) 24.6°C

a 38. Calculate the amount of heat absorbed by 10.0 grams of ice at –15.0°C in converting it to liquid water at 50.0°C. (Sp. heat of $H_2O(s)$ = 2.09 J/g·°C, Sp. heat of $H_2O(\ell)$ = 4.18 J/g·°C, heat of fusion of $H_2O(s)$ = 333 J/g)

(a) 5.73 x 10^3 J   (b) 6.76 x 10^2 J   (c) 1.70 x 10^2 J
(d) 2.83 x 10^3 J   (e) 3.29 x 10^3 J

c 39. Calculate the amount of heat required to convert 10.0 grams of ice at –20.°C to steam at 120.°C. (Sp. heat of $H_2O(s)$ = 2.09 J/g·°C, Sp. heat of $H_2O(\ell)$ = 4.18 J/g·°C, Sp heat of $H_2O(g)$ = 2.03 J/g·°C; heat of fus. of $H_2O(s)$ = 333 J/g, heat of vap. of $H_2O(\ell)$ = 2260 J/g)

(a) 18.6 kJ   (b) 26.3 kJ   (c) 30.9 kJ   (d) 41.2 kJ   (e) 46.4 kJ

e 40. If 10.0 g of ice at –10.0°C is placed in 200. g of water at 80.0°C in an insulated container, what will be the temperature of the system when equilibrium is established? (Sp. heat of $H_2O(s)$ = 2.09 J/g·°C, Sp. heat of $H_2O(\ell)$ = 4.18 J/g·°C, heat of fusion of $H_2O(s)$ = 333 J/g)

(a) 76°C   (b) 65°C   (c) 27°C   (d) 20°C   (e) 72°C

e 41. What will be the final temperature of the liquid water resulting from the mixing of 10.0 grams of steam at 130.°C with 40.0 g of ice at –10.0°C? (Sp. heat of $H_2O(s)$ = 2.09 J/g·°C, Sp. heat of $H_2O(\ell)$ = 4.18 J/g·°C, Sp. heat of $H_2O(g)$ = 2.03 J/g·°C; heat of fusion of $H_2O(s)$ = 333 J/g, heat of vap. of $H_2O(\ell)$ = 2260 J/g)

(a) 80.4°C   (b) 72.6°C   (c) 54.3°C   (d) 46.1°C   (e) 63.3°C

b*42. How much heat would be required to convert 234.3 g of solid benzene, $C_6H_6(s)$, at 5.5°C into benzene vapor, $C_6H_6(g)$, at 100.0°C?

mp of $C_6H_6(s)$ = 5.5°C   molar heat of fusion at 5.5°C = 9.92 kJ/mol
bp of $C_6H_6(\ell)$ = 80.1°C   molar heat of vaporization at 80.1°C = 30.8 kJ/mol
molar heat capacity of $C_6H_6(\ell)$ = 136 J/mol·°C
molar heat capacity of $C_6H_6(g)$ = 81.6 J/mol·°C

(a) 106 kJ   (b) 158 kJ   (c) 53 kJ   (d) 32 kJ   (e) 5049 kJ

Sublimation

b 43. Some solids can be converted directly to the vapor phase by heating. The process is called _____.

(a) fusion   (b) sublimation   (c) vaporization
(d) condensation   (e) distillation

Phase Diagrams (P vs. T)

e 44. A sketch of the phase diagram (not to scale) of water is given below.

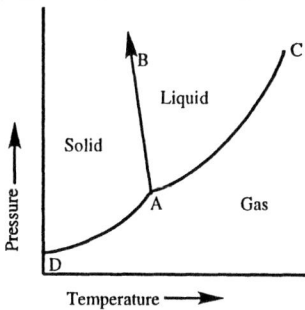

Which statement is **false**?

(a) Line AD is the sublimation curve - solid and vapor are in equilibrium.
(b) Point A is the triple point - solid, liquid, and vapor are at equilibrium.
(c) Line AC is the vapor pressure curve - liquid and gas (vapor) are in equilibrium.
(d) Line AB is the melting curve - solid and liquid are in equilibrium.
(e) The slope of line AB is negative showing that as the liquid is cooled, the molecules get closer and closer together as they solidify.

e 45. A sketch of the phase diagram (not to scale) of water is given below.

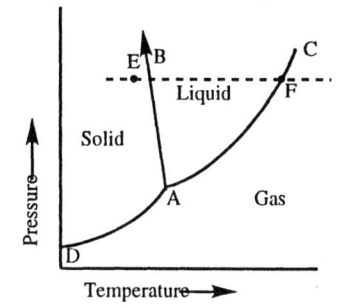

Which statement concerning the path (broken line) is **false**?

(a) At point E the water is all solid (ice).
(b) If heat is added to the ice at point E, the temperature of the ice increases until line AB is reached, then the temperature remains constant until all of the ice is melted into liquid.
(c) Once all the ice has melted (at the intersection of the broken line and line AB) the temperature of the water increases as heat is added until point F is reached.
(d) If more heat is added upon reaching point F, the temperature will remain constant as the liquid water vaporizes.
(e) Another way to vaporize all of the liquid water at point F is to increase the pressure.

Structures of Crystals

b 46. In a body-centered cubic lattice, how many atoms are contained in a unit cell?

(a) one   (b) two   (c) three   (d) four   (e) five

c 47. In a face-centered cubic lattice, how many atoms are contained in a unit cell?

(a) one   (b) two   (c) four   (d) five   (e) six

c 48. In any cubic lattice, an atom lying at the corner of a unit cell is shared equally by how many unit cells?

(a) one   (b) two   (c) eight   (d) four   (e) sixteen

b 49. In a face-centered cubic lattice, an atom laying in a face of a unit cell is shared equally by how many unit cells?

(a) one   (b) two   (c) eight   (d) four   (e) sixteen

d 50. In a cubic lattice, an atom on the edge of a unit cell is shared equally by how many unit cells?

(a) one   (b) two   (c) eight   (d) four   (e) twelve

e 51. A single substance that can crystallize in more than one arrangement is said to be _____.

(a) isomorphous   (b) amorphous   (c) primitive
(d) triclinic   (e) polymorphous

Bonding in Solids

c 52. Which one of the following statements is **not** applicable to **covalent** solids?

(a) The units that occupy the lattice points are atoms.
(b) The binding forces in covalent solids are shared electrons.
(c) Covalent solids have low melting points.
(d) Covalent solids are very hard.
(e) Covalent solids do not conduct electric current well.

b 53. Which one of the following statements is **not** applicable to **metallic** solids?

(a) The units that occupy the lattice points are positive ions.
(b) The binding forces in metallic solids are shared electron pairs.
(c) The melting points of metallic solids vary over a large range.
(d) The hardness of metallic solids varies from quite soft to quite hard.
(e) Metallic solids conduct electric current well.

d 54. Which one of the following statements is **not** applicable to **molecular** solids?

(a) The units that occupy the lattice points are molecules.
(b) The binding forces in molecular solids are dispersion forces or dispersion forces and dipole-dipole interactions.
(c) Molecular solids have relatively low melting points.
(d) Molecular solids are usually excellent conductors of electric current.
(e) Molecular solids are soft compared to covalent solids.

d 55. Which one of the following statements is **not** applicable to **ionic** solids?

(a) The units that occupy the lattice points are ions.
(b) The binding forces in ionic solids are electrostatic attractions.
(c) Most ionic solids are hard and brittle.
(d) Ionic solids are usually excellent conductors of electric current.
(e) Ionic solids have fairly high melting points.

b 56. Which one of the following is **not** a general property of **ionic** solids?

(a) hard and brittle
(b) good electrical conductors in the solid state
(c) relatively high melting points
(d) strongest interparticle attractions are electrostatic
(e) positions of ions define the unit cell (lattice)

d 57. Which statement is **false**?

(a) The size of an ion in a crystal is influenced by its environment, that is, by the other ions surrounding it.
(b) About 26% of the total volume of any cubic close packed crystal lattice is empty space.
(c) Two solids that crystallize in the same kind of crystal lattice are said to be isomorphous.
(d) An example of an amorphous solid is diamond.
(e) Glass window panes flow like liquids, but very very slowly.

d 58. Which one of the following is a covalent solid?

(a) sulfur trioxide (b) nickel (c) ammonium chloride
(d) silicon carbide, SiC (e) sucrose, $C_{12}H_{22}O_{11}$

c 59. Which one of the following is an ionic solid?

(a) graphite (b) nickel (c) ammonium chloride
(d) silicon carbide, SiC (e) sucrose, $C_{12}H_{22}O_{11}$

c 60. Which one of the following is a metallic solid?

(a) graphite (b) sulfur (c) nickel (d) iodine (e) neon

d 61. Which one of the following is classified as a covalent solid?

(a) K_2SO_4 (b) Cr (c) CO_2 (d) C (e) CH_4

d 62. Which one of the following is a molecular solid?

(a) NH_4Cl (b) Pb (c) SiC (d) C_6H_6 (e) $KC_2H_3O_2$

c 63. Which one of the following crystallizes in a metallic lattice?

(a) $C_{10}H_8$ (b) graphite (c) In (d) LiF (e) $KMnO_4$

c 64. Which of the following, in the solid state, would be an example of a molecular crystal?

(a) diamond
(b) copper
(c) phosphorus trichloride
(d) magnesium fluoride
(e) diamond

a 65. Which of the following in the solid state would be an example of a molecular crystal?

(a) carbon dioxide
(b) graphite
(c) cesium fluoride
(d) iron
(e) quartz

d 66. Which one of the following substances is **incorrectly** matched with the kind of solid it forms?

	Substance	Kind of Solid
(a)	sulfur dioxide	molecular
(b)	graphite	covalent
(c)	calcium bromide	ionic
(d)	lithium	ionic
(e)	methane	molecular

d 67. Which one of the following pairs is **incorrectly** matched?

	Substance	Classification
(a)	H$_2$O	molecular solid
(b)	paraffin	molecular solid
(c)	KF	ionic solid
(d)	CsI	covalent solid
(e)	Ni	metallic solid

b 68. Which one of the following pairs is **incorrectly** matched?

	Substance	Classification
(a)	sand	covalent solid
(b)	diamond	molecular solid
(c)	Fe	metallic solid
(d)	CaF$_2$	ionic solid
(e)	quartz	covalent solid

e 69. Elemental silicon exists as a solid with a crystal structure like that of diamond. But silicon is less dense than diamond. Which response contains all the **correct** conclusions that can be drawn?

I. Silicon and diamond are allotropes.
II. The carbon atoms in diamond are more closely spaced than silicon atoms in solid silicon.
III. Silicon is a poor electrical conductor.
IV. Silicon is amorphous in the solid state.
V. One would expect silicon to have a very high melting point.

(a) II, III, and IV
(b) I and III
(c) II and V
(d) III, IV, and V
(e) II, III, and V

e 70. Which statement is **false**?

(a) Molecular solids generally have lower melting points than covalent solids.
(b) Metallic solids exhibit a wide range of melting points.
(c) The lattice of a metallic solid is defined by the position of the metal nuclei, and the valence electrons are distributed over the lattice as a whole.
(d) Most molecular solids melt at lower temperatures than metallic solids.
(e) The interactions among the molecules in molecular solids are generally stronger than those among the particles that define either covalent or ionic lattices.

b 71. Substances have properties that are related to their structures. Which of the following statements regarding properties of solids is **not** expected to be correct?

(a) Solid potassium should be a good conductor of electricity.
(b) Solid $CaSO_4$ should sublime readily.
(c) Molten LiCl should be a good conductor of electricity.
(d) Graphite should have a high melting point.
(e) Solid CO_2 should have a low melting point.

d 72. Substances have properties that are related to their structures. Which of the following statements regarding properties of solids is **not** expected to be correct?

(a) Molten KBr should be a good conductor of electricity.
(b) Diamond should have a high melting point.
(c) Solid sodium should be a good conductor of electricity.
(d) Solid CaF_2 should have a low melting point.
(e) Silicon carbide, SiC, should not sublime readily.

e*73. Which one of the following elements has the lowest melting point?

(a) Li   (b) Na   (c) K   (d) Rb   (e) Cs

a 74. Which one of the following melts at the lowest temperature at a given pressure?

(a) CH_4 (b) LiCl   (c) $C_{10}H_8$ (d) $CaCl_2$ (e) H_2O

a 75. Which of the following compounds would be expected to have the highest melting point?

(a) BaF_2 (b) $BaCl_2$ (c) $BaBr_2$ (d) BaI_2 (e) NaF

e 76. Arrange the following **ionic** compounds in order of increasing melting points.

NaF, MgF_2, AlF_3, NaBr, NaI

Increasing Melting Points →

(a) NaF   <   NaBr   <   NaI   < MgF_2 < AlF_3
(b) NaBr <   NaI   <   NaF   < AlF_3 < MgF_2
(c) MgF_2 < AlF_3 <   NaF   <   NaI   <   NaBr
(d) AlF_3 < MgF_2 <   NaF   <   NaBr   <   NaI
(e) NaI   <   NaBr   <   NaF   < MgF_2 < AlF_3

c 77. Arrange the following in order of increasing melting points.

CsCl, BaCl$_2$, diamond (C), H$_2$, HF

Increasing Melting Points →

(a) H$_2$ < HF < BaCl$_2$ < CsCl < diamond
(b) HF < H$_2$ < CsCl < BaCl$_2$ < diamond
(c) H$_2$ < HF < CsCl < BaCl$_2$ < diamond
(d) H$_2$ < HF < diamond < CsCl < BaCl$_2$
(e) HF < H$_2$ < diamond < CsCl < BaCl$_2$

d 78. Arrange the following in order of increasing melting points.

KCl, He, H$_2$O, HF

(a) He < H$_2$O < HF < KCl
(b) H$_2$O < HF < He < KCl
(c) KCl < H$_2$O < HF < He
(d) He < HF < H$_2$O < KCl
(e) H$_2$O < He < KCl < HF

a 79. For crystal structures that contain only one kind of atom, the nearest neighbors of each atom can be visualized as lying along a line of the unit cell. For a simple cubic structure what is the orientation of that line with respect to the unit cell and how many atomic radii does it contain?

(a) cell edge and 2 atomic radii
(b) face diagonal and 4 atomic radii
(c) body diagonal and 3 atomic radii
(d) face diagonal and 2 atomic radii
(e) body diagonal and 4 atomic radii

e 80. For crystal structures that contain only one kind of atom, the nearest neighbors of each atom can be visualized as lying along a line of the unit cell. For a body-centered cubic structure what is the orientation of that line with respect to the unit cell and how many atomic radii does it contain?

(a) cell edge and 2 atomic radii
(b) face diagonal and 4 atomic radii
(c) body diagonal and 3 atomic radii
(d) face diagonal and 2 atomic radii
(e) body diagonal and 4 atomic radii

b 81. For crystal structures that contain only one kind of atom, the nearest neighbors of each atom can be visualized as lying along a line of the unit cell. For a face-centered cubic structure what is the orientation of that line with respect to the unit cell and how many atomic radii does it contain?

(a) cell edge and 2 atomic radii
(b) face diagonal and 4 atomic radii
(c) body diagonal and 3 atomic radii
(d) face diagonal and 2 atomic radii
(e) body diagonal and 4 atomic radii

e 82. In ideal close-packed structures of metallic solids, how many nearest neighbors does a metal ion have?

(a) 4   (b) 3   (c) 6   (d) 8   (e) 12

d 83. What is the coordination number for each sphere (metal ion) in a body-centered cubic structure?

(a) 4   (b) 3   (c) 6   (d) 8   (e) 12

b 84. Metallic calcium crystallized in a face-centered cubic lattice and the atomic radius of calcium is 1.97Å. Calculate the edge length, a, of a unit cell of calcium.

(a) 4.19Å   (b) 5.57Å   (c) 6.05Å   (d) 6.24Å   (e) 6.83Å

c 85. Potassium bromide, KBr, crystallizes in the NaCl (face-centered cubic) lattice. The ionic radii of K^+ and Br^- ions are 1.33Å and 1.95Å, respectively. Assuming anion-anion contact along the face diagonal and anion-cation contact along the edge of the unit cell, calculate the unit cell edge length, a.

(a) 6.24Å   (b) 6.33Å   (c) 6.56Å   (d) 6.84Å   (e) 7.14Å

c*86. Metallic calcium crystallized in a face-centered cubic lattice, and the atomic radius of calcium is 1.97Å. Calculate the density of calcium.

(a) 1.28 g/cm³   (b) 1.42 g/cm³   (c) 1.54 g/cm³
(d) 1.84 g/cm³   (e) 2.11 g/cm³

e*87. Potassium bromide, KBr, crystallizes in the NaCl (face-centered cubic) lattice. The ionic radii of K^+ and Br^- ions are 1.33Å and 1.95Å, respectively. Assuming anion-anion contact along the face diagonal and anion-cation contact along the edge of the unit cell, calculate the density of such a crystal of KBr.

(a) 2.84 g/cm³   (b) 3.56 g/cm³   (c) 1.46 g/cm³
(d) 4.19 g/cm³   (e) 2.80 g/cm³

b*88. The density of palladium is 12.0 g/cm³. The unit cell of Pd is a face-centered cube. Calculate the atomic radius of Pd.

(a) 1.26Å   (b) 1.37Å   (c) 1.44Å   (d) 1.56Å   (e) 1.64Å

d*89. An unknown metal crystallized in the hexagonal closest-packed structure and has a density of 12.2 g/cm³. Its atomic radius is 1.34Å. There are 4 atoms per unit cell. What is the metal?

(a) Mg   (b) Ba   (c) Fe   (d) Ru   (e) Au

a*90. A certain metal crystallizes in the hexagonal closest-packed structure and has a density of 4.51 g/cm³. Its atomic radius is 1.47Å. There are four atoms per unit cell. Determine its atomic weight. What is the metal?

(a) Ti   (b) Be   (c) Zn   (d) Ir   (e) Sr

Band Theory of Metals

d 91. Consider the metallic bonding in lithium. The interaction of all the 2s orbitals of one mole of lithium atoms produces a band consisting of _____ mole(s) of molecular orbitals (σ_{2s} and $\sigma_{2s}^{\star}$).

(a) one-eighth (b) one-fourth (c) one-half
(d) one (e) two

a 92. The conduction band of magnesium is thought to result from the combination of molecular orbitals resulting from overlap of _____ atomic orbitals.

(a) 3s and 3p (b) 3s and 3d (c) 3d and 4s
(d) 3p and 3d (e) 3s and 4s

c 93. Which one of the following elements is considered a semiconductor?

(a) Li   (b) Fe   (c) Cl   (d) Ni   (e) Si

c 94. The electrical conductivity of a metal __a__ with increasing temperature and that of a semiconductor __b__ with increasing temperature.

	a	b
(a)	increases	increases
(b)	increases	decreases
(c)	decreases	increases
(d)	decreases	decreases
(e)	remains the same	increases or decreases

d 95. Elements that have their highest energy electrons in a **filled** band of molecular orbitals that is separated from the lowest empty band by an energy difference much too large for electrons to jump between bands are called _____.

(a) semiconductors (b) metals (c) conductors
(d) insulators (e) isomorphs

14 Solutions

The Dissolution Process

e 1. The dissolution process is exothermic if the amount of energy released in bringing about ___ __a__ interactions exceeds the sum of the amounts of energy absorbed in overcoming __b__ and __c__ interactions.

	a	b	c
(a)	solute-solute	solvent-solvent	solvent-solute
(b)	solvent-solvent	solute-solute	solvent-solute
(c)	solvent-solute	solute-solute	crystal lattice
(d)	solute-solute	crystal lattice	solvent-solvent
(e)	solvent-solute	solute-solute	solvent-solvent

c 2. The change in energy accompanying the process

$$M^+(g) + X^-(g) \rightarrow MX(s)$$ is the _____ of MX.

(a) heat of hydration (b) heat of solution
(c) crystal lattice energy of MX (d) heat of ionization
(e) heat of dissociation

e 3. Consider the three statements below. Which numbered response contains all the statements that are **true** and no false statements?

I. Hydration is a special case of solvation in which the solvent is water.
II. The oxygen end of water molecules is attracted toward Ca^{2+} ions.
III. The hydrogen end of water molecules is attracted toward Cl^- ions.

(a) I (b) II (c) III (d) I and II (e) I, II, and III

b 4. Consider a series of chloride salts, MCl_2. As the charge-to-radius ratio of M^{2+} __a__, the hydration energy of M^{2+} __b__ in magnitude and the crystal lattice energy of MCl_2 __c__ in magnitude.

	a	b	c
(a)	increases	decreases	increases
(b)	increases	increases	increases
(c)	increases	increases	decreases
(d)	increases	decreases	decreases
(e)	increases	does not change	may increase or decrease

e 5. Which of the following metal ions has the most negative heat of hydration?

(a) Ca^{2+} (b) K^+ (c) Mn^{2+} (d) Li^+ (e) Al^{3+}

a 6. Which one of the following solutes is **most** likely to have low water solubility due to the dissolution process being highly endothermic?

(a) Al_2O_3 (b) RbF (c) CaF_2 (d) AgCl (e) $FeCl_2$

c 7. Which of the following compounds is **not** miscible with water?

(a) CH₃OH (b) CH₃COOH (c) CCl₄
(d) CH₃CN (e) HOCH₂CH₂OH

b 8. Consider the following pairs of liquids. Which numbered response contains all the pairs that are miscible and none that are immiscible?

I. benzene, C₆H₆, and hexane, C₆H₁₄
II. water, H₂O, and methanol, CH₃OH
III. water, H₂O, and hexane, C₆H₁₄

(a) I (b) I and II (c) II and III (d) II (e) I, II, and III

a 9. Select the most appropriate explanation of the following observation on solubility: "Hydrogen chloride, HCl, is very soluble in water."

(a) Water is a polar solvent, and it promotes the ionization of many polar molecules.
(b) Opposites attract, that is, polar solutes dissolve in non-polar solvents and vice-versa.
(c) Water promotes the dissociation of many ionic solids.
(d) Gases that interact only slightly with solvents dissolve freely.
(e) Relatively light molecules are generally more soluble in water than heavier molecules.

d 10. Which one of the following statements is **false**?

(a) All ions are hydrated in aqueous solution.
(b) All cations are hydrated in aqueous solution.
(c) All anions are hydrated in aqueous solution.
(d) Hydration is generally highly endothermic for ionic compounds.
(e) The heats of hydration of cations increases as their charge-to-radius ratios increase.

a 11. Which of the following terms is **not** generally used in describing the dissolution of solids and gases in liquids?

(a) miscibility (b) saturation (c) molality
(d) molarity (e) % solute by mass

b 12. Which one of the following statements is **false**?

(a) The effects of lattice energies and hydration energies oppose each other in the dissolution of solids in liquids.
(b) The solvation process usually absorbs heat.
(c) Nonpolar solids do not dissolve appreciably in polar solvents.
(d) The solubility of a gas that does not react with the solvent decreases as temperature increases.
(e) The separation of solute particles from a crystal requires energy.

e 13. Which numbered response lists all of the following statements that are **true**, and no false statements?

    I. Many solids that dissolve in endothermic processes have solubilities that increase as temperature increases.
    II. The solubility in water of a gas that does not react with water increases as the partial pressure of that gas above the surface of the solution increases.
    III. Most gases that are reasonably soluble in water are polar or else they react with water or ionize in water.

(a) I   (b) II   (c) I and II   (d) II and III   (e) I, II, and III

d 14. Which one of the following statements is **false**?

(a) Carbon tetrachloride, CCl_4, is more miscible with hexane, C_6H_{14}, than it is with a polar solvent like methanol, CH_3OH.
(b) Gases are generally more soluble in water under high pressures than under low pressures.
(c) The solubilities of solids in liquids may either increase or decrease as temperature increases.
(d) Apparent percentage ionization of a polar covalent solute in water increases as concentration of the solute increases.
(e) A supersaturated solution can be described as being metastable.

d 15. If the concentration of CO_2 is 2.90 g of CO_2 per 1.00 L of soft drink when bottled under 2.0 atm of CO_2 pressure, what will be the concentration of the CO_2 in the drink after it has been opened and left to come to equilibrium with the atmosphere which has a CO_2 partial pressure of 3.0 x 10^{-4} atm?

(a) 2.2 x 10^{-3} g CO_2/L   (b) 2.0 x 10^{-4} g CO_2/L
(c) 1.0 x 10^{-4} g CO_2/L   (d) 4.4 x 10^{-4} g CO_2/L
(e) 4.6 x 10^{-2} g CO_2/L

Molality and Mole Fraction

a 16. To calculate the **molality** of a solution, which of the following **must** be measured?

I. mass of solute   II. mass of solvent   III. total volume of solution

(a) I and II   (b) I and III   (c) II and III   (d) only I   (e) I, II and III

c 17. Calculate the molality of a solution that contains 25 g of H_2SO_4 dissolved in 80. g of H_2O.

(a) 1.6 m   (b) 2.2 m   (c) 3.2 m   (d) 6.3 m   (e) 7.0 m

a 18. Calculate the molality of a solution prepared by dissolving 150. grams of sodium dichromate dihydrate, $Na_2Cr_2O_7 \cdot 2H_2O$ in 350. gm of water.

(a) 1.44 m   (b) 1.56 m   (c) 1.79 m   (d) 1.86 m   (e) 2.30 m

d 19. What is the molality of an aqueous solution that is 10.0% ethanol, C_2H_5OH, by mass?

(a) 1.38 m (b) 1.77 m (c) 2.17 m (d) 2.42 m (e) 2.66 m

c 20. Calculate the molality of a 10.0% H_3PO_4 solution in water.

(a) 0.380 m (b) 0.760 m (c) 1.13 m (d) 1.51 m (e) 1.89 m

a 21. How many grams of sucrose, $C_{12}H_{22}O_{11}$, must be dissolved in 750. mL of water to prepare a 0.250 molal solution?

(a) 64.1 g   (b) 114 g   (c) 85.5 g   (d) 78.2 g   (e) 96.4 g

c 22. What mass of water must be used to dissolve 20.0 grams of ethanol, C_2H_5OH, to prepare a 0.0500 molal solution of ethanol?

(a) 3.76 kg   (b) 4.00 kg   (c) 8.70 kg   (d) 6.35 kg   (e) 7.18 kg

b 23. Calculate the molality of a solution that contains 51.2 g of naphthalene, $C_{10}H_8$, in 500. mL of carbon tetrachloride. The density of CCl_4 is 1.60 g/mL.

(a) 0.250 m (b) 0.500 m (c) 0.750 m (d) 0.840 m (e) 1.69 m

b 24. What is the molality of a solution prepared by dissolving 10.0 grams of methylamine, CH_3NH_2, in 50.0 mL of ethanol, C_2H_5OH? The specific gravity of ethanol is 0.789.

(a) 6.30 m (b) 8.18 m (c) 4.26 m (d) 5.16 m (e) 3.87 m

d 25. If the mole fraction of methyl alcohol in a solution (with only water) is 0.28, what is the mole fraction of the water?

(a) 0.28   (b) 1.28   (c) 0.62   (d) 0.72   (e) 0.36

a 26. What is the mole fraction of ethanol, C_2H_5OH, in a solution of 47.5 g of C_2H_5OH in 850. g of water?

(a) 0.021   (b) 0.18   (c) 0.032   (d) 0.98   (e) 0.028

b 27. If 8.32 grams of methanol, CH_3OH, are dissolved in 10.3 grams of water, what is the mole fraction of methanol in the solution?

(a) 0.61   (b) 0.31   (c) 0.11   (d) 0.43   (e) 0.36

c 28. If 41.6 g of acetic acid, CH_3COOH, are dissolved in 65.0 g of water, what is the mole fraction of acetic acid?

(a) 0.200   (b) 0.275   (c) 0.161   (d) 0.192   (e) 0.840

b 29. Calculate the mole fraction of ethyl alcohol, C_2H_5OH, in a solution that contains 230. grams of C_2H_5OH and 312 grams of benzene, C_6H_6.

(a) 0.44   (b) 0.56   (c) 0.57   (d) 1.8   (e) 2.3

a 30. Calculate the mole fraction of C_2H_5OH in a solution that contains 46 grams of ethyl alcohol, C_2H_5OH, and 64 grams of methyl alcohol, CH_3OH.

(a) 0.33   (b) 0.42   (c) 0.50   (d) 0.67   (e) 1.5

e 31. What is the mole fraction of CH_3OH in a 3.50 m aqueous solution of CH_3OH?

(a) 0.0630   (b) 0.0679   (c) 0.650   (d) 0.350   (e) 0.0592

c*32. If the mole fraction of CH_3OH in a solution with only water is 0.0250, what is the molality of the CH_3OH?

(a) 12.4 m (b) 3.82 m (c) 1.42 m (d) 5.76 m (e) 0.0256 m

d 33. What is the mole fraction of methanol, CH_3OH, in an aqueous solution that is 20.0% methanol by mass?

(a) 0.250   (b) 0.544   (c) 0.101   (d) 0.123   (e) 0.308

a*34. Calculate the molarity of a solution that contains 9.0 g of $(COOH)_2$ in 1500. mL of solution.

(a) 0.067 M   (b) 0.10 M   (c) 0.13 M   (d) 0.20 M   (e) 0.26 M

d*35. What is the molarity of 2500. mL of a solution that contains 160. grams of NH_4NO_3?

(a) 0.333 M   (b) 0.450 M   (c) 0.600 M   (d) 0.800 M   (e) 1.00 M

Lowering of Vapor Pressure and Raoult's Law

c 36. Sucrose is a nonvolatile, nonionizing solute in water. Determine the vapor pressure lowering, at 27°C, of a solution of 75.0 grams of sucrose, $C_{12}H_{22}O_{11}$, dissolved in 180. g of water. The vapor pressure of pure water at 27°C is 26.7 torr. Assume the solution is ideal.

(a) 0.585 torr   (b) 0.058 torr   (c) 0.571 torr
(d) 5.62 torr   (e) 0.548 torr

a 37. Calculate the vapor pressure of a solution prepared by dissolving 70.0 g of naphthalene, $C_{10}H_8$ (a nonvolatile nonelectrolyte), in 220.0 g of benzene, C_6H_6, at 20°C. Assume the solution is ideal. The vapor pressure of pure benzene is 74.6 torr at 20°C.

(a) 62.5 torr   (b) 14.5 torr   (c) 40.8 torr   (d) 60.1 torr   (e) 12.1 torr

d 38. A solution consists of 0.450 mole of pentane, C_5H_{12}, and 0.250 mole of cyclopentane, C_5H_{10}. What is the vapor pressure of pentane in this solution at 25°C? The vapor pressure of the pure liquids at 25°C are 451 torr for pentane and 321 torr for cyclopentane. Assume that the solution is an ideal solution.

(a) 80.2 torr   (b) 188 torr   (c) 203 torr   (d) 290 torr   (e) 306 torr a 39. A solution is made by mixing 52.1 g of propyl chloride, C_3H_7Cl, and 38.4 g of propyl bromide, C_3H_7Br. What is the vapor pressure of propyl chloride in the solution at 25°C? The vapor pressure of pure propyl chloride is 347 torr at 25°C and that of pure propyl bromide is 133 torr at 25°C. Assume that the solution is an ideal solution.

(a) 236 torr (b) 128 torr (c) 136 torr (d) 147 torr (e) 155 torr

c 40. At 25°C a solution consists of 0.450 mole of pentane, C_5H_{12}, and 0.250 mole of cyclopentane, C_5H_{10}. What is the **lowering** of the vapor pressure of pentane in this solution? The vapor pressure of the pure liquids at 25°C are 451 torr for pentane and 321 torr for cyclopentane. Assume that the solution is an ideal solution.

(a) 115 torr (b) 138 torr (c) 161 torr (d) 187 torr (e) 206 torr

c 41. A solution is made by mixing 52.1 g of propyl chloride, C_3H_7Cl, and 38.4 g of propyl bromide, C_3H_7Br. What is the **lowering** of the vapor pressure of propyl chloride at 25°C in this solution? The vapor pressure of pure propyl chloride is 347 torr at 25°C and that of pure propyl bromide is 133 torr at 25°C. Assume that the solution is an ideal solution.

(a) 192 torr (b) 200 torr (c) 111 torr (d) 219 torr (e) 171 torr

c 42. At 25°C a solution consists of 0.450 mole of pentane, C_5H_{12}, and 0.250 mole of cyclopentane, C_5H_{10}. What is the mole fraction of pentane in the vapor that is in equilibrium with this solution? The vapor pressure of the pure liquids at 25°C are 451 torr for pentane and 321 torr for cyclopentane. Assume that the solution is an ideal solution.

(a) 0.284 (b) 0.551 (c) 0.716 (d) 0.643 (e) 0.357

a 43. At 25°C a solution consists of 0.450 mole of pentane, C_5H_{12}, and 0.250 mole of cyclopentane, C_5H_{10}. What is the mole fraction of cyclopentane in the vapor that is in equilibrium with this solution? The vapor pressure of the pure liquids at 25°C are 451 torr for pentane and 321 torr for cyclopentane. Assume that the solution is an ideal solution.

(a) 0.284 (b) 0.551 (c) 0.716 (d) 0.643 (e) 0.357

Fractional Distillation

b 44. Crude oil is a complex solution of liquids, and some gases and solids. The components are commonly separated from each other by utilizing a technique called fractional distillation. A physical property of the components upon which this technique is **not** dependent is _____.

(a) condensation point (b) density (c) volatility
(d) boiling point (e) vapor pressure

Boiling Point Elevation

b 45. Which of these aqueous solutions would be expected to have the highest boiling point?

(a) 0.100 m $CaCl_2$ (b) 0.200 m NaOH (c) 0.050 m K_2SO_4
(d) 0.050 m $Al_2(SO_4)_3$ (e) 0.200 m CH_3OH

b 46. If 4.27 grams of sucrose, $C_{12}H_{22}O_{11}$, are dissolved in 15.2 grams of water, what will be the boiling point of the resulting solution? K_b for water = 0.512°C/m.

(a) 101.64°C   (b) 100.42°C   (c) 99.626°C   (d) 100.73°C   (e) 101.42°C

d 47. Calculate the boiling point of a solution prepared by dissolving 70.0 g of naphthalene, $C_{10}H_8$ (a nonvolatile nonelectrolyte), in 220.0 g of benzene, C_6H_6. The K_b for benzene = 2.53°C/m. The boiling point of pure benzene is 80.1°C.

(a) 87.8°C   (b) 73.8°C   (c) 83.2°C   (d) 86.4°C   (e) 106.3°C

Freezing Point Depression

d 48. Calculate the freezing point of a solution that contains 8.0 g of sucrose ($C_{12}H_{22}O_{11}$) in 100. g of H_2O. K_f for H_2O = 1.86°C/m.

(a) –0.044°C   (b) –0.22°C   (c) –0.39°C   (d) –0.44°C   (e) 0.04°C

e 49. Calculate the freezing point of a solution that contains 68.4 g of sucrose (table sugar) in 300. g of water. One mole of sucrose is 342 g. K_f for H_2O = 1.86°C/m.

(a) –0.186°C   (b) –0.372°C   (c) –0.558°C
(d) –0.744°C   (e) –1.24°C

e 50. Calculate the freezing point of a solution that contains 30.0 g of urea, CH_4N_2O, in 200. g of water. Urea is a nonvolatile nonelectrolyte. K_f for H_2O = 1.86°C/m.

(a) –1.86°C   (b) –2.79°C   (c) –3.72°C   (d) –4.23°C   (e) –4.65°C

Determination of Molecular Weight by Freezing Point Depression or Boiling Point Elevation

c 51. The freezing point of a solution made by dissolving 0.200 g of a molecular compound in 100. g of H_2O is –0.0120°C. Calculate the molecular weight of the compound. The compound does not ionize in water. K_f for H_2O = 1.86°C/m.

(a) 155 amu   (b) 215 amu   (c) 310 amu   (d) 395 amu   (e) 450 amu

d 52. When 1.150 grams of an unknown nonelectrolyte dissolves in 10.0 grams of water, the solution freezes at –2.16°C. What is the molecular weight of the unknown compound? K_f for water = 1.86°C/m.

(a) 88.6 g/mol   (b) 116 g/mol   (c) 74.2 g/mol
(d) 99.1 g/mol   (e) 132 g/mol

d 53. A 4.305-gram sample of a nonelectrolyte is dissolved in 105 grams of water. The solution freezes at –1.23°C. Calculate the molecular weight of the solute. K_f for water = 1.86°C/m.

(a) 39.7 g/mol   (b) 58.4 g/mol   (c) 46.2 g/mol
(d) 62.0 g/mol   (e) 74.2 g/mol

a 54. When 20.0 grams of an unknown compound are dissolved in 500. grams of benzene, the freezing point of the resulting solution is 3.77°C. The freezing point of pure benzene is 5.48°C, and the K_f for benzene is 5.12°C/m. What is the molecular weight of the unknown compound?

(a) 120 g/mol (b) 80.0 g/mol (c) 100 g/mol
(d) 140 g/mol (e) 160 g/mol

e 55. The freezing point of a solution of 1.048 g of an unknown nonelectrolyte dissolved in 36.21 g of benzene is 1.39°C. Pure benzene freezes at 5.48°C and its K_f value is 5.12°C/m. What is the molecular weight of the compound?

(a) 59.2 g/mol (b) 54.0 g/mol (c) 61.4 g/mol
(d) 42.4 g/mol (e) 36.3 g/mol

a 56. When 35.0 g of an unknown nonelectrolyte is dissolved in 220.0 g of benzene, the solution boils at 83.2°C. Calculate the molecular weight of the unknown nonelectrolyte. The K_b for benzene = 2.53°C/m. The boiling point of pure benzene is 80.1°C.

(a) 130 g/mol (b) 20.3 g/mol (c) 183 g/mol
(d) 156 g/mol (e) 194 g/mol

b 57. When 27.0 g of an unknown nonelectrolyte is dissolved in 250.0 mL of benzene, the solution boils at 82.5°C. Calculate the molecular weight of the unknown nonelectrolyte. The K_b for benzene = 2.53°C/m. The boiling point of pure benzene is 80.1°C. The density of benzene = 0.876 g/mL.

(a) 60 g/mol (b) 130 g/mol (c) 220 g/mol
(d) 110 g/mol (e) 150 g/mol

Colligative Properties and Dissociation of Electrolytes

b 58. A 0.1000 m aqueous solution of a weak acid, HA, is 1.5% ionized. At what temperature does it freeze? K_f for water = 1.86°C/m.

(a) −0.0764°C (b) −0.189°C (c) −0.372°C
(d) −0.564°C   (e) −0.721°C

b 59. A 0.0100 m solution of HX, a weak acid, freezes at −0.02000°C. Calculate the % ionization of HX. K_f for water = 1.86°C/m.

(a) 2.3% (b) 7.5% (c) 9.5% (d) 11% (e) 18%

a 60. A 0.0490 molal aqueous NaBr solution freezes at −0.173°C. What is its apparent percent dissociation in this solution? K_f = 1.86°C/m for water.

NaBr → Na⁺ + Br⁻

(a) 89.8% (b) 84.2% (c) 96.4% (d) 77.0% (e) 68.9%

Chapter 14   217

c 61.   K₂SO₄ dissolves in water according to

$$K_2SO_4 \rightarrow 2K^+ + SO_4^{2-}$$

A 0.100 m solution of K₂SO₄ freezes at −0.432°C. Calculate the apparent percent dissociation of K₂SO₄ in this solution. K_f for water = 1.86°C/m.

(a) 62%   (b) 64%   (c) 66%   (d) 68%   (e) 83%

c*62.   The compound X₂Y is only partially dissociated in water solution to form X⁺ and Y²⁻ ions. A 0.0100 m solution is found to freeze at −0.040°C. Calculate the apparent percent dissociation of X₂Y. K_f for water = 1.86°C/m.

(a) 42%   (b) 56%   (c) 58%   (d) 79%   (e) 85%

a 63.   A 1.0 g sample of a molecular compound having a molecular weight of 100,000 g/mol is dissolved in 100. g of water. Calculate the osmotic pressure of the solution in torr at a temperature of 27°C. (Assume the volume of the solution is 100. mL.)

(a) 1.9 torr   (b) 2.9 torr   (c) 3.9 torr   (d) 4.9 torr   (e) 5.9 torr

e 64.   Calculate the osmotic pressure of a solution that contains 1.22 g of sucrose ($C_{12}H_{22}O_{11}$) dissolved in 100. g of water at 25°C. (Assume the volume of the solution is 100. mL.)

(a) 6.32 torr   (b) 108 torr   (c) 249 torr   (d) 497 torr   (e) 663 torr

e 65.   Calculate the osmotic pressure associated with 50.0 g of an enzyme of molecular weight 98,000 g/mol dissolved in 2600. mL of benzene at 30.0°C.

(a) 0.484 torr   (b) 1.68 torr   (c) 1.96 torr   (d) 2.48 torr   (e) 3.71 torr

e 66.   Which of the following is **not** an advantage of osmotic pressure over boiling point elevation and freezing point depression for the determination of molecular weights?

(a)   The relatively large change in osmotic pressure is easier to measure accurately than the small boiling point elevation.
(b)   The relatively large change in osmotic pressure can be determined without the special apparatus needed for accurate measurement of the smaller freezing point depression.
(c)   A more dilute solution can be used for osmotic pressure measurements thereby saving money when the molecular weights of expensive substances are to be determined.
(d)   Use of a more dilute solution allows determinations for sparingly soluble high-molecular-weight substances.
(e)   Osmotic pressure can be used to accurately determine the molecular weight of impure substances.

a 67.   Estimate the molecular weight of a biological macromolecule if a 0.100-gram sample dissolved in 50.0 mL of benzene has an osmotic pressure of 9.76 torr at 25.0°C.

(a) 3.8 x 10³ g/mol   (b) 4.2 x 10⁴ g/mol   (c) 5.6 x 10⁴ g/mol
(d) 6.7 x 10⁴ g/mol   (e) 8.3 x 10³ g/mol

e 68. Estimate the molecular weight of a polymer if a 100.-mL solution of 6.5 g of the polymer in toluene has an osmotic pressure of 0.044 atm at 27°C.

(a) 62,000   (b) 45,000   (c) 22,000   (d) 5500   (e) 36,000

Colloids and the Tyndall Effect

b 69. Which one of the following combinations must be a colloid?

	dispersed phase	dispersing medium
(a)	solid	solid
(b)	liquid	gas
(c)	gas	gas
(d)	gas	liquid
(e)	liquid	liquid

c 70. Which one of the following is an example of an emulsion?

(a) shaving cream   (b) fog   (c) mayonnaise
(d) styrofoam   (e) white gold (an alloy)

d 71. The Tyndall effect describes _____.

(a) precipitation of colloidal particles using electrically charged plates
(b) the adsorption of positive ions onto the surface of a hydrophilic solid
(c) hydrophobic interactions between nonpolar molecules
(d) the scattering of light by colloidal particles
(e) reverse osmosis involving saline solutions

Hydrophilic and Hydrophobic Colloids

d 72. Which one of the following statements about soaps and soap molecules is **false**?

(a) They have a polar end.
(b) They have a hydrophobic end.
(c) They are often sodium salts of long chain fatty acids.
(d) The hydrophilic end of a soap molecule is attracted by grease.
(e) They precipitate in water that contains Fe^{3+} ions.

15 Chemical Thermodynamics

The First Law of Thermodynamics

d 1. Which statement is **incorrect**?

    (a) Energy is the capacity to do work or to transfer heat.
    (b) Kinetic energy is the energy of motion.
    (c) Potential energy is the energy that a system possesses by virtue of its position or composition.
    (d) A process that absorbs energy from its surroundings is called exothermic.
    (e) The Law of Conservation of Energy is another statement of the First Law of Thermodynamics.

Some Thermodynamic Terms

a*2. Which one of the following thermodynamic quantities is **not** a state function?

    (a) work   (b) internal energy   (c) free energy
    (d) enthalpy   (e) entropy

e 3. Which statement concerning state functions is **false**?

    (a) A change in a state function describes a difference between two states and is independent of the process by which the change occurs.
    (b) State functions are represented by capital letters.
    (c) Differences in state functions are often described using an arbitrary scale.
    (d) The change in a quantity X is described as $\Delta X = X_{final} - X_{initial}$.
    (e) If the initial value is greater than the final value for $\Delta X = X_{final} - X_{initial}$, then ΔX is positive.

Calorimetry

b 4. If 4.168 kJ of heat is added to a calorimeter containing 75.40 g of water, the temperature of the water and the calorimeter increases from 24.58°C to 35.82°C. Calculate the heat capacity of the calorimeter (in J/°C). The specific heat of water is 4.184 J/g·°C.

    (a) 622 J/°C   (b) 55.34 J/°C   (c) 315.5 J/°C
    (d) 25.31 J/°C   (e) 17.36 J/°C

b 5. A 51.6-mL dilute solution of acid at 23.85°C is mixed with 48.5 mL of a dilute solution of base, also at 23.85°C, in a coffee-cup calorimeter. After the reaction occurs, the temperature of the resulting mixture is 27.25°C. The density of the final solution is 1.03 g/mL. Calculate the amount of heat evolved. Assume the specific heat of the solution is 4.184 J/g·°C. The heat capacity of the calorimeter is 23.9 J/°C.

    (a) 3.05 kJ   (b) 1.55 kJ   (c) 5.49 kJ   (d) 0.837 kJ   (e) 14.6 kJ

Thermochemical Equations

e 6. How much heat is absorbed in the complete reaction of 3.00 grams of SiO_2 with excess carbon in the reaction below? ΔH^0 for the reaction is +624.6 kJ.

$$SiO_2(g) + 3C(s) \rightarrow SiC(s) + 2CO(g)$$

(a) 366 kJ (b) 1.13 x 10^5 kJ (c) 5.06 kJ
(d) 1.33 x 10^4 kJ (e) 31.2 kJ

a 7. The burning of 80.3 g of SiH_4 at constant pressure gives off 3790 kJ of heat. Calculate ΔH for this reaction.

$$SiH_4(g) + 2O_2(g) \rightarrow SiO_2(s) + 2H_2O(\ell)$$

(a) –1520 kJ/mol rxn (b) –47.2 kJ/mol rxn (c) –4340 kJ/mol rxn
(d) –2430 kJ/mol rxn (e) +4340 kJ/mol rxn

d 8. The "roasting" of 48.7 g of ZnS at constant pressure gives off 220. kJ of heat Calculate the ΔH for this reaction.

$$2ZnS(s) + 3O_2(g) \rightarrow 2ZnO(s) + 2SO_2(g)$$

(a) –110 kJ/mol rxn (b) –293 kJ/mol rxn (c) –440. kJ/mol rxn
(d) –881 kJ/mol rxn (e) +440. kJ/mol rxn

Standard States and Standard Enthalpy Changes

b 9. Which of the following statements is **incorrect**?

(a) The thermochemical standard state of a substance is its most stable state under one atmosphere pressure and at some specific temperature (298 K if not specified).
(b) A superscript zero, such as ΔH^0, indicates a specified temperature of 0°C.
(c) For a pure substance in the liquid or solid phase, the standard state is the pure liquid or solid.
(d) For a pure gas, the standard state is the gas at a pressure of one atmosphere.
(e) For a substance in solution, the standard state refers to one-molar concentration.

Standard Molar Enthalpies of Formation, ΔH_f^0

c 10. For which of the following substances does $\Delta H_{f298}^0 = 0$?

(a) ZnO(s) (b) $H_2O(\ell)$ (c) $Hg(\ell)$ (d) $I_2(\ell)$ (e) C(diamond)

b 11. Which response includes all of the following substances that have $\Delta H_{f298}^0 = 0$, and no other substances?

I. HCl(g) II. Na(s) III. HCl(aq) IV. $F_2(g)$

(a) I and II (b) II and IV (c) I, II, and IV
(d) II (e) I, II, III, and IV

a 12. Calculate the amount of heat released in the complete combustion of 8.17 grams of Al to form $Al_2O_3(s)$ at 25°C and 1 atm. ΔH_f^0 for $Al_2O_3(s)$ = 1676 kJ/mol

$$4Al(s) + 3O_2(g) \rightarrow 2Al_2O_3(s)$$

(a) 254 kJ   (b) 203 kJ   (c) 127 kJ   (d) 237 kJ   (e) 101 kJ

a 13. How much heat energy is liberated when 11.0 grams of manganese is converted to Mn_2O_3 at standard state conditions? $\Delta H_{f\,Mn_2O_3(s)}^0$ is –962.3 kJ/mol.

(a) 96.2 kJ   (b) 192 kJ   (c) 289 kJ   (d) 460 kJ   (e) 964 kJ

Hess's Law

b 14. From the following data at 25°C,

$H_2(g) + Cl_2(g) \rightarrow 2HCl(g)$ $\Delta H^0 = -185$ kJ
$2H_2(g) + O_2(g) \rightarrow 2H_2O(g)$ $\Delta H^0 = -483.7$ kJ

Calculate ΔH^0 at 25°C for the reaction below.

$$4HCl(g) + O_2(g) \rightarrow 2Cl_2(g) + 2H_2O(g)$$

(a) +299 kJ   (b) –114 kJ   (c) –299 kJ   (d) +114 kJ   (e) –86.8 kJ

e 15. Given the following at 25°C and 1.00 atm:

ΔH^0
$SO_3(g) + H_2O(\ell) \rightarrow H_2SO_4(\ell)$ –133 kJ
$Pb(s) + PbO_2(s) + 2H_2SO_4(\ell) \rightarrow 2PbSO_4(s) + 2H_2O(\ell)$ –509 kJ

Calculate the ΔH^0 for the reaction below at 25°C.

$$Pb(s) + PbO_2(s) + 2SO_3(g) \rightarrow 2PbSO_4(s)$$

(a) +376 kJ   (b) –376 kJ   (c) – 642 kJ   (d) –243 kJ   (e) –775 kJ

d 16. Given the enthalpy changes for the following reactions, calculate ΔH_f^0 for CO(g).

$C\text{ (graphite)} + O_2(g) \rightarrow CO_2(g)$ $\Delta H_f^0 = -393.5$ kJ
$CO(g) + 1/2 O_2 \rightarrow CO_2(g)$ $\Delta H^0 = -283.0$ kJ

(a) 6.78 x 10^2 kJ   (b) –6.78 x 10^2 kJ   (c) +110.5 kJ
(d) –110.5 kJ   (e) –173 kJ

a 17. Given the standard heats of formation for the following compounds, calculate ΔH_{298}^0 for the following reaction.

$CH_4(g) + H_2O(g) \rightarrow CH_3OH(\ell) + H_2(g)$
ΔH_f^0(kJ/mol)   –75   –242   –238   0

(a) +79 kJ   (b) –79 kJ   (c) +594 kcal
(d) –594 kcal   (e) – 405 kJ

b 18. Calculate ΔH^0 at 25°C for the reaction below.

$$2ZnS(s) + 3O_2(g) \rightarrow 2ZnO(s) + 2SO_2(g)$$
ΔH_f^0 (kJ/mol)   −205.6   0   −348.3   −296.8

(a) −257.1 kJ   (b) −879.0 kJ   (c) +257.1 kJ
(d) −582.2 kJ   (e) +879.0 kJ

c 19. Calculate ΔH^0 for the following reaction at 25.0°C.

$$Fe_3O_4(s) + CO(g) \rightarrow 3FeO(s) + CO_2(g)$$
ΔH_f^0(kJ/mol)   −1118   −110.5   −272   −393.5

(a) −263 kJ   (b) 54 kJ   (c) 19 kJ   (d) −50 kJ   (e) 109 kJ

e 20. Calculate the standard enthalpy change for the reaction below.

$$C\text{ (graphite)} + 4HNO_3(\ell) \rightarrow CO_2(g) + 4NO_2(g) + 2H_2O(\ell)$$
ΔH_f^0 (kJ/mol)   0   −174.1   −393.5   33.2   −285.8

(a) −123.9 kJ   (b) −472.1 kJ   (c) −201.9 kJ
(d) −404.8 kJ   (e) −135.9 kJ

d 21. Evaluate ΔH^0 for the reaction below at 25°C.

$$SiO_2(s) + 4HF(aq) \rightarrow SiF_4(g) + 2H_2O(\ell)$$
ΔH_f^0 (kJ/mol)   −910.9   −320.8   −1615   −285.8

(a) +293.3 kJ   (b) −954.9 kJ   (c) −366.5 kJ
(d) 7.5 kJ   (e) −1781.1 kJ

a 22. Use the data below to calculate ΔH_f^0 for benzene, $C_6H_6(\ell)$, at 25°C and 1 atm.

$$2C_6H_6(\ell) + 15O_2(g) \rightarrow 12CO_2(g) + 6H_2O(\ell) \quad \Delta H^0 = -6535 \text{ kJ}$$

$\Delta H_{f\ CO_2(g)}^0 = -393.5$ kJ/mol, $\Delta H_{fH_2O(\ell)}^0 = -285.8$ kJ/mol

(a) 49.1 kJ/mol   (b) 3.51 x 10^4 kJ/mol   (c) 103 kJ/mol
(d) 1.76 x 10^3 kJ/mol (e) 561 kJ/mol

e 23. Given the following at 25°C, calculate ΔH_f^0 for $HPO_3(s)$ at 25°C.

$$P_4O_{10}(s) + 4HNO_3(\ell) \rightarrow 4HPO_3(s) + N_2O_5(s) \quad \Delta H_{rxn}^0 = -180.6 \text{ kJ}$$

$\Delta H_f^0 = -2984$ kJ/mol for $P_4O_{10}(s)$, −174.1 kJ/mol for $HNO_3(\ell)$, and −43.1 kJ/mol for $N_2O_5(s)$.

(a) −528.0 kJ/mol   (b) −1474 kJ/mol   (c) −948.5 kJ/mol
(d) +1474 kJ/mol   (e) −954.5 kJ/mol

b 24. Given the following, calculate ΔH_f^0 at 25°C for CuO(s).

$$2NH_3(g) + 3CuO(s) \rightarrow N_2(g) + 3Cu(s) + 3H_2O(g) \quad \Delta H^0_{298} = -162.2 \text{ kJ}$$

Substance	ΔH_f^0 (kJ / mol)
$NH_3(g)$	-46.11
$N_2(g)$	0
$Cu(s)$	0
$H_2O(g)$	-241.8

(a) –79.1 kJ (b) –157 kJ (c) –101 kJ (d) +60.2 kJ (e) +120 kJ

b 25. Given that ΔH^0 for the oxidation of sucrose, $C_{12}H_{22}O_{11}(s)$, is –5648 kJ per mole of sucrose at 25°C, evaluate ΔH_f^0 for sucrose.

$$C_{12}H_{22}O_{11}(s) + 12O_2(g) \rightarrow 12CO_2(g) + 11H_2O(\ell)$$
ΔH_f^0 (kJ/mol)   ?   0   –393.5   –285.8

(a) –1676 kJ/mol   (b) –2218 kJ/mol   (c) –1431 kJ/mol
(d) –1067 kJ/mol   (e) –2640 kJ/mol

a 26. Calculate ΔH_f^0 at 25°C for CO(g), given that ΔH^0 at 25°C for the reaction below is –809.9 kJ.

$$2CH_4(g) + O_2(g) + 4Cl_2(g) \rightarrow 8HCl(g) + 2CO(g)$$
ΔH_f^0 (kJ/mol)   –74.81   0   0   –92.31   ?

(a) –110.5 kJ/mol   (b) –177.5 kJ/mol   (c) –160.0 kJ/mol
(d) –437.7 kJ/mol   (e) –486.6 kJ/mol

a 27. How much heat is released when 6.38 grams of Ag(s) reacts by the equation shown below at standard state conditions?

$$4Ag(s) + 2H_2S(g) + O_2(g) \rightarrow 2Ag_2S(s) + 2H_2O(\ell)$$

Substance	ΔH_f^0 (kJ / mol)
$Ag(s)$	0
$H_2S(g)$	-20.6
$O_2(g)$	0
$Ag_2S(s)$	-32.6
$H_2O(\ell)$	-285.8

(a) 8.80 kJ (b) 69.9 kJ (c) 22.1 kJ (d) 90.8 kJ (e) 40.5 kJ

b 28. How much heat is evolved in the formation of 35.0 grams of $Fe_2O_3(s)$ at 25°C and 1.00 atm pressure by the following reaction?

$$4Fe(s) + 3O_2(g) \rightarrow 2Fe_2O_3(s)$$
ΔH_f^0 (kJ/mol)   0   0   –824.2

(a) 90.4 kJ (b) 180.7 kJ (c) 151 kJ (d) 360.1 kJ (e) 243. 9 kJ a*29. How much heat is released or absorbed in the reaction of 10.0 grams of SiO_2 (quartz) with excess hydrofluoric acid?

$$SiO_2(s) + 4HF(aq) \rightarrow SiF_4(g) + 2H_2O(\ell)$$
ΔH_f^0 (kJ/mol)   −910.9   −320.8   −1615   −285.8

(a) 1.25 kJ absorbed (b) 1.25 kJ released (c) 11.3 kJ absorbed
(d) 11.3 kJ released (e) 6.56 kJ released

c 30. Calculate the standard enthalpy change at 25°C accompanying the reaction of 114 grams of sulfur trioxide with a stoichiometric quantity of water.

$$SO_3(g) + H_2O(\ell) \rightarrow H_2SO_4(\ell)$$
ΔH_f^0 (kJ/mol)   −395.6   −285.8   −814.0

(a) −251 kJ   (b) −442 kJ   (c) −189 kJ
(d) −118 kJ   (e) 1.17 x 10³ kJ

c 31. How much heat would be released if 12.0 g of methane, CH_4, was completely burned in oxygen to form carbon dioxide and water at standard state conditions? ΔH_f^0 for $CH_4(g)$ = −74.81 kJ/mol, for $CO_2(g)$ = −393.5 kJ/mol and for $H_2O(\ell)$ = −285.8 kJ/mol

(a) 77.5 kJ   (b) 453 kJ   (c) 668 kJ   (d) 190. kJ   (e) 890. kJ

d 32. Calculate the amount of heat absorbed if 10.0 g of methane, CH_4, reacted with steam to produce methanol, $CH_3OH(\ell)$, and H_2. ΔH_f^0 for $CH_4(g)$ = −74.81 kJ/mol, ΔH_f^0 for $H_2O(g)$ = −241.8 kJ/mol and ΔH_f^0 for $CH_3OH(\ell)$ = −238.4 kJ/mol

(a) 78.2 kJ   (b) 253 kJ   (c) 347 kJ   (d) 48.9 kJ   (e) 125 kJ

a*33 Calculate the standard heat of vaporization, ΔH_{vap}^0, for tin(IV) chloride, $SnCl_4$, **in kJ per mole**. ΔH_f^0 = −511.3 kJ/mol for $SnCl_4(\ell)$ and −471.5 kJ/mol for $SnCl_4(g)$.

(a) 39.8   (b) 16.4   (c) 26.4   (d) 44.8   (e) 53.2

Bond Energies

c 34. The heat of reaction of one of the following reactions is the average bond energy for the N-H bond in NH_3. Which one?

(a) $2NH_3(g) \rightarrow N_2(g) + 3H_2(g)$ (b) $NH_3(g) \rightarrow 1/2 N_2(g) + 3/2 H_2(g)$
(c) $1/3 NH_3(g) \rightarrow 1/3 N(g) + H(g)$ (d) $2/3 NH_3(g) \rightarrow 1/3 N_2(g) + H_2(g)$
(e) $1/3 N(g) + H(g) \rightarrow 1/3 NH_3(g)$

c*35. Evaluate ΔH^0 for the following reaction from the given bond energies.

$$2HBr(g) \rightarrow H_2(g) + Br_2(g)$$
ΔH_{H-H} = 436 kJ/mol, ΔH_{Br-Br} = 193 kJ/mol, ΔH_{H-Br} = 366 kJ/mol

(a) −103 kJ   (b) −143 kJ   (c) +103 kJ   (d) +142 kJ   (e) 259 kJ

a 36. Estimate the enthalpy change for the reaction below from the average bond energies given. There are two Cl-Cl and two C-H bonds in CH_2Cl_2. Remember that energy is absorbed when bonds are broken and released when they are formed.

$$CH_4(g) + 2Cl_2(g) \rightarrow CH_2Cl_2(g) + 2HCl(g)$$

Average Bond Energies
C-H	413 kJ/mol	Cl-Cl	242 kJ/mol
H-Cl	432 kJ/mol	C-Cl	339 kJ/mol

(a) –232 kJ/mol   (b) +578 kJ/mol   (c) +232 kJ/mol
(d) –578 kJ/mol   (e) +541 kJ/mol

a 37. Estimate the heat of reaction at 298 K for the reaction shown, given the average bond energies below.

$$Br_2(g) + 3F_2(g) \rightarrow 2BrF_3(g)$$

Bond	Bond Energy
Br-Br	193 kJ/mol
F-F	155 kJ/mol
Br-F	249 kJ/mol

(a) –836 kJ   (b) –150 kJ   (c) –89 kJ   (d) –665 kJ   (e) –1222 kJ

b*38. Calculate the enthalpy change for the following reaction using the given bond energies.

$$C_2H_4(g) + H_2O(g) \rightarrow C_2H_5OH(g)$$

C=C 602 kJ/mol   C–H 413 kJ/mol   O–H 463 kJ/mol
C–O 358 kJ/mol   C–C 346 kJ/mol

(a) –550. kJ   (b) –52 kJ   (c) –654 kJ   (d) 35 kJ   (e) 361 kJ

c 39. Calculate the average N-H bond energy in $NH_3(g)$. ΔH_f^0 for $NH_3(g) = -46.11$ kJ/mol, ΔH_f^0 for $N(g) = 472.7$ kJ/mol, ΔH_f^0 for $H(g) = 218.0$ kJ/mol, ΔH_f^0 for $N_2(g) = 0$ kJ/mol, ΔH_f^0 for $H_2(g) = 0$ kJ/mol.

(a) –46.11 kJ   (b) –15.4 kJ   (c) –390.9 kJ   (d) 264.5 kJ   (e) 1173 kJ

d 40. Calculate the average S-F bond energy in SF_6. ΔH_f^0 for $SF_6(g) = -1209$ kJ/mol, for $S(g) = 278.8$ kJ/mol, and for $F(g) = 78.99$ kJ/mol.

(a) 1962 kJ   (b) 1209 kJ   (c) 200.8 kJ   (d) 327.0 kJ   (e) 1565 kJ

d 41. Determine the average P-H bond energy in phosphine, PH_3.
ΔH_f^0 for $PH_3(g) = 5.4$ kJ/mol ΔH_f^0 for $P_4(s) = 0$ kJ/mol ΔH_f^0 for $H_2(g) = 0$ kJ/mol

$P_4(s) \rightarrow 4P(g)$ $\Delta H^0 = 235.6$ kJ
$H_2(g) \rightarrow 2H(g)$ $\Delta H^0 = 436$ kJ

(a) 685 kJ   (b) 369 kJ   (c) 529 kJ   (d) 236 kJ   (e) 69.5 kJ

b 42. Given the following bond energies and standard enthalpy of formation for $NF_3(g)$:
F-F   155 kJ/mol,   N≡N   945 kJ/mol
$1/2 N_2(g) + 3/2 F_2(g) \rightarrow NF_3(g)$ $\Delta H^0 = -103$ kJ/mol
Evaluate the average N-F bond energy in $NF_3(g)$.

(a) 113 kJ/mol   (b) 269 kJ/mol   (c) 317 kJ/mol
(d) 66 kJ/mol   (e) 328 kJ/mol

a 43. Given: H-H bond energy = 435 kJ, Cl-Cl bond energy = 243 kJ, and the standard heat of formation of HCl(g) is –92 kJ/mol, calculate the H-Cl bond energy.

(a) 431 kJ   (b) 247 kJ   (c) 180 kJ   (d) 4.6 kJ   (e) 326 kJ

c 44. Calculate the average bond energy in kJ per mol of bonds for the C-H bond from the following data:
$C(graphite) + 2H_2(g) \rightarrow CH_4(g)$ $\Delta H^0_{rxn} = -74.81$ kJ
ΔH^0_f for H(g) = 218.0 kJ ΔH^0_f for C(g) = 716.7 kJ

(a) 590.4 kJ/mol   (b) 1011 kJ/mol   (c) 415.9 kJ/mol
(d) 1665 kJ/mol   (e) 1229 kJ/mol

d 45. The ΔH^0_f for gaseous acetylene, H—C≡C—H, is 227 kJ/mol. What is the C≡C bond energy? The bond energies are 423 kJ/mol for C–H and 436 kJ/mol for H–H. The heat of sublimation for carbon is 717 kJ/mol.

(a) 98 kJ/mol   (b) 348 kJ/mol   (c) 986 kJ/mol
(d) 817 kJ/mol   (e) 1251 kJ/mol

Changes in Internal Energy, ΔE

c 46. Consider the following reaction at constant pressure. Which response is **true**?
$N_2(g) + O_2(g) \rightarrow 2NO(g)$

(a) Work is done on the system as it occurs.
(b) Work is done by the system as it occurs.
(c) No work is done as the reaction occurs.
(d) Work may be done on or by the system as the reaction occurs, depending upon the temperature.
(e) The amount of work depends on the pressure.

d 47. Which statement concerning sign conventions for $\Delta E = q + w$ is **false**?

(a) For heat absorbed by the system, q is positive.
(b) For work done by the system, w is negative.
(c) When energy is released by the reacting system, ΔE is negative.
(d) If ΔE is positive, energy can be written as a product in the equation for the reaction.
(e) For an expansion, w is negative.

e 48. A 1.00-g sample of hexane, C_6H_{14}, undergoes complete combustion with excess O_2 in a bomb calorimeter. The temperature of the 1500. g of water surrounding the bomb rises from 22.64°C to 29.30°C. The heat capacity of the calorimeter is 4.04 kJ/°C. What is ΔE for the reaction in kJ/mol of C_6H_{14}. The specific heat of water is 4.184 J/g·°C.

(a) -9.96×10^3 kJ/mol (b) -4.52×10^3 kJ/mol (c) -1.15×10^4 kJ/mol
(d) -7.40×10^4 kJ/mol (e) -5.91×10^3 kJ/mol

e 49. A 0.900-g sample of toluene, C_7H_8, was completely burned in a bomb calorimeter containing 4560. g of water which increased in temperature from 23.800°C to 25.718°C. What is ΔE for the reaction in kJ/mol C_7H_8? The heat capacity of the calorimeter was 780. J/°C. The specific heat of water is 4.184 J/g·°C.

(a) -4520 kJ/mol (b) $+3500$ kJ/mol (c) -38.1 kJ/mol
(d) -2220 kJ/mol (e) -3900 kJ/mol

b 50. The temperature of the water and calorimeter changes from 25.023°C to 28.263°C, when 0.902 g of benzoic acid (C_6H_5COOH) is burned in a bomb calorimeter with excess oxygen. Calculate ΔE (in kJ/g) for this reaction. The value of the "calorimeter constant", the heat capacity of the calorimeter plus the water, is 7.706 kJ/°C.

(a) -19.1 kJ/g (b) -27.7 kJ/g (c) -22.5 kJ/g
(d) -15.6 kJ/g (e) -13.9 kJ/g

Relationship Between ΔH and ΔE

c 51. Which one of the following statements is **false**?

(a) The change in internal energy, ΔE, for a process is equal to the amount of heat absorbed at constant volume, q_v.
(b) The change in enthalpy, ΔH, for a process is equal to the amount of heat absorbed at constant pressure, q_p.
(c) A bomb calorimeter measures ΔH directly.
(d) If q_p for a process is negative, the process must be exothermic.
(e) The work done in a process occurring at constant pressure is zero if Δn_{gases} is zero.

c 52. Which one of the following statements is **false**?

(a) The amount of heat absorbed by a system at **constant volume**, q_v, is ΔE for the process.
(b) The amount of heat absorbed by a system at **constant pressure**, q_p, is ΔH for the process.
(c) In the relationship $\Delta E = q + w$, as applied to a typical chemical reaction, w is usually much larger than q.
(d) At constant temperature and pressure, the work done by a system involving gases is $-\Delta n_{gases}(RT)$ where $\Delta n_{gas} = n_{product\ gases} - n_{reactant\ gases}$ for the process of interest.
(e) At constant pressure, the work done in a process by a system involving gases can be expressed as $-P\Delta V$.

b 53. Which one of the following statements is **false**? For a reaction carried out at constant temperature and **constant pressure** in an open container, _____.

(a) the work done by the system can be set equal to $-P\Delta V$
(b) the work done by the system can be set equal to $V\Delta P$
(c) the work done by the system can be set equal to $-\Delta nRT$ where Δn is the number of moles of gaseous products minus the number of moles of gaseous reactants
(d) the heat absorbed by the system can be called q_p
(e) the heat absorbed by the system can be called ΔH

e 54. Which statement is **false**?

(a) The thermodynamic quantity most easily measured in a "coffee cup" calorimeter is ΔH.
(b) No work is done in a reaction occurring in a bomb calorimeter.
(c) ΔH is sometimes exactly equal to ΔE.
(d) ΔH is often nearly equal to ΔE.
(e) ΔH is equal to ΔE for the process: $2H_2(g) + O_2(g) \rightarrow 2H_2O(g)$.

b 55. Assuming the gases are ideal, calculate the amount of work done, in joules, for the conversion of 1.00 mole of Ni to $Ni(CO)_4$ at 75°C in the reaction below. The value of R is 8.314 J/mol·K.

$$Ni(s) + 4CO(g) \rightarrow Ni(CO)_4(g)$$

(a) 1.80×10^3 J (b) 8.68×10^3 J (c) -1.80×10^3 J
(d) -8.68×10^3 J (e) -494 J

b 56. The reaction of 1.00 mole of $H_2(g)$ with 0.500 mole of $O_2(g)$ to produce 1.00 mole of steam, $H_2O(g)$, at 100°C and 1.00 atm pressure **evolves** 242 kJ of heat. Calculate ΔE per mole of $H_2O(g)$ produced. The universal gas constant is 8.314 J/mol·K.

(a) +240 kJ (b) –240 kJ (c) +242 kJ (d) –242 kJ (e) –238 kJ

Entropy

a 57. What is the entropy change of the reaction below at 298 K and 1 atm pressure?

$$N_2(g) + 3H_2(g) \rightarrow 2NH_3(g)$$
S^0_{298} (J/mol·K)   191.5   130.6   192.3

(a) –198.7 J/K (b) 76.32 J/K (c) –129.7 J/K
(d) 303.2 J/K (e) 384.7 J/K

c 58. Evaluate ΔS^0 for the reaction below at 25°C and 1 atm.

$$3NO_2(g) + H_2O(\ell) \rightarrow 2HNO_3(aq) + NO(g)$$
S^0 (J/mol·K)   240   69.91   146   210.7

(a) $+1.37 \times 10^3$ J/K (b) +287.2 J/K (c) –287.2 J/K
(d) $+1.37 \times 10^3$ J/K (e) –531.4 J/K

b 59. Given the following data at 298 K, calculate ΔS^0 for the given reaction.

$$2Ag_2O(s) \rightarrow 4Ag(s) + O_2(g)$$

ΔS^0 (J / mol • K)

Ag(s) 42.55   Ag$_2$O(s) 121.7   O$_2$(g) 205.0

(a) 68.62 J/K   (b) 131.8 J/K   (c) 117.7 J/K
(d) 93.76 J/K   (e) 606.7 J/K

b 60. Calculate ΔS^0 for the reaction below at 25°C. S^0 for SiH$_4$ = 204.5 J/mol•K, for O$_2$(g) = 205.0 J/mol•K, for SiO$_2$(s) = 41.84 J/mol•K, for H$_2$O(ℓ) = 69.91 J/mol•K.

$$SiH_4(g) + 2O_2(g) \rightarrow SiO_2(s) + 2H_2O(\ell)$$

(a) –353.5 J/K   (b) – 432.8 J/K   (c) 595.0 J/K
(d) –677.0 J/K   (e) –880.3 J/K

c 61. If the entropy change for the reaction below at 298 K and 1 atm pressure is 137 J/K and S^0 = 205 J/mol•K for O$_2$(g), what is S^0 for O$_3$(g)?

$$2O_3(g) \rightarrow 3O_2(g)$$

(a) 364 J/mol•K   (b) 478 J/mol•K   (c) 239 J/mol•K
(d) –117 J/mol•K   (e) –59 J/mol•K

d 62. The heat of vaporization of methanol, CH$_3$OH, is 35.20 kJ/mol. Its boiling point is 64.6°C. What is the change in entropy for the vaporization of methanol?

(a) –17.0 J/mol•K   (b) 3.25 J/mol•K   (c) 17.0 J/mol•K
(d) 104 J/mol•K   (e) 543 J/mol•K

c 63. Which of following would have the **lowest** value of absolute entropy per mole?

(a) water at 50°C   (b) steam at 100°C   (c) ice at 0°C
(d) superheated steam at 200°C   (e) a salt-ice-water bath for freezing ice cream

e 64. Which one of the following reactions has a **positive** entropy change?

(a) H$_2$O(g) $\rightarrow$ H$_2$O(ℓ)   (b) BF$_3$(g) + NH$_3$(g) $\rightarrow$ F$_3$BNH$_3$(s)
(c) 2SO$_2$(g) + O$_2$(g) $\rightarrow$ 2SO$_3$(g)   (d) N$_2$(g) + 3H$_2$(g) $\rightarrow$ 2NH$_3$(g)
(e) 2NH$_4$NO$_3$(s) $\rightarrow$ 2N$_2$(g) + 4H$_2$O(g) + O$_2$(g)

d 65. Based on the relationship of entropy to the degree of disorder in a system, which response includes all the following changes that represent an **increase** in entropy, and no others?

I.   the freezing of water   II.   the condensation of steam
III. sublimation (vaporization) of dry ice, solid CO$_2$
IV.   the extraction of salts and pure water from seawater

(a) I and IV   (b) II and IV   (c) I, and II   (d) III   (e) I and III

c 66. Which response includes all the following processes that are accompanied by an **increase** in entropy, and only those processes?

    I.   boiling water   II.   freezing water
    III. $N_2(g) + 3H_2(g) \rightarrow 2NH_3(g)$
    IV. $Br_2(\ell) \rightarrow Br_2(g)$

(a) I and II   (b) III and IV   (c) I and IV
(d) II, III, and IV   (e) another one or another combination

a 67. Which response contains all the processes below that occur with an **increase** in entropy and no others?

    I.   The evaporation of CCl_4.
    II.   The precipitation of white silver chloride, AgCl, from a solution containing silver ions and chloride ions.
    III.   The reaction $PCl_3(g) + Cl_2(g) \rightarrow PCl_5(g)$.
    IV.   Thirty-five pennies are removed from a bag and all are placed heads up on a table.

(a) I   (b) II and IV   (c) I, III, and IV
(d) II and III   (e) another one or another combination

d 68. Based on the relationship of entropy to the degree of disorder in a system, which of the following changes are accompanied by an **increase** in entropy?

    I.   The sublimation of iodine.
    II.   The freezing of ethyl alcohol.
    III.   The extraction of metals from their ores.
    IV.   The burning of a **solid** rocket fuel.
    V.   The thermal decomposition of ammonium nitrate.
            $2NH_4NO_3(s) \rightarrow 2N_2(g) + 4H_2O(g) + O_2(g)$

(a) II and III   (b) I and IV   (c) III, IV, and V
(d) I, IV, and V   (e) II, III, and V

d 69. Which response contains all the processes below for which ΔS is **positive** and none for which ΔS is negative?

    I. $CCl_4(\ell) \rightarrow CCl_4(g)$
    II. $N_2(g) + 3H_2(g) \rightarrow 2NH_3(g)$
    III. $2N_2O(g) \rightarrow 2N_2(g) + O_2(g)$
    IV. $I_2(g) \rightarrow I_2(s)$

(a) I and II   (b) III and IV   (c) II and IV
(d) I and III   (e) I, II, and IV

Free Energy Change, ΔG, and Spontaneity

d 70. Consider the conversion of a substance from solid to liquid.
$$\text{Solid} \rightleftharpoons \text{Liquid}$$
At one atmosphere pressure and at the melting point of the substance, _____.

(a) $\Delta H = 0$ for the process
(b) $\Delta S = 0$ for the process
(b) $\Delta E = 0$ for the process
(d) $\Delta G = 0$ for the process
(e) both $\Delta H = 0$ and $\Delta E = 0$ for the process

d 71. Evaluate ΔG^0 for the reaction below at 25°C.

$$2C_2H_2(g) + 5O_2(g) \rightarrow 4CO_2(g) + 2H_2O(\ell)$$
ΔG_f^0 (kJ/mol) 209.2   0   −394.4   −237.2

(a) −1409 kJ (b) −2599 kJ (c) −1643 kJ
(d) −2470 kJ (e) −766 kJ

c 72. Calculate the standard Gibbs free energy change for the following reaction at 25°C.

$$CaCO_3(s) + 2HCl(g) \rightarrow CaCl_2(s) + CO_2(g) + H_2O(\ell)$$
ΔG_f^0 (kJ/mol) −1129   −95.3   −750.2   −394.4   −237.2

(a) −41 kJ (b) −158 kJ (c) −62 kJ (d) −87 kJ (e) −104 kJ

c 73. Calculate the ΔG_f^0 at 298 K for $PbCl_2(s)$ from the following information. ΔG^0 for the reaction below is −58.4 kJ at 298 K.

$$PbS(s) + 2HCl(g) \rightarrow PbCl_2(s) + H_2S(g)$$
ΔG_f^0 (kJ/mol) −98.7   −95.30   ?   −33.6

(a) −16.0 kJ/mol (b) −47.6 kJ/mol (c) −314.1 kJ/mol
(d) −36.2 kJ/mol (e) −52.3 kJ/mol

d 74. Calculate ΔG^0 at 298 K for the reaction below.

$$Fe_2O_3(s) + 13CO(g) \rightarrow 2Fe(CO)_5(g) + 3CO_2(g)$$
ΔH_f^0 (kJ/mol) −824.2   −110.5   −733.8   −393.5
S^0 (J/mol·K) 87.4   197.6   445.2   213.6

(a) +63.6 kJ (b) +26.8 kJ (c) −243.1 kJ
(d) −52.2 kJ (e) −193.3 kJ

d 75. Evaluate ΔG_{298}^0 for the following reaction at 25°C.

$$2ZnS(s) + 3O_2(g) \rightarrow 2ZnO(s) + 2SO_2(g)$$
ΔH_f^0 (kJ/mol) −205.6   0   −348.3   −296.8
S^0 (J/mol·K) 57.7   205.0   43.64   248.1

(a) −951.1 kJ (b) −922.6 kJ (c) −704.2 kJ
(d) −835.2 kJ (e) −1902 kJ

c 76. For the following reaction at 25°C, $\Delta H^0 = -26.88$ kJ and $\Delta S^0 = 11.2$ J/K. Calculate ΔG^0 for the reaction at 25°C in kilojoules.

$$I_2(g) + Cl_2(g) \rightarrow 2ICl(g)$$

(a) –102 kJ (b) +50.6 kJ (c) –30.2 kJ (d) –50.6 kJ (e) +77.0 kJ

e 77. Evaluate ΔG^0 for the reaction below at 25°C.

$$P_4O_{10}(s) + 6H_2O(\ell) \rightarrow 4H_3PO_4(s)$$

	$P_4O_{10}(s)$	$6H_2O(\ell)$	$4H_3PO_4(s)$
ΔH_f^0 (kJ/mol)	–2984	–285.8	–1281
S^0 (J/mol·K)	228.9	69.91	110.5

(a) –50.33 kJ (b) –172.0 kJ (c) –282.5 kJ
(d) –304.8 kJ (e) –363.7 kJ

b 78. Evaluate ΔG^0_{rxn} for the following reaction at 25°C.

$$2N_2(g) + 3O_2(g) \rightarrow 2N_2O_3(g)$$

	ΔH_f^0	S^0
$N_2(g)$		191.5 J/mol·K
$O_2(g)$		205.0 J/mol·K
$N_2O_3(g)$	83.72 kJ/mol	312.2 J/mol·K

(a) +540.0 kJ (b) +278.8 kJ (c) –540.0 kJ
(d) –56.1 kJ (e) +56.1 kJ

c 79. The standard molar enthalpy of formation of $NO_2(g)$ is 33.2 kJ/mol at 25°C and that of $N_2O_4(g)$ is 9.16 kJ/mol. At 25°C their absolute entropies are 240.0 and 304.2 J/mol·K, respectively. Use the above data to calculate the standard Gibbs free energy change for the following reaction at 25°C. Express your answer in the form $\Delta G^0 =$ _____ kJ.

$$N_2O_4(g) \rightarrow 2NO_2(g)$$

(a) 4.1 kJ (b) 21.3 kJ (c) 4.8 kJ (d) 41.5 kJ (e) 11.4 kJ

The Temperature Dependence of Spontaneity

c 80. **Estimate** the temperature at which $\Delta G = 0$ for the following reaction.

$$NH_3(g) + HCl(g) \rightarrow NH_4Cl(s)$$
$\Delta H = -176$ kJ; $\Delta S = -284.5$ J/K

(a) 467 K (b) 582 K (c) 619 K (d) 634 K (e) 680 K

e*81. **Estimate** the boiling point of tin(IV) chloride, $SnCl_4$, at one atmosphere pressure. For $SnCl_4(\ell)$, $\Delta H_f^0 = -511.3$ kJ/mol, $S^0 = 258.6$ J/mol·K and for $SnCl_4(g)$, $\Delta H_f^0 = -471.5$ kJ/mol, $S^0 = 366$ J/mol·K at 298 K.

(a) 266°C (b) 213°C (c) 186°C (d) 154°C (e) 99°C

b*82. **Estimate** the boiling point of hydrogen peroxide, H_2O_2.
For $H_2O_2(\ell)$, $\Delta H_f^0 = -187.8$ kJ/mol and $S^0 = 109.6$ J/mol·K.
For $H_2O_2(g)$, $\Delta H_f^0 = -136.3$ kJ/mol and $S^0 = 233$ J/mol·K.

(a) 67.6°C   (b) 145°C   (c) 26.2°C   (d) –11.4°C   (e) 118°C

d 83. **Estimate** the boiling point of water if for $H_2O(\ell)$, $\Delta H_f^0 = -285.8$ kJ/mol and $S^0 = 69.91$ J/mol·K and for $H_2O(g)$, $\Delta H_f^0 = -241.8$ kJ/mol and $S^0 = 188.7$ J/mol·K.

(a) 101 K   (b) 387 K   (c) 398 K   (d) 370 K   (e) 274 K

d 84. A process **cannot** be spontaneous (product-favored) if _____.

(a) it is exothermic, and there is an increase in disorder
(b) it is endothermic, and there is an increase in disorder
(c) it is exothermic, and there is a decrease in disorder
(d) it is endothermic, and there is a decrease in disorder
(e) the entropy of the universe increases

d 85. Which one of the following statements is **not** correct?

(a) When ΔG for a reaction is negative, the reaction is spontaneous.
(b) When ΔG for a reaction is positive, the reaction is nonspontaneous.
(c) When ΔG for a reaction is zero, the system is at equilibrium.
(d) When ΔH for a reaction is negative, the reaction is never spontaneous.
(e) When ΔH for a reaction is very positive, the reaction is not expected to be spontaneous.

d 86. Which statement below is **false**?

(a) For reactions that release heat to the surroundings, ΔH is negative.
(b) For reactions in which the reacting system becomes more disordered, ΔS is positive.
(c) If the free energy change of reaction is positive, the reaction cannot occur to give predominantly products under the given conditions.
(d) The entropy of a system increases when order increases.
(e) Endothermic reactions may be spontaneous.

a 87. For the reaction given below, $\Delta H^0 = -1516$ kJ at 25°C and $\Delta S^0 = -432.8$ J/K at 25°C. This reaction is spontaneous _____.

$$SiH_4(g) + 2O_2(g) \rightarrow SiO_2(s) + 2H_2O(\ell)$$

(a) only below a certain temperature
(b) only above a certain temperature
(c) at all temperatures
(d) at no temperatures
(e) cannot tell from the information available

c 88. Consider the reaction below at 25°C for which $\Delta S^0 = 16.1$ J/K.

$$CH_4(g) + N_2(g) + 163.8 \text{ kJ} \rightarrow HCN(g) + NH_3(g)$$

Which one of the following statements describes the reaction?

(a) Spontaneous at all temperatures
(b) Spontaneous at relatively low temperatures only
(c) Spontaneous at relatively high temperatures only
(d) Nonspontaneous at all temperatures
(e) Insufficient information to estimate temperature range of spontaneity

b 89. Consider the following equation carefully, and determine the sign of ΔS^0 for the reaction it describes.

$$NH_4Br(s) \rightarrow NH_3(g) + HBr(g) \quad \Delta H^0_{rxn} = +188.3 \text{ kJ}$$

Which response describes the thermodynamic spontaneity of the reaction?

(a) The reaction is spontaneous at all temperatures.
(b) The reaction is spontaneous only at relatively high temperatures.
(c) The reaction is spontaneous only at relatively low temperatures.
(d) The reaction is not spontaneous at any temperatures.
(e) We cannot tell from information given.

c 90. For which set of values of ΔH and ΔS will a reaction be spontaneous (product-favored) at all temperatures?

(a) $\Delta H = +10$ kJ, $\Delta S = -5$ J/K
(b) $\Delta H = -10$ kJ, $\Delta S = -5$ J/K
(c) $\Delta H = -10$ kJ, $\Delta S = +5$ J/K
(d) $\Delta H = +10$ kJ, $\Delta S = +5$ J/K
(e) no such values exist

a 91. For which set of values of ΔH and ΔS will a reaction be nonspontaneous (reactant-favored) at all temperatures?

(a) $\Delta H = +10$ kJ, $\Delta S = -5$ J/K
(b) $\Delta H = -10$ kJ, $\Delta S = -5$ J/K
(c) $\Delta H = -10$ kJ, $\Delta S = +5$ J/K
(d) $\Delta H = +10$ kJ, $\Delta S = +5$ J/K
(e) no such values exist

c 92. Joseph Priestley prepared oxygen by heating mercury(II) oxide. The compound HgO is stable at room temperature but decomposes into its elements (Hg and O_2) at high temperatures. What conclusions can be drawn concerning ΔH and ΔS for this decomposition reaction?

(a) ΔH is negative and ΔS is positive.
(b) ΔH is negative and ΔS is negative.
(c) ΔH is positive and ΔS is positive.
(d) ΔH is positive and ΔS is negative.
(e) ΔH becomes negative at high temperatures.

c 93. For a certain process at 127°C, $\Delta G = -16.20$ kJ and $\Delta H = -17.0$ kJ. What is the entropy change for this process at this temperature? Express your answer in the form, $\Delta S = \underline{\hspace{2cm}}$ J/K.

(a) –6.3 J/K (b) +6.3 J/K (c) –2.0 J/K (d) +2.0 J/K (e) –8.1 J/K

Chapter 15   235

b 94. For a certain process at 27°C, $\Delta G = +210.6$ kJ and $\Delta H = -168.2$ kJ. What is the entropy change for this process at this temperature? Express your answer in the form, $\Delta S = $ _____ J/K.

(a) 1.26 x 10³ J/K   (b) –1.26 x 10³ J/K   (c) –141.3 J/K
(d) +141.3 J/K   (e) +628.3 J/K

d 95. Consider the following reaction and its ΔH^0_{rxn} and ΔG^0_{rxn} values at 25°C. Evaluate ΔS^0_{rxn} at 25°C.

$$2C_2H_2(g) + 5O_2(g) \rightarrow 4CO_2(g) + 2H_2O(\ell)$$

$\Delta H^0_{rxn} = -2599$ kJ, $\Delta G^0_{rxn} = -2470$ kJ

(a) +340 J/K   (b) –340 J/K   (c) +386 J/K
(d) –433 J/K   (e) –386 J/K

c 96. Evaluate ΔS^0 for the reaction below at 25°C.

	$CH_4(g)$ +	$2Cl_2(g)$	→	$CCl_4(\ell)$ +	$2H_2(g)$
ΔH^0_f (kJ/mol)	–74.81	0		–135.4	0
ΔG^0_f (kJ/mol)	–50.75	0		–65.27	0

(a) –360 J/K   (b) –66.9 J/K   (c) –155 J/K
(d) –487 J/K   (e) –387 J/K

c 97. Calculate ΔS^0 at 25°C for the reaction below.

	PbS(s) +	2HCl(g)	→	$PbCl_2(s)$ +	$H_2S(g)$
ΔH^0_f (kJ/mol)	–100.4	–92.31		–359.4	–20.6
ΔG^0_f (kJ/mol)	–98.7	–95.30		–314.1	–33.6

(a) 686 J/K   (b) –741 J/K   (c) –123 J/K
(d) 1.33 x 10³ J/K   (e) 515 J/K

a 98. The ΔG at 298 K for the reaction below is +115 kJ/mol. Calculate S^0 for $Mn_3O_4(s)$.

	$3MnO_2(s)$	→	$Mn_3O_4(s)$ +	$O_2(g)$
ΔH^0_f (kJ/mol)	–520.		–1387	0
S^0 (J/mol·K)	53		?	205

(a) 149 J/mol·K   (b) 534 J/mol·K   (c) 43 J/mol·K
(d) 339 J/mol·K   (e) 3440 J/mol·K

c 99. Estimate the temperature above which this reaction is spontaneous. $\Delta S^0 = 16.1$ J/K.

$$CH_4(g) + N_2(g) + 163.8 \text{ kJ} \rightarrow HCN(g) + NH_3(g)$$

(a) 9.91°C   (b) 1045 K   (c) 9.90 x 10³°C   (d) 10.7 K   (e) 10.1°C

d 100. Priestley prepared oxygen by heating mercury(II) oxide. From the data given below estimate the temperature above which this reaction will become spontaneous.

$$HgO(s) \rightarrow Hg(\ell) + \tfrac{1}{2}O_2(g) \qquad \Delta H^0 = 90.83 \text{ kJ}$$

$S^0(Hg) = 76.02$ J/mol·K $S^0(HgO) = 70.29$ J/mol·K $S^0(O_2) = 205.0$ J/mol·K

(a) 108 K   (b) 566 K   (c) 430 K   (d) 840 K   (e) 739 K

c*101. At 25°C $\Delta H = 128.9$ kJ and $\Delta G = 33.5$ kJ for a reaction. Above what minimum temperature will this reaction become spontaneous?

(a) 298 K   (b) 332 K   (c) 403 K   (d) 530 K   (e) 1150 K

e*102. Estimate the temperature above which this reaction is nonspontaneous.

$$PbS(s) + 2HCl(g) \rightarrow PbCl_2(s) + H_2S(g)$$

	PbS(s)	2HCl(g)	PbCl₂(s)	H₂S(g)
ΔH_f^0 (kJ/mol)	−100.4	−92.31	−359.4	−20.6
ΔG_f^0 (kJ/mol)	−98.7	−95.30	−314.1	−33.6

(a) −144°C   (b) 88°C   (c) 16°C
(d) −42°C   (e) 499°C

16 Chemical Kinetics

The Rate of a Reaction

d 1. Which of the following is a kinetic quantity?

(a) enthalpy (b) internal energy (c) free energy
(d) rate of reaction (e) entropy

c 2. Of the following questions, which ones are **thermodynamic** rather than kinetic concepts?

I. Can substances react when they are put together?
II. If a reaction occurs, how fast will it occur?
III. What is the mechanism by which the reaction occurs?
IV. If substances react, what energy changes are associated with the reaction?

(a) I and III (b) II and IV (c) I and IV (d) II and III (e) I, III, and IV

e 3. One of the reactions that is used to produce gaseous hydrogen commercially follows. A proper expression for the rate of this reaction **could be** _____.

$$H_2O(g) + CO(g) \rightarrow H_2(g) + CO_2(g)$$

(a) $\dfrac{-\Delta[CO_2]}{\Delta t}$ (b) $\dfrac{-\Delta[H_2]}{\Delta t}$ (c) k (d) $\dfrac{\Delta[CO]}{\Delta t}$ (e) $\dfrac{-\Delta[H_2O]}{\Delta t}$

b 4. At some time, the rate of formation of C is observed to be 0.036 mol/L·s for the reaction below. **In order**, what is the rate of change of A, the rate of change of B, the rate of change of D and the rate of reaction at this time (all given in mol/L·s)?

$$2A + 3B \rightarrow 4C + 2D$$

(a) 0.018, 0.027, 0.018, 0.0090
(b) −0.018, −0.027, 0.018, 0.0090
(c) −0.072, −0.048, 0.072, 0.144
(d) −0.036, −0.036, 0.036, 0.0090
(e) −0.018, −0.012, −0.018, 0.018

Factors That Affect Reaction Rates

d 5. Four of the following factors can affect the forward rate of a chemical reaction. Which one **cannot** affect this **rate**? (Note: the question refers to **reaction rate**, not equilibrium.)

(a) temperature
(b) presence of a catalyst
(c) concentration of reactants of the forward reaction
(d) removal of some of the products of the forward reaction
(e) physical state or state of subdivision of solid reactants

e 6. Suppose a reaction A + B → C occurs at some initial rate at 25°C. Which response includes all of the changes below that **could** increase the rate of this reaction?

    I.   lowering the temperature   II.   adding a catalyst
    III.   increasing the initial concentration of B

(a) I   (b) II   (c) III   (d) I and II   (e) II and III

Nature of the Reactants

c 7. Which of the following reactions would be expected to be the slowest?

(a) $Ag^+(aq) + Cl^-(aq) \rightarrow AgCl(s)$
(b) $H^+(aq) + OH^-(aq) \rightarrow H_2O(\ell)$
(c) $CH_4(g) + 2O_2(g) \rightarrow CO_2(g) + 2H_2O(g)$
(d) $Pb^{2+}(aq) + CrO_4^{2-}(aq) \rightarrow PbCrO_4(s)$
(e) $H^+(aq) + CN^-(aq) \rightarrow HCN(aq)$

e 8. Which is **not** an example of the effect of subdivision of the reactant on the rate of **chemical** reaction?

(a) Violent explosions that occur in grain elevators.
(b) A container of flammable liquid will burn on the surface but allowed to vaporize will burn explosively.
(c) A chunk of iron takes months to rust completely while iron wool will rust in days.
(d) Some metals may be fused (welded) with minimal loss while their powders will burn in a flame.
(e) The Grand Canyon was created by dissolution by water over millions of years.

Concentrations of Reactants: the Rate-Law Expression

c 9. For a given reaction, the rate-law expression is _____.

(a) a constant of proportionality between reaction rate and the concentrations of reactants
(b) the sum of the powers to which reactant concentrations appear
(c) an equation in which reaction rate is equal to a mathematical expression involving, or related to, concentrations of reactants involved in the rate-determining step
(d) an equation that gives the additional energy that reactants must obtain in order to react
(e) 55 miles per hour

b 10. What can be said about the stoichiometric coefficients of a balanced chemical equation for a reaction and the powers to which the concentrations are raised in the rate law expression?

(a) There is an exact relationship between the two.
(b) Not much can be said except that there is no necessary relationship.
(c) The powers can be equal to the number of molecules that are formed.
(d) The powers are equal to the number of molecules that must collide and react in the fastest step.
(e) The powers equal the coefficients of the equation for the reaction at 298 K.

a 11. The rate expression for the following reaction is found to be: rate = $k[N_2O_5]$. What is the overall reaction order?

$$2N_2O_5(g) \rightarrow 4NO_2(g) + O_2(g)$$

(a) first   (b) second   (c) third   (d) fourth   (e) none of these

c 12. The gas phase reaction A + B → C has a reaction rate which is experimentally observed to follow the relationship rate = $k[A]^2[B]$. The overall order of the reaction

(a) is first.   (b) is second.   (c) is third.   (d) is zero.   (e) is one-half.

d 13. The gas phase reaction A + B → C has a reaction rate which is experimentally observed to follow the relationship rate = $k[A]^2$. The reaction is _____ order in B.

(a) first   (b) second   (c) third   (d) zero   (e) one-half

a 14. The gas phase reaction A + B → C has a reaction rate which is experimentally observed to follow the relationship rate = $k[A]^2[B]$. Which one of the following would affect the value of the specific rate constant, k?

(a) decreasing the temperature
(b) changing the concentration of A
(c) changing the concentration of B
(d) changing the concentration of C
(e) letting the reaction go on for a long time

a 15. The units of the rate constant for a second order reaction could be _____.

(a) $M^{-1} \cdot s^{-1}$ (b) M (c) $M \cdot s^{-1}$ (d) s^{-1} (e) $M^2 \cdot s^{-1}$

d 16. The units of the rate constant for a first order reaction could be _____.

(a) $M^{-1} \cdot min^{-1}$ (b) M (c) $M \cdot min^{-1}$ (d) min^{-1} (e) $M^2 \cdot min^{-1}$

b 17. A hypothetical reaction X + 2Y → Products is found to be first order in X and second order in Y. What are the **units** of k, the specific rate constant, if reaction rate is expressed in units of moles per liter per second?

(a) $M \cdot s^{-1}$ (b) $M^{-2} \cdot s^{-1}$ (c) $M^{-3} \cdot s$ (d) $M^2 \cdot s^{-1}$ (e) $M^{-1} \cdot s$

a 18. A reaction A + 2B → C is found to be first order in A and first order in B. What are the **units** of the rate constant, k, if the rate is expressed in units of moles per liter per minute?

(a) $M^{-1} \cdot min^{-1}$ (b) M (c) $M \cdot min^{-1}$ (d) min^{-1} (e) $M^2 \cdot min^{-1}$

e 19. Consider the following rate law expression: rate = $k[A]^2[B]$. Which of the following is not true about the reaction having this expression?

(a)   The reaction is first order in B.
(b)   The reaction is overall third order.
(c)   The reaction is second order in A.
(d)   A and B must both be reactants.
(e)   Doubling the concentration of A doubles the rate.

e 20. Consider the following rate law expression: rate = k[A][B]. Which of the following is not true about the reaction having this expression?

(a) The reaction is first order in B.
(b) The reaction is overall second order.
(c) The reaction is first order in A.
(d) A and B must both be reactants.
(e) Doubling the concentrations of A and of B doubles the rate.

d 21. The gas phase reaction A + B → C has a reaction rate which is experimentally observed to follow the relationship rate = k[A]²[B]. If the concentration of A is tripled and the concentration of B is doubled, the reaction rate would be increased by a factor of _____.

(a) 6 (b) 9 (c) 12 (d) 18 (e) 36

a 22. A reaction is first order in X and second order in Y. Tripling the initial concentration of X and cutting the initial concentration of Y to three-fourths of its previous concentration at constant temperature causes the initial rate to _____ by a factor of _____.

(a) increase, 1.69 (b) decrease, 0.19 (c) increase, 1.33
(d) decrease, 1.25 (e) increase, 2.25

a 23. A reaction is second order in X and zero order in Y. Doubling the initial concentration of X and halving the initial concentration of Y at constant temperature causes the initial rate to

(a) increase by a factor of 4. (b) decrease by a factor of 2.
(c) remain unchanged. (d) increase by a factor of 2.
(e) be undeterminable without the balanced equation.

b 24. Consider the following rate data for the reaction below at a particular temperature.

2A + 3B → Products

Experiment	Initial [A]	Initial [B]	Initial Rate of Loss of A
1	0.10 M	0.30 M	7.20 x 10^{-5} M·s^{-1}
2	0.10 M	0.60 M	1.44 x 10^{-4} M·s^{-1}
3	0.20 M	0.90 M	8.64 x 10^{-4} M·s^{-1}

The reaction is _____ order in A and _____ order in B.

(a) first, first (b) second, first (c) first, second
(d) second, second (e) third, first

c 25. Given the following data for the NH_4^+ + NO_2^- → N_2 + $2H_2O$ reaction

Trial	[NH_4^+]	[NO_2^-]	Rate
1	0.010 M	0.020 M	0.020 M/s
2	0.015	0.020	0.030
3	0.010	0.010	0.005

The rate law for the reaction is

(a) rate = k[NH_4^+][NO_2^-] (b) rate = k[NH_4^+]²[NO_2^-] (c) rate = k[NH_4^+][NO_2^-]²
(d) rate = k[NH_4^+]²[NO_2^-]² (e) None of the above

d 26. Determine the rate-law expression for the reaction below.

$$2A + B_2 + C \rightarrow A_2B + BC$$

Trial	Initial [A]	Initial [B$_2$]	Initial [C]	Initial Rate of Formation of BC
1	0.20 M	0.20 M	0.20 M	2.4 x 10^{-6} M·min^{-1}
2	0.40 M	0.30 M	0.20 M	9.6 x 10^{-6} M·min^{-1}
3	0.20 M	0.30 M	0.20 M	2.4 x 10^{-6} M·min^{-1}
4	0.20 M	0.40 M	0.40 M	4.8 x 10^{-6} M·min^{-1}

(a) rate = $k[A]^2[B_2][C]$ (b) rate = $k[B_2]^2[C]^2$ (c) rate = $k[A][C]^2$
(d) rate = $k[A]^2[C]$ (e) rate = $k[A][B_2][C]$

c 27. Rate data have been determined at a particular temperature for the overall reaction
$2NO + 2H_2 \rightarrow N_2 + 2H_2O$ in which all reactants and products are gases.

Trial Run	Initial [NO]	Initial [H$_2$]	Initial Rate (M·s^{-1})
1	0.10 M	0.20 M	0.0150
2	0.10 M	0.30 M	0.0225
3	0.20 M	0.20 M	0.0600

The rate-law expression is _____.

(a) rate = $k[NO]^2[H_2]^2$ (b) rate = $k[NO][H_2]^2$ (c) rate = $k[NO]^2[H_2]$
(d) rate = $k[NO][H_2]$ (e) None of the preceding answers is correct.

d 28. NO reacts with chlorine in a gas phase reaction to form nitrosyl chloride, NOCl. From the following experimental data, determine the form of the equation that describes the relationship of reaction rate to initial concentrations of reactants.

$$2NO + Cl_2 \rightarrow 2NOCl$$

Run	Initial [NO]	Initial [Cl$_2$]	Initial Rate of Formation of NOCl
1	0.50 M	0.50 M	1.14 M/hr
2	1.00 M	1.00 M	9.12 M/hr
3	1.00 M	0.50 M	4.56 M/hr

(a) rate = $k[NO]$ (b) rate = $k[NO][Cl_2]$ (c) rate = $k[NO]^2$
(d) rate = $k[NO]^2[Cl_2]$ (e) rate = $k[NO]^2[Cl_2]^2$

e 29. The following data were collected for the following reaction at a particular temperature. What is the rate-law expression for this reaction? rate = _____.

$$A + B \rightarrow C$$

Experiment	Initial [A]	Initial [B]	Initial Rate of Formation of C
1	0.10 M	0.10 M	4.0 x 10^{-4} M/min
2	0.20 M	0.20 M	3.2 x 10^{-3} M/min
3	0.10 M	0.20 M	1.6 x 10^{-3} M/min

(a) $k[A]$ (b) $k[A]^2$ (c) $k[A][B]$ (d) $k[B]$ (e) $k[A][B]^2$

d 30. Determine the rate-law expression for the reaction below at the temperature at which the tabulated initial rate data were obtained.   rate = _____

$$A + 2B + 3C \rightarrow \text{Products}$$

Experiment	Initial [A]	Initial [B]	Initial [C]	Initial Rate of Loss of A
1	0.10 M	0.20 M	0.10 M	4.0 x 10^{-2} M·min^{-1}
2	0.40 M	0.20 M	0.10 M	4.0 x 10^{-2} M·min^{-1}
3	0.20 M	0.20 M	0.25 M	1.0 x 10^{-1} M·min^{-1}
4	0.20 M	0.40 M	0.10 M	1.6 x 10^{-1} M·min^{-1}

(a) k[A][B]   (b) k[A]²[C]   (c) k[C]²   (d) k[B]²[C]   (e) none of these

e 31. The following rate data apply to the reaction $3X + Y + 2Z \rightarrow$ Products   at a particular temperature. The form of the rate law expression is rate = _____ .

Experiment	Initial [X]	Initial [Y]	Initial [Z]	Initial Rate of Loss of X
1	0.300 M	0.200 M	0.100 M	6.40 x 10^{-3} M·min^{-1}
2	0.300	0.300	0.250	4.00 x 10^{-2}
3	0.600	0.300	0.250	4.00 x 10^{-2}
4	0.900	0.200	0.400	1.02 x 10^{-1}
5	0.800	0.400	0.100	6.40 x 10^{-3}

(a) k[X][Y][Z]   (b) k[X]³[Y][Z]²   (c) k[X]²[Z]
(d) k[X][Y]²   (e) k[Z]²

c 32. Evaluate the specific rate constant for this reaction at 800°C. The rate-law expression is rate = k[NO]²[H₂]. (Choose the closest answer.)

$$2NO(g) + 2H_2(g) \rightarrow N_2(g) + 2H_2O(g)$$

Experiment	Initial [NO]	Initial [H₂]	Initial Rate of Reaction (M·s^{-1})
1	0.0010 M	0.0060 M	7.9 x 10^{-7}
2	0.0040 M	0.0060 M	1.3 x 10^{-5}
3	0.0040 M	0.0030 M	6.4 x 10^{-6}

(a) 22 $M^{-2}·s^{-1}$ (b) 4.6 $M^{-2}·s^{-1}$ (c) 1.3 x 10^2 $M^{-2}·s^{-1}$
(d) 0.82 $M^{-2}·s^{-1}$ (e) 0.024 $M^{-2}·s^{-1}$

c 33. Evaluate the specific rate constant for the reaction at the temperature for which the data were obtained. The rate-law expression is rate = k[A][B]².

$$A + B \rightarrow C$$

Experiment	Initial [A]	Initial [B]	Initial Rate of Formation of C
1	0.10 M	0.10 M	4.0 x 10^{-4} M/min
2	0.20 M	0.20 M	3.2 x 10^{-3} M/min
3	0.10 M	0.20 M	1.6 x 10^{-3} M/min

(a) 1.2 x 10^{-2} M^{-2}·min^{-1} (b) 3.6 x 10^{-2} M^{-2}·min^{-1} (c) 4.0 x 10^{-1} M^{-2}·min^{-1}
(d) 6.2 x 10^{-1} M^{-2}·min^{-1} (e) 7.0 x 10^{-3} M^{-2}·min^{-1}

e 34. Evaluate the specific rate constant at the temperature at which the data were collected. The rate-law expression is rate = $k[NO]^2[H_2]$.

$$H_2(g) + NO(g) \rightarrow N_2O(g) + H_2O(g)$$

Experiment	Initial [NO] (M)	Initial [H$_2$] (M)	Initial Rate (M·s^{-1})
1	0.30	0.35	2.835 x 10^{-3}
2	0.60	0.35	1.134 x 10^{-2}
3	0.60	0.70	2.268 x 10^{-2}

(a) 9.4 x 10^{-3} M^{-2}·s^{-1}
(b) 2.7 x 10^{-2} M^{-2}·s^{-1}
(c) 1.6 x 10^{-4} M^{-2}·s^{-1}
(d) 8.1 x 10^{-3} M^{-2}·s^{-1}
(e) 9.0 x 10^{-2} M^{-2}·s^{-1}

d 35. Evaluate the specific rate constant, k, at the temperature of reaction. The rate-law expression is rate = $k[B]^2[C]$.   A + 2B + 3C → Products

Experiment	Initial [A]	Initial [B]	Initial [C]	Initial Rate of Loss of A
1	0.10 M	0.20 M	0.10 M	4.0 x 10^{-2} M·min^{-1}
2	0.40 M	0.20 M	0.10 M	4.0 x 10^{-2} M·min^{-1}
3	0.20 M	0.20 M	0.25 M	1.0 x 10^{-1} M·min^{-1}
4	0.20 M	0.40 M	0.10 M	1.6 x 10^{-1} M·min^{-1}

(a) 1.0 M^{-2}·min^{-1}
(b) 0.48 M^{-2}·min^{-1}
(c) 6.1 M^{-2}·min^{-1}
(d) 10 M^{-2}·min^{-1}
(e) 14 M^{-2}·min^{-1}

b*36. Rate data have been determined at a particular temperature for the overall reaction

$$2NO + 2H_2 \rightarrow N_2 + 2H_2O$$ in which all reactants and products are gases. The value of the specific rate constant at this temperature is _____.

Trial Run	Initial [NO]	Initial [H$_2$]	Initial Rate (M·s^{-1})
1	0.10 M	0.20 M	0.0150
2	0.10 M	0.30 M	0.0225
3	0.20 M	0.20 M	0.0600

(a) 0.75 M^{-1}·s^{-1}
(b) 7.5 M^{-2}·s^{-1}
(c) 3.0 x 10^{-3} M^{-2}·s^{-1}
(d) 3.0 x 10^{-4} M^{-1}·s^{-1}
(e) 375 M^{-2}·s^{-1}

c*37. The following rate data apply to the reaction   3X + Y + 2Z → Products   at a particular temperature. What is the value of k at this temperature?

Experiment	Initial [X]	Initial [Y]	Initial [Z]	Initial Rate of Loss of X
1	0.300 M	0.200 M	0.100 M	6.40 x 10^{-3} M·min^{-1}
2	0.300	0.300	0.250	4.00 x 10^{-2}
3	0.600	0.300	0.250	4.00 x 10^{-2}
4	0.900	0.200	0.400	1.02 x 10^{-1}
5	0.800	0.400	0.100	6.40 x 10^{-3}

(a) 0.274 M^{-1}·min^{-1}
(b) 0.341 M^{-1}·min^{-1}
(c) 0.640 M^{-1}·min^{-1}
(d) 0.890 M^{-1}·min^{-1}
(e) 0.914 M^{-1}·min^{-1}

d 38. Consider a chemical reaction involving compounds A and B, which is found to be first order in A and second order in B. At what rate will the reaction occur in experiment 2?

Experiment	Rate ($M\cdot s^{-1}$)	Initial [A]	Initial [B]
1	0.10	1.0 M	0.20 M
2	?	2.0 M	0.60 M

(a) 1.2 $M\cdot s^{-1}$ (b) 0.20 $M\cdot s^{-1}$ (c) 0.60 $M\cdot s^{-1}$
(d) 1.8 $M\cdot s^{-1}$ (e) 0.36 $M\cdot s^{-1}$

a 39. The oxidation of NO by O_3 is first order in each of the reactants, and its rate constant is $1.5 \times 10^7 \, M^{-1}\cdot s^{-1}$. If the concentrations of NO and O_3 are each 5.0×10^{-7} M, what is the rate of oxidation of NO in $M\cdot s^{-1}$?

(a) 3.8×10^{-6} (b) 2.5×10^{-14} (c) 7.5×10^{-7}
(d) 15   (e) 7.5

e*40. Consider the hypothetical reaction and rate data below. Determine the form of the rate-law expression (i.e., determine the values of a and b in rate = $k[A]^a[B]^b$) and also the value of the specific rate constant, k. Which of the answers below would be the initial rate of reaction for $[A]_{initial} = 0.40$ M and $[B]_{initial} = 0.10$ M?

3A + 2B → Products

Run	$[A]_{initial}$	$[B]_{initial}$	Initial Rate of Reaction (moles per liter per second)
1	0.10 M	0.10 M	4.0×10^{-4}
2	0.20 M	0.30 M	4.8×10^{-3}
3	0.30 M	0.10 M	3.6×10^{-3}

(a) $1.6 \times 10^{-4} \, M\cdot s^{-1}$ (b) $3.4 \times 10^{-3} \, M\cdot s^{-1}$ (c) $1.2 \times 10^{-3} \, M\cdot s^{-1}$
(d) $4.8 \times 10^{-4} \, M\cdot s^{-1}$ (e) $6.4 \times 10^{-3} \, M\cdot s^{-1}$

a*41. Rate data have been determined at a particular temperature for the overall reaction
$$2NO + 2H_2 \rightarrow N_2 + 2H_2O$$
in which all reactants and products are gases.

Trial Run	Initial [NO]	Initial [H_2]	Initial Rate ($M\cdot s^{-1}$)
1	0.10 M	0.20 M	0.0150
2	0.10 M	0.30 M	0.0225
3	0.20 M	0.20 M	0.0600

What would be the initial rate of the reaction if the initial molar concentration of NO = 0.30 M and the initial molar concentration of H_2 = 0.10 M?

(a) 0.068 $M\cdot s^{-1}$ (b) 0.22 $M\cdot s^{-1}$ (c) 0.022 $M\cdot s^{-1}$
(d) 0.040 $M\cdot s^{-1}$ (e) 0.10 $M\cdot s^{-1}$

a*42. What will be the initial rate of this reaction when [A] = 0.60 M, [B] = 0.30 M, and [C] = 0.10 M?   A + 2B + 3C → Products

Experiment	Initial [A]	Initial [B]	Initial [C]	Initial Rate of Loss of A
1	0.10 M	0.20 M	0.10 M	4.0 x 10^{-2} M·min^{-1}
2	0.40 M	0.20 M	0.10 M	4.0 x 10^{-2} M·min^{-1}
3	0.20 M	0.20 M	0.25 M	1.0 x 10^{-1} M·min^{-1}
4	0.20 M	0.40 M	0.10 M	1.6 x 10^{-1} M·min^{-1}

(a) 0.090 M·min^{-1} (b) 4.0 M·min^{-1} (c) 0.30 M·min^{-1}
(d) 0.40 M·min^{-1} (e) 0.016 M·min^{-1}

Concentration Versus Time: The Integrated Rate Equation

c 43. The half-life for the reactant A in the first order reaction A → B is 36.2 seconds. What is the rate constant for this reaction at the same temperature?

(a) 52.2 s^{-1} (b) 0.0276 s^{-1} (c) 0.0191 s^{-1} (d) 18.1 s^{-1} (e) 0.00832 s^{-1}

a 44. The rate constant for the first order reaction A → B + C is k = 3.3 x 10^{-2} min^{-1} at 57 K. What is the half-life for this reaction at 57 K?

(a) 21 min   (b) 30 min   (c) 61 min   (d) 9.1 min   (e) 1200 min

c 45. The decomposition of dimethylether at 504°C is first order with a half-life of 1570. seconds. What fraction of an initial amount of dimethylether remains after 4710. seconds?

(a) 1/3   (b) 1/6   (c) 1/8   (d) 1/16   (e) 1/32

b 46. The gas phase reaction 3C + 2D → E + F obeys the rate-law expression rate = k[D] and has a half-life of 0.860 s. If 2.00 mole of D is injected into a 1.00-L container with excess C, what concentration of D remains after 1.50 seconds?

(a) 0.48 M   (b) 0.60 M   (c) 1.68 M   (d) 1.40 M   (e) 1.06 M

c 47. The decomposition of dinitrogen pentoxide obeys the rate-law expression rate = 0.080 min^{-1}[N_2O_5]. If the initial concentration of N_2O_5 is 0.30 M, what is the concentration after 2.6 minutes? N_2O_5 → N_2O_3 + O_2

(a) 0.38 M   (b) 0.028 M   (c) 0.24 M   (d) 0.13 M   (e) 0.32 M

a 48. A chemical reaction A → B + C is first order in A and has a rate constant of 1.2 x 10^{-3} min^{-1}. If the initial concentration of A is 0.40 M., how much time must pass in order to reduce the concentration of A to 0.22 M?

(a) 5.0 x 10^2 min   (b) 3.0 x 10^2 min   (c) 7.4 x 10^{-3} min
(d) 4.3 x 10^{-4} min   (e) 2.2 x 10^2 min

d 49. The gas phase reaction below obeys the rate-law expression rate = $k[SO_2Cl_2]$. At 593 K the specific rate constant is 2.2×10^{-5} s^{-1}. A 2.0-g sample of SO_2Cl_2 is introduced into a closed 4.0-L container.

$$SO_2Cl_2 \rightarrow SO_2 + Cl_2$$

How much time must pass in order to reduce the amount of SO_2Cl_2 present to 1.8 grams?

(a) 7.4×10^3 seconds   (b) 2.1×10^2 seconds   (c) 3.5×10^2 seconds
(d) 4.8×10^3 seconds   (e) 5.8×10^4 seconds

c 50. Cyclopropane rearranges to form propene in a reaction that is first order. If the rate constant is 2.74×10^{-3} s^{-1}, how long would it take for 85.6% of the cyclopropane to rearrange if the initial concentration was 0.460 M?

(a) 51.0 s   (b) 62.0 s   (c) 707 s   (d) 2.74×10^{-3} s   (e) 3.83×10^{-4} s

b 51. At 300 K the reaction below obeys the rate law Rate $=k[NOCl]^2$ where $k = 2.8 \times 10^{-5}$ $M^{-1} \cdot s^{-1}$.

$$2NOCl \rightarrow 2NO + Cl_2$$

Suppose 1.0 mole of NOCl is introduced into a 2.0-liter container at 300 K. Evaluate the half-life of the reaction.

(a) 2.6×10^3 seconds   (b) 3.6×10^4 seconds   (c) 2.4×10^4 seconds
(d) 1.1×10^3 seconds   (e) 4.0×10^4 seconds

d 52. Compounds A and B react to form C and D in a reaction that is found to be second-order overall and second-order in B. The rate constant at 50.°C is 2.48 liter per mole per minute. What is the half-life of B (in min) if 0.822 M B reacts with excess A?

$$A + B \rightarrow C + D$$

(a) 0.0139   (b) 12.0   (c) 1.39   (d) 0.491   (e) 5.88

e 53. At 300 K the reaction below obeys the rate law rate $= k[NOCl]^2$ where $k = 2.8 \times 10^{-5}$ $M^{-1} \cdot s^{-1}$.

$$2NOCl \rightarrow 2NO + Cl_2$$

Suppose 1.0 mole of NOCl is introduced into a 2.0-liter container at 300 K. How much NOCl will remain after 30 minutes?

(a) 0.77 mol   (b) 0.84 mol   (c) 0.87 mol
(d) 0.63 mol   (e) 0.95 mol

e 54. At a certain temperature the reaction $2B \rightarrow C + D$ obeys the rate-law expression rate $= (1.14 \times 10^{-3}\ M^{-1} \cdot s^{-1})[B]^2$. If 5.00 mol of B is initially present in a 1.00-L container at that temperature, how long would it take for 2.00 mol of B to be consumed at constant temperature?

(a) 224 s   (b) 87.5 s   (c) 46.0 s   (d) 73.0 s   (e) 58.5 s

e 55. Compounds A and B react to form C and D in a reaction that is found to be second-order overall and second-order in B. The rate constant at 30°C is 0.622 liter per mole per minute.

$$A + B \rightarrow C + D$$

How many minutes does it take 4.0×10^{-2} M B (mixed with excess A) to be reduced to 3.3×10^{-2} M B?

(a) 1.4 min   (b) 3.6 min   (c) 5.0 min   (d) 6.4 min   (e) 8.5 min

b 56. Which statement is **incorrect**?

(a) The reaction rate for a zero-order reaction is independent of concentrations.
(b) The specific rate constant for a second-order reaction is independent of temperature.
(c) The half-life for a first-order reaction is independent of initial concentrations.
(d) The rate law expression relates rate and concentration.
(e) The integrated rate equation relates time and concentration.

c*57. Which of the following statements concerning graphical methods for determining reaction order is **false**?

(a) For a first-order reaction the plot of ln [A] vs. time gives a straight line.
(b) For a first-order reaction the slope of the straight-line graph equals $-ak$.
(c) For a second-order reaction the plot of $[A]^2$ vs. time gives a straight line.
(d) For a first-order reaction the intercept of the straight-line graph equals ln $[A]_0$.
(e) For a zero-order reaction the plot of [A] vs. time gives a straight line.

Collision Theory and Transition State Theory

d 58. Which one of the following statements is **false**?

(a) In order for a reaction to occur, reactant molecules must collide with each other.
(b) A catalyst alters the rate of a reaction and is neither a product nor a reactant in the overall equation.
(c) According to collision theory a three-body collision is less likely than a two-body collision.
(d) In reactions that are second order in one reactant and first order in another, the slow step generally involves a three-body collision of these reactants.
(e) The transition state is a short-lived, high energy state, intermediate between reactants and products.

d 59. Which statement is **false**?

(a) If a reaction is thermodynamically spontaneous, it may occur rapidly.
(b) A fast reaction may be thermodynamically spontaneous.
(c) If a reaction is thermodynamically spontaneous, it may occur slowly.
(d) If a reaction is thermodynamically spontaneous, it must have a low activation energy.
(e) Rate of reaction is a kinetic quantity rather than a thermodynamic quantity.

a 60. A reaction has an activation energy of 40 kJ and an overall energy change of reaction of –100 kJ. In each of the following potential energy diagrams, the horizontal axis is the reaction coordinate and the vertical axis is potential energy in kJ. Which potential energy diagram best describes this reaction?

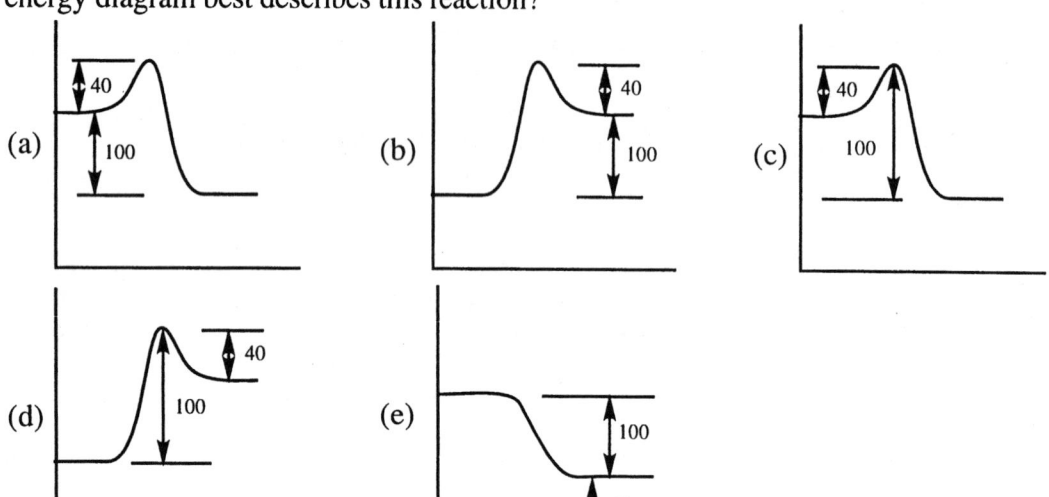

(a) (b) (c) (d) (e)

a 61. Given the following potential energy diagram for the one-step reaction
    X + Y → Z + R   The reaction _____.

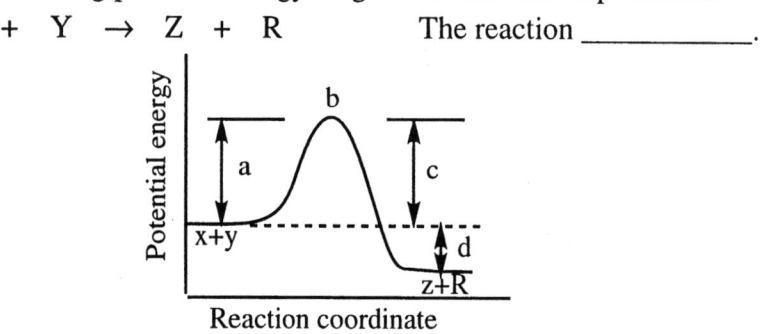

(a) releases energy (b) absorbs energy (c) is impossible
(d) occurs without a net change in energy (e) may either absorb or release energy

e 62. Given the following potential energy diagram for the one-step reaction
    X + Y → Z + R   The arrow "d" represents the _____.

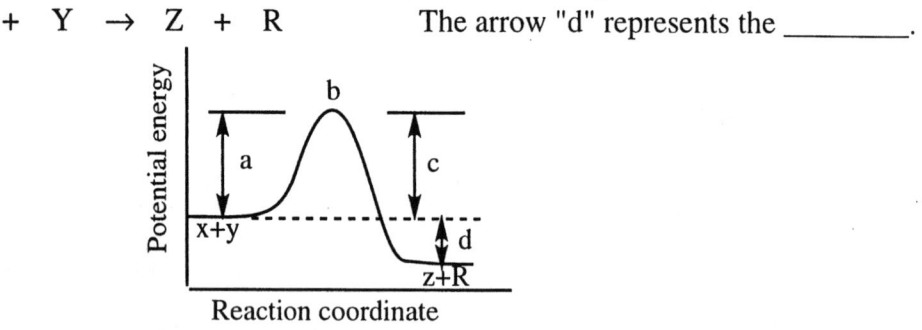

(a) energy content of products (b) activation energy for the forward reaction
(c) energy content of reactants (d) activation energy for the reverse reaction
(e) the net change in energy for the reaction

b 63. Given the following potential energy diagram for the one-step reaction
X + Y → Z + R   The point "b" represents _____.

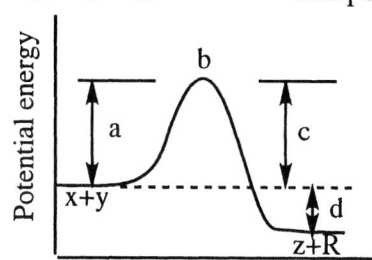

(a) the energy of the mixture when half of the reactants have been converted to products
(b) the energy of the transition state
(c) the number of moles of transition state that must be formed
(d) the energy of the forward reaction
(e) the energy of the reverse reaction

d 64. Given the following potential energy diagram for the one-step reaction
X + Y → Z + R   The arrow "c" represents the _____.

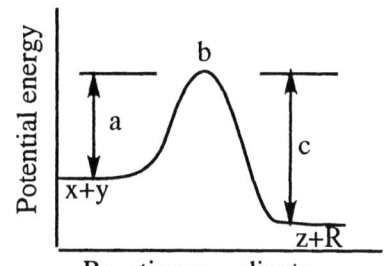

(a) net energy of reaction for the forward reaction
(b) activation energy for the forward reaction
(c) net energy of reaction for the reverse reaction
(d) activation energy for the reverse reaction
(e) energy content for the reaction

Reaction Mechanisms and the Rate-Law Expression

b 65. A reaction mechanism will usually be

(a) the only possible explanation for the reaction.
(b) difficult to verify experimentally.
(c) proven experimentally to be the balanced chemical equation.
(d) obvious from a consideration of the balanced chemical equation.
(e) obvious from a consideration of the reaction rate data.

d 66. Which of the following is a kinetics concept?

(a) free energy   (b) enthalpy   (c) spontaneity
(d) reaction mechanism (e) entropy

b 67. Consider the hypothetical reaction shown below.

$$A + 2B \rightarrow AB_2$$

Assume that the following proposed mechanism is consistent with the rate data.

B + B	$\rightarrow$	B_2		slow
B_2 + A	$\rightarrow$	AB + B		fast
B + AB	$\rightarrow$	AB_2		fast
A + 2B	$\rightarrow$	AB_2		overall

Which one of the following statements must be true? The reaction is _____.

(a) first order in A, second order in B, and third order overall
(b) second order in B and second order overall
(c) first order in A and first order overall
(d) second order in B, zero order in A, and third order overall
(e) second order in A and second order overall

c*68. Consider the following proposed mechanism. If this mechanism for the overall reaction were correct, and if k_1 were much less than k_2, then the observed rate law would be

$$2A \xrightarrow{k_1} C + I$$
$$I + B \xrightarrow{k_2} C + D$$

(a) rate = $k_1[A]$ (b) rate = $k_2[I][B]$ (c) rate = $k_1[A]^2$
(d) rate = $k_1[A]^2 - k_2[C][D]$ (e) rate = $k_1k_2[A]^2[I][B]$

a 69. Suppose the reaction

$$2AB + C_2 \rightarrow A_2C + B_2C$$

occurs by the following mechanism.

Step 1	AB	+	C_2	$\rightarrow$	AC_2 + B		slow
Step 2	B	+	AB	$\rightarrow$	AB_2		fast
Step 3	AC_2	+	AB_2	$\rightarrow$	A_2C_2 + B_2		fast
Step 4	A_2C_2	+	B_2	$\rightarrow$	A_2C + B_2C		fast
Overall	2 AB	+	C_2	$\rightarrow$	A_2C + B_2C		

The rate law expression must be rate = _____.

(a) $k[AB][C_2]$ (b) $k[AB]^2[C_2]$ (c) $k[AB]^2$
(d) $k[AB]$ (e) $k[C_2]$

a 70. The disproportionation (auto-oxidation-reduction) of the hypochlorite ion to chlorate and chloride ions occurs in aqueous solution. The rate expression is found to be rate = $k[ClO^-]^2$. Which one of the following (possible) mechanisms is consistent with this information?

$$3ClO^- \rightarrow ClO_3^- + 2Cl^-$$

(a) $ClO^- + ClO^- \rightarrow ClO_2^- + Cl^-$ (slow)
    $ClO^- + ClO_2^- \rightarrow ClO_3^- + Cl^-$ (fast)

(b) $ClO^- \rightarrow Cl + O^-$ (slow)
    $O^- + ClO^- \rightarrow Cl^- + O_2^-$ (fast)
    $O_2^- + Cl \rightarrow Cl^- + O_2$ (fast)
    $O_2 + ClO^- \rightarrow ClO_3^-$ (fast)

(c) $3ClO^- \rightarrow ClO_3^- + 2Cl^-$ (one step)

(d) $ClO^- \rightarrow Cl^- + O$ (slow)
    $O + ClO^- \rightarrow O_2 + Cl^-$ (fast)
    $ClO^- + O_2 \rightarrow ClO_3^-$ (fast)

(e) $3ClO^- \rightarrow ClO_2^- + Cl_2O^{2-}$ (slow)
    $Cl_2O^{2-} \rightarrow Cl^- + ClO^-$ (fast)
    $ClO^- + ClO_2^- \rightarrow Cl^- + ClO_3^-$ (fast)

a*71. The reaction A + B → C + D, obeys the rate law expression rate =$k[A][B]$. Which response lists all the proposed mechanisms below that **are consistent** with this information and none that are inconsistent?

I.   A + B → C + D   one step

II. A + B ⇌ E   fast, equilibrium
     E → C + D   slow

III. A + B ⇌ E   fast, equilibrium
     A + E → C + B   slow

(a) I and II   (b) II   (c) III
(d) II and III   (e) I, II, and III

a 72. Consider the reaction below and its observed rate law expression. Which proposed mechanisms **are consistent** with the rate law expression?

$$2NO_2 \rightarrow 2NO + O_2 \qquad \text{rate} = k[NO_2]^2$$

I. $NO_2 + NO_2 \rightarrow N_2O_4$ slow
    $N_2O_4 \rightarrow N_2 + 2O_2$ fast
    $N_2 + O_2 \rightarrow 2NO$ fast
    $2NO_2 \rightarrow 2NO + O_2$ overall

II. $NO_2 \rightarrow N + O_2$ slow
    $NO_2 + N \rightarrow N_2O_2$ fast
    $N_2O_2 \rightarrow 2NO$ fast
    $2NO_2 \rightarrow 2NO + O_2$ overall

III. $NO_2 \rightarrow NO + O$ slow
     $O + NO_2 \rightarrow NO + O_2$ fast
     $2NO_2 \rightarrow 2NO + O_2$ overall

(a) I (b) II (c) III (d) I and III
(e) another one or another combination

b*73. At 300 K the following reaction is found to obey the rate law rate $=k[NOCl]^2$.

$$2NOCl \rightarrow 2NO + Cl_2$$

Consider the three postulated mechanisms given below. Then choose the response that lists all those that are **possibly** correct and no others.

I. $NOCl \rightarrow NO + Cl$ slow
    $Cl + NOCl \rightarrow NOCl_2$ fast
    $NOCl_2 + NO \rightarrow 2NO + Cl_2$ fast
    $2NOCl \rightarrow 2NO + Cl_2$

II. $2NOCl \rightarrow NOCl_2 + NO$ slow
    $NOCl_2 \rightarrow NO + Cl_2$ fast
    $2NOCl \rightarrow 2NO + Cl_2$

III. $NOCl \rightarrow NO + Cl$ fast
     $NOCl + 2Cl \rightarrow NO + Cl_2$ slow
     $2NOCl \rightarrow 2NO + Cl_2$

(a) I (b) II (c) III (d) I and II (e) II and III

e 74. Which statement concerning a possible mechanism for a reaction is **false**?

(a) A possible mechanism must be consistent with the experimental data.
(b) Each elementary step is represented by a balanced equation.
(c) The elementary steps must add to give the equation for the overall reaction.
(d) The speed of the slow step limits the rate at which the overall reaction occurs.
(e) For all reactions the experimentally determined reaction orders of the reactants indicate the number of molecules of those reactants involved in the slow step of the mechanism.

e 75. Which of the following statements concerning a reaction and its mechanism is **false**?

(a) For a reactant, more is consumed than is formed.
(b) For a product, more is formed than is consumed.
(c) A reaction intermediate is formed in early steps and completely consumed in later steps.
(d) For a multi-step mechanism, the slowest step has the highest activation energy.
(e) Reactions involving simultaneous trimolecular collisions are very common in gases.

Temperature: The Arrhenius Equation

d 76. The **principal** reason for the increase in reaction rate with increasing temperature is

(a) molecules collide more frequently at high temperatures.
(b) the pressure exerted by reactant molecules increases with increasing temperature.
(c) the activation energy increases with increasing temperature.
(d) the fraction of high energy molecules increases with increasing temperature.
(e) the activation energy for the forward reaction decreases while the activation energy for the reverse reaction increases.

b 77. Which response contains all the following statements that are **true** and no false statements?

I. Reactions with more negative values of ΔG^0 are more spontaneous and proceed at higher rates than those with less negative values of ΔG^0.
II. The activation energy, E_a, is usually about the same as ΔE for a reaction.
III. The activation energy for a reaction does not change as temperature changes.
IV. Reactions generally occur at faster rates at higher temperatures.

(a) I and II (b) III and IV (c) I, II, and III
(d) II, III, and IV (e) I, II, III, and IV

d 78. Suppose the activation energy of a certain reaction is 250 kJ/mol. If the rate constant at $T_1 = 300$ K is k_1, and the rate constant at $T_2 = 320$ K is k_2, then $k_2/k_1 =$ _____. (The universal gas constant = 8.314 J/mol·K.)

$$\ln\left(\frac{k_2}{k_1}\right) = \frac{E_a}{R}\left(\frac{1}{T_1} - \frac{1}{T_2}\right)$$

(a) 3×10^{-29} (b) 0.067   (c) 15.0   (d) 525   (e) 3×10^{28}

d 79. Suppose the activation energy of a certain reaction is 250 kJ/mol. If the rate constant at $T_1 = 300$ K is k_1, and the rate constant at $T_2 = 320$ K is k_2, then $k_2/k_1 =$ _____. (The universal gas constant = 8.314 J/mol·K.)

(a) 3×10^{-29} (b) 0.067   (c) 15.0   (d) 525   (e) 3×10^{28}

c 80. The specific rate constant, k, for a reaction is 2.64×10^{-2} s^{-1} at 25°C, and the activation energy is 74.0 kJ/mol. Calculate k at 50°C. (The universal gas constant = 8.314 J/mol·K.)

$$\ln\left(\frac{k_2}{k_1}\right) = \frac{E_a}{R}\left(\frac{1}{T_1} - \frac{1}{T_2}\right)$$

(a) 0.832 s^{-1} (b) 71.9 s^{-1} (c) 0.266 s^{-1} (d) 1.08 s^{-1} (e) 0.0265 s^{-1}

c 81. The specific rate constant, k, for a reaction is 2.64×10^{-2} s^{-1} at 25°C, and the activation energy is 74.0 kJ/mol. Calculate k at 50°C.

(a) 0.832 s^{-1} (b) 71.9 s^{-1} (c) 0.266 s^{-1} (d) 1.08 s^{-1} (e) 0.0265 s^{-1}

a 82. The rate constant, k, for a first-order reaction is 1.20×10^2 s^{-1} at 45°C and the activation energy is 98.2 kJ/mol. Calculate the rate constant for this reaction at 95°C.

(a) 1.87×10^4 s^{-1} (b) 7.72×10^{-1} s^{-1} (c) 6.06×10^{-2} s^{-1}
(d) 1.40×10^3 s^{-1} (e) 2.02×10^5 s^{-1}

b 83. The rate constant, k, for a first-order reaction is 1.36×10^3 s^{-1} at 90.°C and the activation energy is 78.4 kJ/mol. Calculate the rate constant for this reaction at 50.°C.

(a) 60.4 s^{-1} (b) 54.5 s^{-1} (c) 23.8 s^{-1} (d) 4.85 s^{-1} (e) 1.78×10^3 s^{-1}

d 84. Calculate the activation energy of a reaction if the rate constant is 0.75 s^{-1} at 25°C and 11.5 s^{-1} at 75°C. (The universal gas constant = 8.314 J/mol·K.)

$$\ln\left(\frac{k_2}{k_1}\right) = \frac{E_a}{R}\left(\frac{1}{T_1} - \frac{1}{T_2}\right)$$

(a) 681 J/mol   (b) 20.4 kJ/mol   (c) 15.8 kJ/mol
(d) 47.1 kJ/mol   (e) 31.4 kJ/mol

d 85. Calculate the activation energy of a reaction if the rate constant is 0.75 s^{-1} at 25°C and 11.5 s^{-1} at 75°C.

(a) 681 J/mol   (b) 20.4 kJ/mol   (c) 15.8 kJ/mol
(d) 47.1 kJ/mol   (e) 31.4 kJ/mol

c 86. Calculate the activation energy of a reaction if its rate constant is 2.8×10^6 s^{-1} at 24°C and 1.5×10^7 s^{-1} at 48°C.

(a) 26 kJ/mol   (b) 80. kJ/mol   (c) 55 kJ/mol
(d) 67 J/mol   (e) 3.51 J/mol

d 87. What would be the activation energy of a reaction if its rate constant at 35°C was double the value of its rate constant at 25°C?

(a) 63.8 kJ/mol (b) 75.1 kJ/mol (c) 8.12 kJ/mol
(d) 52.9 kJ/mol (e) 68.3 J/mol

Catalysts

e 88. A catalyst _____.

(a) is used up in a chemical reaction
(b) changes the value of ΔG^0 of the reaction
(c) is always a solid
(d) does not influence the reaction in any way
(e) changes the activation energy of the reaction

c 89. A catalyst

(a) increases the amount of products present at equilibrium.
(b) increases the rate at which equilibrium is reached but decreases the equilibrium constant.
(c) increases the rate at which equilibrium is reached without changing the equilibrium constant.
(d) increases ΔH for the process.
(e) lowers ΔS for the process.

d 90. Which response includes all the true statements and no false statements? A catalyst can _____.

I. make a nonspontaneous reaction spontaneous
II. speed up a forward reaction and slow down the reverse reaction
III. lower the activation energy of the forward reaction
IV. lower the activation energy of the reverse reaction

(a) I and III (b) II and III (c) I and IV
(d) III and IV (e) another one or another combination

d 91. Which of the following is **not** an example of an important, useful reaction catalyzed by transition metals and/or their oxides?

(a) the Haber process for the production of ammonia
(b) the contact process for the production of sulfur trioxide in producing sulfuric acid
(c) the hydrogenation of unsaturated hydrocarbons
(d) the reaction of leaded fuels with the catalysts in catalytic converters
(e) the chlorination of benzene

d 92. The catalytic converters installed in newer models of automobiles are designed to catalyze certain kinds of favorable reactions. Unfortunately, other unfavorable reactions also are catalyzed. Which one of those listed below, all of which are catalyzed in such mufflers, is an unfavorable reaction?

(a) $2CO(g) + O_2(g) \rightarrow 2CO_2(g)$
(b) $2C_8H_{18}(g) + 25O_2(g) \rightarrow 16CO_2(g) + 18H_2O(g)$
(c) $C(s) + O_2(g) \rightarrow CO_2(g)$
(d) $2SO_2(g) + O_2(g) \rightarrow 2SO_3(g)$
(e) $2NO(g) \rightarrow N_2(g) + O_2(g)$

c 93. Which statement concerning biological catalysts is **false**?

(a) Enzymes are proteins that act as catalysts for specific biochemical reactions.
(b) The reactants in enzyme-catalyzed reactions are called substrates.
(c) Each enzyme catalyzes many different reactions in a living system.
(d) Enzyme-catalyzed reactions are important examples of zero-order reactions.
(e) Discovery or synthesis of catalysts that mimic the efficiency of naturally occurring enzymes would saved on the costs of using high temperature and high pressure in commercial processes.

c 94. The reaction, $A + 2B \rightarrow B_2 + A$, proceeds by the following mechanism: (A is a catalyst.)

$A + B \rightarrow AB$ (slow)
$AB + B \rightarrow B_2 + A$ (fast)

What is the rate law expression for this reaction?

(a) Rate $=k[A]$
(d) Rate $=k[A][B]^2$
(b) Rate $=k[B]$
(e) Rate $=k[A]^2[B]$
(c) Rate $=k[A][B]$

17 Chemical Equilibrium

Basic Concepts and the Equilibrium Constant

e 1. Which of the statements concerning equilibrium is **false**?

(a) A system that is disturbed from an equilibrium condition responds in a manner to restore equilibrium.
(b) Equilibrium in molecular systems is dynamic, with two opposing processes balancing one another.
(c) The value of the equilibrium constant for a given reaction mixture is the same regardless of the direction from which equilibrium was attained.
(d) A system moves spontaneously toward a state of equilibrium.
(e) The equilibrium constant usually is independent of temperature.

d 2. When the system A + B $\rightleftharpoons$ C + D is at equilibrium,

(a) the forward reaction has stopped.
(b) the reverse reaction has stopped.
(c) both the forward and the reverse reactions have stopped.
(d) neither the forward nor the reverse reaction has stopped.
(e) the sum of the concentrations of A and B must equal the sum of the concentrations of C and D.

e 3. Consider the **gas phase** system below at a high temperature. The form of the expression for the equilibrium constant, K_c, _____.

$$4NH_3 + 5O_2 \rightleftharpoons 4NO + 6H_2O$$

(a) cannot be determined without rate data

(b) is $K_c = \dfrac{[NH_3][O_2]}{[NO][H_2O]}$

(c) is $K_c = \dfrac{[NH_3]^4[O_2]^5}{[NO]^4[H_2O]^6}$

(d) is $K_c = \dfrac{[NO]^4}{[NH_3][O_2]}$

(e) is $K_c = \dfrac{[NO]^4[H_2O]^6}{[NH_3]^4[O_2]^5}$

c 4. Consider the following reaction. What would be the equilibrium constant expression?
$$4Br_2(g) + CH_4(g) \rightleftharpoons 4HBr(g) + CBr_4(g)$$

(a) $K_c = \dfrac{[CBr_4][HBr]}{[Br_2][CH_4]}$

(b) $K_c = \dfrac{[CH_4][Br_2]^4}{[HBr]^4[CBr_4]}$

(c) $K_c = \dfrac{[CBr_4][HBr]^4}{[Br_2]^4[CH_4]}$

(d) $K_c = \dfrac{[CH_4][Br_2]}{[HBr][CBr_4]}$

(e) $K_c = \dfrac{[HBr]^4}{[Br_2]^4[CH_4]}$

b 5. Consider the following reaction involving a solid.
$$2NH_4NO_3(s) \rightleftharpoons 2NH_3(g) + 2NO(g) + H_2(g) + 2O_2(g)$$
The appropriate equilibrium constant expression would be:

(a) $K_c = [NH_3][NO][H_2][O_2]$
(b) $K_c = [NH_3]^2[NO]^2[H_2][O_2]^2$

(c) $K_c = \dfrac{[NH_3][NO][H_2][O_2]}{[NH_4NO_3]}$
(d) $K_c = \dfrac{[NH_3]^2[NO]^2[H_2][O_2]^2}{[NH_4NO_3]^2}$

(e) $K_c = \dfrac{[NH_3]^2[NO]^2[H_2][O_2]^2}{2}$

a 6. The equilibrium constant expression for the reaction
$$NH_4Cl(s) \rightleftharpoons NH_3(g) + HCl(g)$$
would be written

(a) $K_c = [NH_3][HCl]$
(b) $K_c = \dfrac{[NH_4Cl][NH_3]}{[HCl]}$
(c) $K_c = \dfrac{[NH_3][HCl]}{[NH_4Cl]}$

(d) $K_c = \dfrac{[NH_4Cl]}{[NH_3][HCl]}$
(e) $K_c = \dfrac{[NH_3]}{[HCl]}$

c 7. For a reversible reaction with a one-step mechanism, A $\rightleftharpoons$ B, the rate of the forward reaction is $\text{rate}_f = 3.2 \times 10^{-6}\ \text{s}^{-1}\ [A]$ and the rate of the reverse reaction is $\text{rate}_r = 4.6 \times 10^{-4}\ \text{s}^{-1}\ [B]$. What is the value of K_c for this reaction?

(a) 2.5×10^{-11} (b) 7.0×10^{9} (c) 7.0×10^{-3}
(d) 1.4×10^{2} (e) 1.5×10^{-9}

c 8. For the reaction $2A + B \rightleftharpoons C + 2D$ at 35°C, the value of k_f is $3.0 \times 10^{-3}\ M^{-2}\cdot s^{-1}$ and the value of $k_r = 1.5 \times 10^{-2}\ M^{-2}\cdot s^{-1}$. Calculate the value of K_c for this reaction.

(a) 2.0   (b) 0.5   (c) 0.20   (d) 5.0   (e) 0.012

d 9. What is the value of K_c for the reaction $2A(g) + 3B(g) \rightleftharpoons 2C(g) + D(g)$ if at equilibrium $[A] = 0.60\ M$, $[B] = 0.30\ M$, $[C] = 0.10\ M$ and $[D] = 0.50\ M$?

(a) 1.9   (b) 0.15   (c) 2.4   (d) 0.51   (e) 0.088

b 10. In a 1.0-liter container there are, at equilibrium, 0.20 mole of I_2, 0.30 mole of H_2, and 0.20 mole of HI. What is the value of K_c for the reaction?
$$H_2(g) + I_2(g) \rightleftharpoons 2HI(g)$$

(a) 0.33   (b) 0.67   (c) 2.7   (d) 1.3   (e) 1.5

c 11. In a 1.0-liter container there are, at equilibrium, 0.10 mole H_2, 0.20 mole N_2, and 0.40 mole NH_3. What is the value of K_c for this reaction at this temperature?
$$N_2(g) + 3H_2(g) \rightleftharpoons 2NH_3(g)$$

(a) 0.0012   (b) 20   (c) 8.0×10^{2} (d) 0.050   (e) 8.0

b 12. Consider the following reversible reaction. In a 3.00-liter container, the following amounts are found in equilibrium at 400°C: 0.0420 mole N_2, 0.516 mole H_2, and 0.0357 mole NH_3. Evaluate K_c.
$$N_2(g) + 3H_2(g) \rightleftharpoons 2NH_3(g)$$

(a) 0.202 (b) 1.99 (c) 16.0 (d) 4.94 (e) 0.503

a 13. Some nitrogen and hydrogen gases are pumped into an empty 5.00-liter vessel at 500°C. When equilibrium was established, 3.00 moles of N_2, 2.10 moles of H_2, and 0.298 moles of NH_3 were present. Evaluate K_c at 500°C.
$$N_2(g) + 3H_2(g) \rightleftharpoons 2NH_3(g)$$

(a) 0.080 (b) 0.63 (c) 0.96 (d) 0.022 (e) 0.0067

a 14. At equilibrium, the following amounts are found at a certain temperature in a 3.0-liter container: 2.0 mole of Cl_2, 0.80 mol of H_2O (steam), 0.0030 mol of HCl, and 0.0045 mol of O_2. Evaluate K_c at that temperature.
$$2Cl_2(g) + 2H_2O(g) \rightleftharpoons 4HCl(g) + O_2(g)$$

(a) 4.7 x 10^{-14} (b) 8.4 x 10^{-7} (c) 1.4 x 10^{-13}
(d) 2.2 x 10^{13} (e) 7.1 x 10^{12}

b*15. The reversible reaction $\quad 2SO_2(g) + O_2(g) \rightleftharpoons 2SO_3(g)$
has come to equilibrium in a vessel of specific volume at a given temperature. **Before** the reaction began, the concentrations of the reactants were 0.060 mol/L of SO_2 and 0.050 mol/L of O_2. After equilibrium is reached, the concentration of SO_3 is 0.040 mol/L. What is the value of K_c?

(a) 2.7 (b) 1.3 x 10^2 (c) 7.5 x 10^{-3} (d) 0.38 (e) 40.

a*16. Consider the following reaction in which all reactants and products are gases. 1.00 mol of A and 2.00 mol of B are placed in a 5.0-liter container. After equilibrium has been established, 0.50 mol of D is present in the container. Calculate the equilibrium constant, K_c, for the reaction.
$$A + 2B \rightleftharpoons 2C + D$$

(a) 1.0 (b) 0.15 (c) 0.33 (d) 3.0 (e) 5.0

d*17. Given: $A(g) + 3B(g) \rightleftharpoons C(g) + 2D(g)$
One (1.0) mole of A and 1.0 mole of B are placed in a 5.0-liter container. After equilibrium has been established, 0.50 mole of D is present in the container. Calculate the equilibrium constant, K_c, for the reaction.

(a) 1.2 (b) 0.68 (c) 12 (d) 27 (e) 1.4 x 10^2

c 18. Nitrosyl chloride, NOCl, dissociates on heating. When a 1.50-gram sample of NOCl is heated to 350°C in a 1.00-liter container, the percent dissociation is 57.2%. Calculate K_c for the reaction as given.
$$NOCl(g) \rightleftharpoons NO(g) + \tfrac{1}{2}Cl_2(g)$$

(a) 0.0421 (b) 0.876 (c) 0.108 (d) 9.26 (e) 1.75 x 10^{-4}

Variation of K_c with the Form of the Balanced Equation

d 19. Given the following reaction and its equilibrium constant at a certain temperature,
$$N_2(g) + 3H_2(g) \rightleftharpoons 2NH_3(g) \qquad K_c = 3.6 \times 10^8$$
calculate the numerical value of the equilibrium constant for the following reaction at the same temperature.
$$NH_3(g) \rightleftharpoons \tfrac{1}{2}N_2(g) + \tfrac{3}{2}H_2(g)$$

(a) 2.8×10^{-9} (b) 1.9×10^4 (c) 1.3×10^7 (d) 5.3×10^{-5} (e) 7.7×10^{-18}

d 20. If the equilibrium constant at a certain temperature is 2.1×10^{13} for the following reaction,
$$4HCl(g) + O_2(g) \rightleftharpoons 2Cl_2(g) + 2H_2O(g)$$
calculate the value of the equilibrium constant at the same temperature for
$$\tfrac{1}{2}Cl_2(g) + \tfrac{1}{2}H_2O(g) \rightleftharpoons HCl(g) + \tfrac{1}{4}O_2(g).$$

(a) 2.2×10^{-7} (b) 3.8×10^3 (c) 5.3×10^{12} (d) 4.7×10^{-4} (e) 1.2×10^{-14}

The Reaction Quotient

d 21. The equilibrium constant for the following gas phase reaction is 0.50 at 600°C. A mixture of HCHO, H_2, and CO is introduced into a flask at 600°C. After a short time, analysis of a small amount of the reaction mixture shows the concentrations to be $[HCHO] = 1.5\ M$, $[H_2] = 0.5\ M$, and $[CO] = 1.0\ M$. Which of the following statements about this reaction mixture is true?
$$HCHO \rightleftharpoons H_2 + CO$$

(a) The reaction mixture is at equilibrium.
(b) The reaction mixture is not at equilibrium, but no further reaction will occur.
(c) The reaction mixture is not at equilibrium, but will move toward equilibrium by forming more HCHO.
(d) The reaction mixture is not at equilibrium, but will move toward equilibrium by using up more HCHO.
(e) The forward rate of this reaction is the same as the reverse rate.

c 22. The equilibrium constant, K_c, for the following gas phase reaction is 0.50 at 600°C. A mixture of HCHO, H_2, and CO is introduced into a flask at 600°C. After a short time, analysis of a small amount of the reaction mixture shows the concentrations to be $[HCHO] = 1.5\ M$, $[H_2] = 1.2\ M$, and $[CO] = 1.0\ M$. Which of the following statements about this reaction mixture is true?
$$HCHO \rightleftharpoons H_2 + CO$$

(a) The reaction mixture is at equilibrium.
(b) The reaction mixture is not at equilibrium, but no further reaction will occur.
(c) The reaction mixture is not at equilibrium, and will move toward equilibrium by forming more HCHO.
(d) The reaction mixture is not at equilibrium, and will move toward equilibrium by using up more HCHO.
(e) The forward rate of this reaction is the same as the reverse rate.

c 23. The equilibrium constant, K_c, for the following reaction is 0.0154 at a high temperature. A mixture in a container at this temperature has the concentrations : $[H_2] = 1.11\ M$, $[I_2] = 1.30\ M$ and $[HI] = 0.181\ M$. Which of the following statements concerning the reaction and the reaction quotient, Q, is true?

$$H_2(g) + I_2(g) \rightleftharpoons 2HI(g)$$

(a) $Q = K_c$
(b) $Q > K_c$; more HI will be produced.
(c) $Q > K_c$; more H_2 and I_2 will be produced.
(d) $Q < K_c$; more HI will be produced.
(e) $Q < K_c$; more H_2 and I_2 will be produced.

a 24. At a certain temperature $K_c = 25$ and a reaction vessel contains a mixture with the following concentrations : $[H_2] = 0.10\ M$, $[Br_2] = 0.10\ M$ and $[HBr] = 0.50\ M$. Which of the following statements concerning the reaction and the reaction quotient, Q, is true?

$$H_2(g) + Br_2(g) \rightleftharpoons 2HBr(g)$$

(a) $Q = K_c$
(b) $Q < K_c$; more HBr will be produced.
(c) $Q < K_c$; more H_2 and Br_2 will be produced.
(d) $Q > K_c$; more HBr will be produced.
(e) $Q > K_c$; more H_2 and Br_2 will be produced.

e 25. At 990°C K_c is 1.6 for the following reaction. If 4.0 mol of CO, 3.0 mol of H_2O, 2.0 mol of H_2, and 1.0 mol of CO_2 are placed in a 5.0-liter container and allowed to reach equilibrium at 990°C, which response includes all of the following statements that are correct, and no others?

$$CO_2(g) + H_2(g) \rightleftharpoons H_2O(g) + CO(g)$$

I. The concentration of H_2 will be greater than 0.40 mol/L.
II. The concentration of H_2O will be less than 0.60 mol/L.
III. The concentration of CO will be less than 0.80 mol/L.
IV. The concentration of CO_2 will be greater than 0.20 mol/L.

(a) I, II, and IV   (b) I and II   (c) III
(d) I and IV   (e) I, II, III, and IV

Uses of the Equilibrium Constant, K_c

d*26. For the gas phase reaction $N_2O_4 \rightleftharpoons 2NO_2$, $K_c = 8$. If 2 moles of N_2O_4 is introduced into a 1-liter vessel, the number of **moles** (not concentration) of NO_2 at equilibrium, x, can be calculated from the equation:

(a) $8 = \dfrac{x}{2 - \frac{x}{2}}$

(b) $8 = \dfrac{4x^2}{2 - x}$

(c) $8 = \dfrac{x^2}{2}$

(d) $8 = \dfrac{x^2}{2 - \frac{x}{2}}$

(e) $8 = \dfrac{x^2}{2 - x}$

c*27. Nitrosyl chloride, NOCl, dissociates on heating as shown below. When a 1.50-gram sample of pure NOCl is heated at 350°C in a volume of 1.00 liter, the percent dissociation is found to be 57.2%. Calculate the equilibrium concentration of NOCl.

$$NOCl(g) \rightleftharpoons NO(g) + \tfrac{1}{2}Cl_2(g)$$

(a) 8.76×10^{-1} M (b) 9.26 M (c) 9.80×10^{-3} M
(d) 1.31×10^{-2} M (e) 1.75×10^{-4} M

b 28. The numerical value of the equilibrium constant, K_c, for the following gas phase reaction is 0.50 at a certain temperature. When a certain reaction mixture reaches equilibrium, the concentration of O_2 is found to be 2.0 M, while the concentration of SO_3 is found to be 10 M. What is the equilibrium concentration of SO_2 in this mixture?

$$2SO_2 + O_2 \rightleftharpoons 2SO_3$$

(a) 0.50 M (b) 10 M (c) 0.10 M (d) 5.0 M (e) 1.0 M

e*29. Consider the following system. The equilibrium constant for the reaction at 900°C is 0.082. A reaction is initiated with only $COCl_2$ in a 2.0-liter vessel. At equilibrium 2.36×10^{-3} mol of Cl_2 is present. How many moles of $COCl_2$ are present at equilibrium?

$$COCl_2(g) \rightleftharpoons CO(g) + Cl_2(g)$$

(a) 1.4×10^{-4} mol (b) 1.4×10^{-3} mol (c) 6.1×10^{-5} mol
(d) 5.8×10^{-4} mol (e) 3.4×10^{-5} mol

b 30. A quantity of HI was sealed in a tube, heated to 425°C, and held at this temperature until equilibrium was reached. The concentration of HI in the tube at equilibrium was found to be 0.0706 moles/liter. Calculate the equilibrium concentration of H_2 (and I_2).
For the reaction $H_2(g) + I_2(g) \rightleftharpoons 2HI(g)$, $K_c = 54.6$ at 425°C.

(a) 4.78×10^{-3} M (b) 9.55×10^{-3} M (c) 2.34×10^{-3} M
(d) 1.17×10^{-3} M (e) 1.85×10^{-4} M

c 31. The equilibrium constant for the following gas phase reaction is 4.0 at a certain temperature. A reaction is carried out at this temperature starting with 2.0 mol/L of CO and 2.0 mol/L of H_2O. What will be the equilibrium concentration of H_2?

$$CO + H_2O \rightleftharpoons CO_2 + H_2$$

(a) 2.0 M (b) 0.75 M (c) 1.3 M (d) 0.67 M (e) 1.5 M

d 32. For the following reaction, K_c is 144 at 200°C. If 0.400 mol of both A and B are placed in a 2.00-liter container at that temperature, what will be the concentration of C at equilibrium? $A(g) + B(g) \rightleftharpoons C(g) + D(g)$

(a) 0.015 M (b) 1.64 M (c) 0.200 M (d) 0.185 M (e) 1.13 M

a 33. For the following reaction, K_c is 144 at 200°C. If 0.400 mol of both A and B are placed in a 2.00-liter container at that temperature, what will be the concentration of B at equilibrium? $A(g) + B(g) \rightleftharpoons C(g) + D(g)$

(a) 0.015 M (b) 1.64 M (c) 0.200 M (d) 0.185 M (e) 1.13 M

e*34. The equilibrium constant for the reaction
$N_2(g) + O_2(g) \rightleftharpoons 2NO(g)$ has the value 4.00×10^{-2}. We introduce 28.0 g (1.00 mole) of N_2 and 32.0 g (1.00 mole) of O_2 into a 250-mL vessel and allow the reaction to reach equilibrium. What will be the equilibrium concentrations of all substances present? (In the answers, all concentrations are in moles/liter.)

	[N_2]	[O_2]	[NO]
(a)	0.91	0.91	0.18
(b)	3.02	3.02	0.98
(c)	1.82	1.82	0.36
(d)	0.36	0.36	0.72
(e)	3.6	3.6	0.72

b 35. At 25°C, K_c for the following reaction is 4.66×10^{-3}. If 0.800 mol of N_2O_4 is injected into a closed 1.00-liter glass container at 25°C, what will be the equilibrium concentration of N_2O_4?

$$N_2O_4(g) \rightleftharpoons 2NO_2(g)$$

(a) 0.016 M (b) 0.770 M (c) 0.088 M (d) 0.236 M (e) 0.667 M

d 36. The equilibrium constant, K_c, is 0.022 at 25°C for the reaction below. What is the concentration of PCl_5 at equilibrium if a reaction is initiated with 0.80 mole of PCl_5 in a 1.00-liter container?

$$PCl_5(g) \rightleftharpoons PCl_3(g) + Cl_2(g)$$

(a) 0.080 M (b) 0.12 M (c) 0.54 M (d) 0.68 M (e) 0.76 M

c 37. Phosgene, $COCl_2$, is a poisonous gas that decomposes into carbon monoxide and chlorine according to the following equation with $K_c = 0.083$ at 900°C. If the reaction is initiated with 0.600 mole of $COCl_2$ at 900°C in a 5.00-liter container, what concentration of CO will be present after equilibrium is established?

$$COCl_2(g) \rightleftharpoons CO(g) + Cl_2(g)$$

(a) 0.072 M (b) 0.048 M (c) 0.067 M (d) 0.012 M (e) 0.090 M

b 38. $K_c = 0.040$ for the system below at 450°C. If a reaction is initiated with 0.20 mole of Cl_2 and 0.20 mole of PCl_3 in a 1.0-liter container, what concentration of PCl_5 will be present at equilibrium?

$$PCl_5(g) \rightleftharpoons PCl_3(g) + Cl_2(g)$$

(a) 0.09 M (b) 0.13 M (c) 0.22 M (d) 0.31 M (e) 0.16 M

a 39. $K_c = 0.040$ for the system below at 450°C. If a reaction is initiated with 0.20 mole of Cl_2 and 0.20 mole of PCl_3 in a 1.0-liter container, what is the equilibrium concentration of Cl_2 in the same system?

$$PCl_5(g) \rightleftharpoons PCl_3(g) + Cl_2(g)$$

(a) 0.07 M (b) 0.16 M (c) 0.11 M (d) 0.04 M (e) 0.26 M

d 40. For the following system, the equilibrium constant at 445°C is 51.0. If a reaction is initiated with the following initial concentrations, [H$_2$] = 2.06 x 10^{-2} M, [I$_2$] = 1.45 x 10^{-2} M, and [HI] = 0, what will be the equilibrium concentration of HI?

$$H_2(g) + I_2(g) \rightleftharpoons 2HI(g)$$

(a) 1.8 x 10^{-2} M (b) 1.4 x 10^{-1} M (c) 2.7 x 10^{-1} M
(d) 2.6 x 10^{-2} M (e) 3.7 x 10^{-3} M

d*41. A reaction begins with 0.600 mole of A and 0.200 mole of B in a 2.00-L container at a certain temperature. What will be the equilibrium concentration of C?

$$A(g) + B(g) \rightleftharpoons C(g) \qquad K_c = 23.5$$

(a) 0.0684 M (b) 0.0200 M (c) 0.044 M
(d) 0.0836 M (e) 0.105 M

b 42. For the following reaction, K_c is 144 at 200°C. If the reaction were initiated with 0.600 mol of A and 0.200 mol of B in a 2.00-liter container, what would be the equilibrium concentration A?

$$A(g) + B(g) \rightleftharpoons C(g) + D(g)$$

(a) 0.684 M (b) 0.200 M (c) 0.444 M (d) 0.982 M (e) 1.05 M

e*43. At 990°C, K_c = 1.6 for the following reaction. How many moles of H$_2$O(g) are present in an equilibrium mixture resulting from the addition of 1.00 mole of H$_2$, 2.00 mol of CO$_2$, 0.75 mol of H$_2$O, and 1.00 mol of CO to a 5.00-liter container at 990°C?

$$H_2(g) + CO_2(g) \rightleftharpoons H_2O(g) + CO(g)$$

(a) 0.60 mol (b) 0.80 mol (c) 1.02 mol (d) 1.45 mol (e) 1.14 mol

Factors that Affect Equilibria

c 44. For the system H$_2$(g) + CO$_2$(g) $\rightleftharpoons$ H$_2$O(g) + CO(g) at equilibrium, the addition of H$_2$(g) would cause (according to LeChatelier's principle)

(a) only more H$_2$O(g) to form.
(b) only more CO(g) to form.
(c) more H$_2$O(g) and CO(g) to form.
(d) only more CO$_2$(g) to form.
(e) no change in amounts of products or reactants.

d 45. For the system H$_2$(g) + CO$_2$(g) $\rightleftharpoons$ H$_2$O(g) + CO(g) at equilibrium, the removal of some of the H$_2$O(g) would cause (according to LeChatelier's principle)

(a) more H$_2$(g) to be form.
(b) more CO$_2$(g) to be form.
(c) no change in the amounts of products or reactants.
(d) more CO(g) to be form.
(e) the amount of CO(g) to remain constant while the amount of H$_2$O(g) increases to the original equilibrium concentration.

d 46. If the system below is at equilibrium in a closed vessel and a small amount of nitrous acid is added, what would be expected to happen?

$HN_3(\ell) + 2H_2O(\ell) \rightleftharpoons N_2H_4(\ell) + HNO_2(\ell)$ $\Delta H^0 = +641$ kJ
hydrazoic   hydrazine   nitrous
  acid   acid

(a) Some HN_3 would be used up in re-establishing equilibrium.
(b) Some HNO_2 would be formed in re-establishing equilibrium.
(c) Some HNO_2 would be formed, and some N_2H_4 would be lost.
(d) More HN_3 and H_2O would be formed.
(e) The temperature would decrease, and the forward reaction would be favored.

c 47. Consider the following systems at equilibrium. Which response includes **all** the stresses listed that would shift the equilibrium to the right (favor the forward reaction)?

	Equilibrium	Stress
I.	$CO(g) + Cl_2(g) \rightleftharpoons COCl_2(g)$	add Cl_2
II.	$CO(g) + Cl_2(g) \rightleftharpoons COCl_2(g)$	remove $COCl_2$
III.	$PCl_3(g) + Cl_2(g) \rightleftharpoons PCl_5(g)$	remove PCl_3

(a) I   (b) III   (c) I and II   (d) II and III   (e) I, II and III

b 48. Suppose we let the reaction below come to equilibrium. Then we decrease the total pressure, by increasing the volume of the container. What will be the effect on the net amount of $SO_3(g)$ present? $2SO_2(g) + O_2(g) \rightleftharpoons 2SO_3(g)$

(a) It increases.   (b) It decreases.   (c) It does not change.
(d) The question cannot be answered without knowing the value of K.
(e) The question cannot be answered without knowing the value of ΔH^0.

c 49. Suppose we let the reaction below come to equilibrium. Then we decrease the total pressure, by increasing the volume of the container. What will be the effect on the numerical value of the equilibrium constant, K?
$2SO_2(g) + O_2(g) \rightleftharpoons 2SO_3(g) + $ heat

(a) It increases.   (b) It decreases.   (c) It does not change.
(d) The question cannot be answered without knowing the initial value of K.
(e) The question cannot be answered without knowing the value of ΔH^0.

c 50. Suppose the following reaction is at equilibrium at a given temperature and pressure. The pressure is then increased at constant temperature, by compressing the reaction mixture, and the mixture is allowed to re-establish equilibrium. At the new equilibrium _____.

$H_2(g) + Cl_2(g) \rightleftharpoons 2HCl(g)$

(a) there is more hydrogen chloride than there was originally
(b) there is less hydrogen chloride than there was originally
(c) there is the same amount of hydrogen chloride as there was originally
(d) the hydrogen and chloride are completely used up
(e) the amount of hydrogen chloride may be either larger or smaller than it was originally, depending on the value of K

e 51. Decreasing the volume of the container (at constant temperature) after the system below has reached equilibrium would be expected to _____.
$$HN_3(\ell) + 2H_2O(\ell) \rightleftharpoons N_2H_4(\ell) + HNO_2(\ell)$$

(a) produce more HN_3 and raise the temperature
(b) produce more H_2O and lower the temperature
(c) produce more N_2H_4 and raise the temperature
(d) produce more HNO_2 and lower the temperature
(e) have no effect on this equilibrium

b 52. Suppose we let this reaction come to equilibrium. Then we increase the temperature of the reaction mixture. What will be the effect on the net amount of $SO_3(g)$ present?
$$2SO_2(g) + O_2(g) \rightleftharpoons 2SO_3(g) + heat$$

(a) It increases.   (b) It decreases.   (c) It does not change.
(d) The question cannot be answered without knowing the value of K.
(e) The question cannot be answered without knowing the value of ΔH^0.

b 53. Suppose we let this exothermic reaction come to equilibrium. Then we increase the temperature of the reaction mixture. What will be the effect on the numerical value of the equilibrium constant, K?
$$2SO_2(g) + O_2(g) \rightleftharpoons 2SO_3(g)$$

(a) It increases.   (b) It decreases.   (c) It does not change.
(d) The question cannot be answered without knowing the initial value of K.
(e) The question cannot be answered without knowing the value of ΔH^0.

a 54. Consider the system below at equilibrium at 200°C.
$$2Cl_2(g) + 2H_2O(g) + heat \rightleftharpoons 4HCl(g) + O_2(g)$$
Which response contains all the stresses listed that will result in a shift of the equilibrium so that more HCl is produced when equilibrium is re-established, and no stresses that will not?

I.   adding some Cl_2
II. raising the temperature at constant pressure
III. decreasing the volume at constant temperature

(a) I and II   (b) II and III   (c) III
(d) I and II   (e) another one or another combination

d 55. Which of the numbered responses lists all of the following stresses that would shift the equilibrium to the left (favor the reverse reaction), and no other stresses?
$$2NOCl(g) + 75 kJ \rightleftharpoons 2NO(g) + Cl_2(g)$$

I.   Add a catalyst.
II. Heat the mixture.
III. Decrease the volume at constant temperature.
IV. Increase the partial pressure of NOCl by adding NOCl.

(a) I, II, and IV   (b) II, III, and IV   (c) II and III
(d) III   (e) a different one or a different combination

a 56. Which of the numbered responses lists all the following stresses that would shift the equilibrium to the right (favor the forward reaction), and no other stresses?

$$2NOCl(g) + 75 \text{ kJ} \rightleftharpoons 2NO(g) + Cl_2(g)$$

I. Add more NOCl.   II. Remove some Cl_2.
III. Lower the temperature.   IV. Add more NO.

(a) I and II   (b) I, II, and III   (c) I and III
(d) II and IV   (e) II, III, and IV

e 57. Consider the reaction below at equilibrium at a certain temperature.

$$2SO_2(g) + O_2(g) \rightleftharpoons 2SO_3(g) + \text{heat}$$

Which response contains **all** the stresses that would shift the equilibrium so as to produce more SO_3 (to the right), and **only** those stresses?

I. increase temperature at constant pressure
II. decrease the volume of the system at constant pressure
III. remove SO_2
IV. add O_2
V. increase the partial pressure of SO_2

(a) II and III   (b) I, II, and IV   (c) I and V
(d) I and III   (e) II, IV, and V

c 58. Consider the following system at equilibrium.

$$H_2(g) + I_2(g) \rightleftharpoons 2HI(g) + \text{heat}$$

Which response includes all the following that will shift the equilibrium to the left, and no others?

I. increasing the temperature   II. decreasing the temperature
III. decreasing the volume   IV. increasing the volume
V. removing some HI   VI. adding some HI
VII. removing some I_2 VIII. adding some I_2

(a) II   (b) II, V, and VIII   (c) I, VI, and VII
(d) I, III, V, and VII   (e) II, IV, VII, and VIII

b 59. For the **gas phase** reaction $SO_2 + \frac{1}{2}O_2 \rightleftharpoons SO_3$ $\Delta H^0 = -1.6 \times 10^2$ kJ for the forward reaction. In order to increase the yield of SO_3, the reaction should be run

(a) at high P, high T.   (b) at high P, low T.   (c) at low P, high T.
(d) at low P, low T.   (e) at high P, but is independent of T.

e 60. Which of the following will require the least time for a reaction to reach equilibrium?

(a) K_c is a very small number.
(b) K_c is a very large number.
(c) K_c is approximately one.
(d) Cannot tell without knowing the value of K_c.
(e) Cannot tell, since the time required to reach equilibrium does not depend on K_c.

The Haber Process

a 61. Suppose the following reaction is at equilibrium at a given temperature and pressure. The pressure is then increased at constant temperature, by compressing the reaction mixture, and the mixture is allowed to re-establish equilibrium. At the new equilibrium _____.

$$N_2(g) + 3H_2(g) \rightleftharpoons 2NH_3(g)$$

(a) there is more ammonia than there was originally
(b) there is less ammonia than there was originally
(c) there is the same amount of ammonia present as there was originally
(d) the nitrogen is used up completely
(e) the amount of ammonia may be either larger or smaller than it was originally, depending on the value of K

e 62. Consider the following reversible reaction at equilibrium.

$$N_2(g) + 3H_2(g) \rightleftharpoons 2NH_3(g) + 9.22 \times 10^4 \text{ J}$$

Which response contains all the choices below that refer to changes that would shift the equilibrium to the right, and no other choices?

I. add H_2 II. remove N_2 III. add an iron catalyst
IV. increase the temperature at constant pressure
V. increase the pressure at constant temperature by decreasing the volume

(a) III, IV and V (b) I, IV and V (c) IV and V
(d) I, II and III (e) I and V

e 63. Consider the following system in equilibrium.

$$N_2(g) + 3H_2(g) \rightleftharpoons 2NH_3(g) + 92.24 \text{ kJ}$$

Which response includes all of the following that will shift the equilibrium to the right, and no others?

I. increasing the temperature II. decreasing the temperature
III. decreasing the volume IV. increasing the volume
V. removing some NH_3 VI. adding some NH_3
VII. removing some N_2 VIII. adding some N_2

(a) I, IV, VI, and VII (b) II, V, and VIII (c) I, VI, and VII
(d) I, III, V, and VII (e) II, III, V, and VIII

d 64. The Haber process demonstrates that commercial processes do not maximize each factor that affects equilibria but use them in combination to get maximum production. Which of the following statements about the Haber process is **false**?

$$N_2(g) + 3H_2(g) \rightleftharpoons 2NH_3(g) + 92.24 \text{ kJ}$$

(a) Although K_c is much greater at 25°C, the reaction is carried out at 450°C.
(b) A catalyst is used to increase reaction speed without using **very** high temperatures.
(c) Although the reaction is exothermic, it is carried out at high (450°C) temperature.
(d) Although high pressure is unfavorable to the production of the $NH_3(g)$, the reaction is carried out under a pressure of 200 - 1000 atmospheres.
(e) The emerging reaction mixture is cooled to remove the NH_3 and the N_2 and H_2 recycled.

Harcourt, Inc

Application of Stress to a System at Equilibrium

d 65. A system at equilibrium in a 1.0-liter container was found to contain 0.20 mol of A, 0.20 mol of B, 0.40 mol of C, and 0.40 mol of D. If 0.15 mol of A and 0.15 mol of B are added to this system, what will be the new equilibrium concentration of A?
$$A(g) + B(g) \rightleftharpoons C(g) + D(g)$$

(a) 0.050 M (b) 0.10 M (c) 0.20 M (d) 0.25 M (e) 0.30 M

d 66. At equilibrium a 1.0-liter container was found to contain 0.20 mol of A, 0.20 mol of B, 0.40 mol of C, and 0.40 mol of D. If 0.10 mol of A and 0.10 mol of B are added to this system, what will be the new equilibrium concentration of A?
$$A(g) + B(g) \rightleftharpoons C(g) + D(g)$$

(a) 0.37 M (b) 0.47 M (c) 0.87 M (d) 0.23 M (e) 0.067 M

e 67. A system at equilibrium in a 1.0-liter container was found to contain 0.20 mol of A, 0.20 mol of B, 0.40 mol of C, and 0.40 mol of D. If 0.12 mol of A and 0.12 mol of B are added to this system, what will be the new equilibrium concentration of C?
$$A(g) + B(g) \rightleftharpoons C(g) + D(g)$$

(a) 0.32 M (b) 0.13 M (c) 0.24 M (d) 0.40 M (e) 0.48 M

a*68. A 1.00-liter vessel contains the following equilibrium concentrations at 400°C: N_2, 1.00 M; H_2, 0.50 M; and NH_3, 0.50 M. How many moles of hydrogen must be removed from the vessel in order to increase the concentration of nitrogen to 1.20 M?
$$N_2(g) + 3H_2(g) \rightleftharpoons 2NH_3(g)$$

(a) 0.94 mol (b) 1.5 mol (c) 0.33 mol (d) 0.76 mol (e) 1.1 mol

b 69. Given: $A(g) \rightleftharpoons B(g) + C(g)$
When the system is at equilibrium at 200°C, the concentrations are found to be: [A] = 0.20 M, [B] = 0.30 M, [C] = 0.30 M. If the volume of the container is suddenly doubled at 200°C, what will be the new equilibrium concentration of C?

(a) 0.060 M (b) 0.18 M (c) 0.24 M (d) 0.29 M (e) 0.35 M

c 70. Given: $A(g) \rightleftharpoons B(g) + C(g)$
When the above system is at equilibrium at 200°C, the concentrations are found to be: [A] = 0.20 M, [B] = [C] = 0.30 M. If the volume of the container is suddenly doubled at 200°C, what will be the new equilibrium concentration of A?

(a) 0.03 M (b) 0.05 M (c) 0.07 M (d) 0.09 M (e) 0.11 M

b 71. Given: $PCl_5(g) \rightleftharpoons PCl_3(g) + Cl_2(g)$ $K_c = 0.040$ at 450°C
What would be the equilibrium concentration of $PCl_5(g)$ if 0.20 mole of $PCl_5(g)$ was placed in a 1.00-L container at this temperature? What would be the new equilibrium concentration of $PCl_5(g)$ if the volume were halved at this same temperature?

(a) 0.070 M, 0.29 M (b) 0.13 M, 0.29 M (c) 0.13 M, 0.11 M
(d) 0.070 M, 0.13 M (e) 0.060 M, 0.14 M

270

a*72. Given: $N_2(g) + O_2(g) \rightleftharpoons 2NO(g)$ $K_c = 0.16$
What would be the equilibrium concentration of $N_2(g)$ if 0.12 mole of NO(g) was placed in a 1.00-L container at this temperature? What would be the new equilibrium concentration of $N_2(g)$ if the volume were halved at this same temperature?

(a) 0.050 M, 0.10 M (b) 0.050 M, 0.12 M (c) 0.070 M, 0.12 M
(d) 0.10 M, 0.10 M   (e) 0.50 M, 1.0 M

Partial Pressures and the Equilibrium Constant

e 73. When equilibrium is established for the following reaction at 298 K, the concentration of N_2O_4 is 1.34 mol/L, and the concentration of NO_2 is 0.448 mol/L. Convert these concentrations to partial pressures. What is the **total** pressure of the system at equilibrium?
$$N_2O_4(g) \rightleftharpoons 2NO_2(g)$$

(a) 2.54 atm   (b) 11.6 atm   (c) 16.7 atm
(d) 34.4 atm   (e) 43.8 atm

e 74. A mixture of 0.40 mol of N_2, 0.60 mol of Ar, and 0.30 mol of O_2 is confined in a 300-liter vessel at 50.0°C. What is the total pressure of the system?

(a) 0.32 atm   (b) 0.016 atm   (c) 0.049 atm
(d) 0.086 atm   (e) 0.11 atm

a 75. The equilibrium constant expression for the reaction,
$$2CO(g) + O_2(g) \rightleftharpoons 2CO_2(g) \text{ is given by } K_p =$$

(a) $\dfrac{(P_{CO_2})^2}{(P_{CO})^2 P_{O_2}}$ (b) $\dfrac{P_{CO_2}}{P_{CO} P_{O_2}}$ (c) $2P_{CO} + P_{O_2}$

(d) $\dfrac{2P_{CO_2}}{2P_{CO} + P_{O_2}}$ (e) $\dfrac{(2P_{CO_2})^2}{(2P_{CO})^2 P_{O_2}}$

a 76. Consider the reversible reaction at equilibrium at 392°C.
$$2A(g) + B(g) \rightleftharpoons C(g)$$
The partial pressures are found to be: A: 6.70 atm, B: 10.1 atm, C: 3.60 atm. Evaluate K_p for this reaction.

(a) 7.94 x 10^{-3} (b) 1.46 x 10^{-1} (c) 5.32 x 10^{-2}
(d) 54.5   (e) 121

c 77. For the following reaction at equilibrium at 445°C the partial pressures were found to be $[H_2] = 0.45$ atm, $[I_2] = 0.10$ atm and $[HI] = 1.53$ atm. Calculate K_p for this reaction.
$$H_2(g) + I_2(g) \rightleftharpoons 2HI(g)$$

(a) 150   (b) 34   (c) 52   (d) 76   (e) 4.4

e 78. Phosphorus pentafluoride, $PF_5(g)$, partially decomposes to $PF_3(g)$ and $F_2(g)$ at a temperature of 600 K. A 2.0-liter container is filled with pure PF_5. At a temperature of 600 K, the initial pressure of PF_5 is 2.0 atm. When equilibrium is reached, half of the original amount of PF_5 has decomposed. The equilibrium constant, K_p, for
$PF_5(g) \rightleftharpoons PF_3(g) + F_2(g)$ is _____ .

(a) 0.0025   (b) 0.12   (c) 0.25   (d) 0.50   (e) 1.0

d 79. If the equilibrium constant, K_p, at a certain temperature is 8.6×10^{11} for the following reaction, $4HCl(g) + O_2(g) \rightleftharpoons 2Cl_2(g) + 2H_2O(g)$
calculate the value of the equilibrium constant, K_p, at the same temperature for
$\frac{1}{2}Cl_2(g) + \frac{1}{2}H_2O(g) \rightleftharpoons HCl(g) + \frac{1}{4}O_2(g)$.

(a) 4.9×10^{-7} (b) 8.5×10^3 (c) 1.2×10^{13} (d) 1.0×10^{-3} (e) 2.7×10^{-14}

Relationship Between K_p and K_c

b 80. The partial pressure of a certain gas in a system at equilibrium at 10.°C is experimentally determined to be 0.340 atmosphere. What is its concentration in moles per liter?

(a) $4.14 \times 10^{-1} M$ (b) $1.46 \times 10^{-2} M$ (c) $68.2 M$
(d) $2.79 \times 10^{-1} M$ (e) $7.90 M$

d 81. Consider the reversible reaction at equilibrium at 392°C.
$2A(g) + B(g) \rightleftharpoons C(g)$
The partial pressures are found to be: A: 6.70 atm, B: 10.1 atm, C: 3.60 atm. What is the concentration of B at equilibrium?

(a) 0.015 M   (b) 1.64 M   (c) 0.200 M   (d) 0.185 M   (e) 1.13 M

b 82. $K_c = 4.6 \times 10^{-3}$ for the reaction below at 25°C. Evaluate K_p at 25°C.
$N_2O_4(g) \rightleftharpoons 2NO_2(g)$

(a) 0.086   (b) 0.11   (c) 6.2×10^{-3}
(d) 1.9×10^{-4} (e) 3.2×10^{-1}

c 83. $K_c = 0.040$ for the system below at 450.°C. Evaluate K_p for the reaction at 450.°C.
$PCl_5(g) \rightleftharpoons PCl_3(g) + Cl_2(g)$

(a) 0.40   (b) 0.64   (c) 2.4   (d) 5.2×10^{-2} (e) 6.7×10^{-4}

d 84. Given: $PCl_5(g) \rightleftharpoons PCl_3(g) + Cl_2(g)$
At 250.°C a sample of PCl_5 was placed in a 24-liter evacuated reaction vessel and allowed to come to equilibrium. Analysis showed that at equilibrium 0.42 mole of PCl_5, 0.64 mole of PCl_3, and 0.64 mole of Cl_2 were present in the vessel. Calculate K_p for the reaction at 250.°C.

(a) 1.2   (b) 1.3   (c) 1.5   (d) 1.8   (e) 2.2

d 85. Certain amounts of the hypothetical substances A_2 and B are mixed at 300. K. When equilibrium is established for the reaction below, the following amounts were present in a 3.00-liter container: 0.200 mole of A_2, 0.400 mole of B, 0.200 mole of D, and 0.100 mole of E. What is K_p, the equilibrium constant, in terms of partial pressures (atm), for this reaction?

$$A_2(g) + 3B(g) \rightleftharpoons 2D(g) + 3E(g)$$

(a) 16.4
(b) 0.084
(c) 3.81 x 10^{-2}
(d) 2.57 x 10^{-2}
(e) 1.46 x 10^{-2}

c 86. Certain amounts of the hypothetical substances A_2 and B are mixed in a 3.00-liter container at 300. K. When equilibrium is established for the reaction the following amounts are present: 0.200 mol of A_2, 0.400 mol of B, 0.200 mol of D, and 0.100 mol of E. What is K_p, the equilibrium constant in terms of partial pressures, for this reaction?

$$A_2(g) + 3B(g) \rightleftharpoons 2D(g) + E(g)$$

(a) 16.4   (b) 0.084   (c) 3.81 x 10^{-2} (d) 1.42 x 10^{-2} (e) 2.50 x 10^{-1}

a 87. For the reaction below, K_p = 6.70 x 10^{-3} at 25°C.

$$COCl_2(g) \rightleftharpoons CO(g) + Cl_2(g)$$

A sample of $COCl_2$ is placed in a closed 15.0-liter vessel at 25°C, and it exerts a pressure of 4.65 atm before decomposition begins. What will be the partial pressure of Cl_2 at equilibrium?

(a) 0.173 atm   (b) 0.206 atm   (c) 0.566 atm   (d) 2.14 atm   (e) 1.16 atm

d 88. The equilibrium constant, K_p, for the following reaction is 280. at 150.°C. Suppose that a quantity of IBr is placed in a closed reaction vessel and the system is allowed to come to equilibrium at 150.°C. When equilibrium is established, the pressure of IBr is 0.200 atm. What is the pressure of I_2 at equilibrium?

$$I_2(g) + Br_2(g) \rightleftharpoons 2IBr(g) + 11.7 \text{ kJ}$$

(a) 0.168 atm   (b) 0.096 atm   (c) 0.067 atm   (d) 0.012 atm   (e) 0.00014 atm

e 89. For the reaction below, K_p = 6.70 x 10^{-3} at 25°C.

$$COCl_2(g) \rightleftharpoons CO(g) + Cl_2(g)$$

A sample of $COCl_2$ is placed in a closed 15.0-liter vessel at 25°C, and it exerts a pressure of 4.65 atm before decomposition begins. What will be the total pressure at equilibrium?

(a) 1.68 atm   (b) 3.80 atm   (c) 5.03 atm   (d) 4.65 atm   (e) 4.83 atm

c 90. The equilibrium constant, K_p, for the following reaction is 280. at 150.°C. Suppose that a quantity of IBr is placed in a closed reaction vessel and the system is allowed to come to equilibrium at 150.°C. When equilibrium is established, the pressure of IBr is 0.200 atm. What is the total pressure inside the system at equilibrium?

$$I_2(g) + Br_2(g) \rightleftharpoons 2IBr(g) + 11.7 \text{ kJ}$$

(a) 0.176 atm
(b) 0.212 atm
(c) 0.224 atm
(d) 0.334 atm
(e) 0.536 atm

Chapter 17   273

b 91. The gas phase system below is at equilibrium at 200°C with the following partial pressures: 0.20 atm A, 0.20 atm B, 0.10 atm C, and 0.40 atm D. If an additional 0.30 atm C is introduced into the reaction vessel, what will be the partial pressure of C at 200°C when equilibrium is re-established?

$$A(g) + B(g) \rightleftharpoons C(g) + D(g)$$

(a) 0.60 atm   (b) 0.30 atm   (c) 0.50 atm   (d) 0.75 atm   (e) 0.45 atm

c 92. The gas phase system below is at equilibrium at 250°C with the following partial pressures: 0.50 atm A, 0.50 atm B, 0.30 atm C, and 0.30 atm D. If an additional 0.20 atm D is introduced into the reaction vessel, what will be the partial pressure of A at 250°C when equilibrium is re-established?

$$A(g) + B(g) \rightleftharpoons C(g) + D(g)$$

(a) 0.57 atm   (b) 0.37 atm   (c) 0.55 atm   (d) 0.52 atm   (e) 0.46 atm

c 93. The gas phase system below is at equilibrium at 200°C with the following partial pressures: 0.20 atm A, 0.20 atm B, 0.10 atm C, and 0.40 atm D. If an additional 0.30 atm C is introduced into the reaction vessel, what will be the total pressure of the system at 200°C when equilibrium is re-established?

$$A(g) + B(g) \rightleftharpoons C(g) + D(g)$$

(a) 0.10 atm   (b) 0.30 atm   (c) 1.20 atm   (d) 0.60 atm   (e) 1.00 atm

Heterogeneous Equilibria

d 94. Given the equilibrium reaction $ZnCO_3(s) \rightleftharpoons ZnO(s) + CO_2(g)$. Which one of the following statements is true?

(a) Equal concentrations of $ZnO(s)$ and $CO_2(g)$ would result from the decomposition of a given amount of $ZnCO_3(s)$.
(b) The same equilibrium condition would result if we started with **only** pure $ZnCO_3(s)$ in a closed container as if we started with **only** pure $ZnO(s)$ in a closed container.
(c) Introducing 1.0 atm pressure of $N_2(g)$ into the system at equilibrium in a closed container would result in more $ZnCO_3(s)$ being formed.
(d) Decreasing the volume of the closed system initially at equilibrium, at constant temperature, would result in more $ZnCO_3(s)$ being formed.
(e) At equilibrium conditions, the forward and reverse reactions have stopped.

c 95. Which one of the following would force the forward reaction to completion?

$$CaCO_3(s) + 2H_3O^+(aq) \rightleftharpoons Ca^{2+}(aq) + 3H_2O(\ell) + CO_2(g)$$

(a) Removing some H_3O^+ from the reaction mixture, by neutralizing it with base.
(b) Adding more Ca^{2+} to the mixture.
(c) Removing CO_2 as it is formed.
(d) Adding CO_2 to the reaction mixture.
(e) None of the preceding would have any effect on the amount of $CaCO_3$ consumed.

d 96. Which response includes all of the following equilibria (and only those) that would be shifted to the right (forward reaction favored) by increasing the volume of the container?

I.	$2CO(g) + O_2(g) \rightleftharpoons 2CO_2(g)$
II.	$2NO(g) \rightleftharpoons N_2(g) + O_2(g)$
III.	$N_2O_4(g) \rightleftharpoons 2NO_2(g)$
IV.	$Ni(s) + 4CO(g) \rightleftharpoons Ni(CO)_4(g)$
V.	$N_2(g) + 3H_2(g) \rightleftharpoons 2NH_3(g)$

(a) II and III (b) I and II (c) I, IV, and V
(d) III (e) another one or another combination

e 97. Consider the following systems at equilibrium. Which response includes all the stresses (at constant volume) listed that would shift the equilibrium to the right (favor the forward reaction)?

I.	$PCl_5(g) \rightleftharpoons PCl_3(g) + Cl_2(g)$	$\Delta H = 92.5$ kJ	raise temperature
II.	$2SO_2(g) + O_2(g) \rightleftharpoons 2SO_3(g) + 47.3$ kcal		lower temperature
III.	$C(s) + H_2O(g) + 131$ kJ $\rightleftharpoons CO(g) + H_2(g)$		lower temperature

(a) II and III (b) I (c) I, II, and III
(d) I and III (e) another one or another combination

d 98. Consider the following system at equilibrium.
$$A(g) + 2B(\ell) \rightleftharpoons 2C(g) + D(\ell) + \text{heat}$$
Which response includes all of the stresses listed that can result in new equilibrium concentrations of A that are **less** than the equilibrium concentrations before the stress was applied, and no others?

I.	increase pressure	II.	decrease volume
III.	decrease temperature	IV.	add more A
V.	remove some C		

(a) I, II, and IV (b) II, IV, and V (c) I and III
(d) III and V (e) none of these

c 99. A sample of only solid ammonium chloride was heated in a closed container
$NH_4Cl(s) \rightleftharpoons NH_3(g) + HCl(g)$. At equilibrium, the pressure of $NH_3(g)$ was found to be 1.75 atm. What is the equilibrium constant, K_p, for the decomposition at this temperature?

(a) 1.46 (b) 1.75 (c) 3.06 (d) 3.72 (e) 4.14

a*100. A sample of only solid ammonium chloride was heated in a 1.00-L container at 500.°C
$NH_4Cl(s) \rightleftharpoons NH_3(g) + HCl(g)$. At equilibrium, the pressure of $NH_3(g)$ was found to be 1.75 atm. What is the equilibrium constant, K_c, for the decomposition at this temperature?

(a) 7.6×10^{-4} (b) 1.2×10^4 (c) 4.8×10^{-2} (d) 1.9×10^2 (e) 1.8×10^{-3}

a*101. At 1470 K the value of $K_p = 6.0 \times 10^{-4}$ for the reaction
$$2CO(g) \rightleftharpoons C\text{ (graphite)} + CO_2(g)$$
CO(g) initially at 2.00 atm is in contact with graphite until equilibrium is reached. What is the partial pressure of $CO_2(g)$ at equilibrium?

(a) 0.0024 atm (b) 0.00030 atm (c) 0.0012 atm
(d) 0.00060 atm (e) 0.00015 atm

Relationship Between ΔG^0 and the Equilibrium Constant

d 102. Calculate the thermodynamic equilibrium constant at 25°C for a reaction for which $\Delta G^0 = -25.60$ kJ per mol of reaction. R = 8.314 J/mol·K

(a) 5.14×10^1 (b) 11.2 (c) 6.12×10^2
(d) 3.07×10^4 (e) 4.32×10^6

a 103. For the following reaction, ΔG^0_{298} is –277 kJ. Calculate the thermodynamic equilibrium constant, K, at 25°C. R = 8.314 J/mol·K
$$SiCl_4(\ell) + 4H_2O(\ell) \rightleftharpoons H_4SiO_4(s) + 4HCl(aq)$$

(a) 3.6×10^{48} (b) 1.6×10^{27} (c) 2.8×10^{-49}
(d) 4.2×10^{-16} (e) 3.6×10^{-64}

c 104. Consider the following reaction at 25°C for which ΔH^0 is –26.9 kJ and ΔS^0 is 11.4 J/K. Evaluate the equilibrium constant, K_p, for the reaction at 25°C. R = 8.314 J/mol·K
$$I_2(g) + Cl_2(g) \rightleftharpoons 2ICl(g)$$

(a) 3.6×10^6 (b) 4.2×10^3 (c) 2.0×10^5
(d) 6.7×10^8 (e) 4.9×10^{-6}

d 105. Evaluate ΔG^0 at 718°C for a gas phase reaction for which $K_p = 7.4 \times 10^{-6}$ at 718°C. R = 8.314 J/mol·K

(a) 68.6 kJ (b) 365 kJ (c) 427 kJ (d) 97.3 kJ (e) 168 kJ

a 106. K_p for a gas phase reaction is 4.7×10^{-2} at 660.°C. Evaluate ΔG^0 at 660.°C. R = 8.314 J/mol·K

(a) 23.7 kJ (b) 109 kJ (c) 2.46 kJ (d) 31.5 kJ (e) 16.1 kJ

b 107. Evaluate ΔG^0 for a gas phase reaction for which $K_p = 3.1 \times 10^{-5}$ at 718°C.

(a) 62.4 kJ (b) 85.6 kJ (c) 98.2 kJ (d) 110 kJ (e) 138 kJ

a*108. For the reversible reaction below at 1500. K, $K_c = 0.150$ at 1500. K. Evaluate ΔG^0 at 1500. K. R = 8.314 J/mol·K = 0.0821 L·atm/mol·K
$$2SO_2(g) + O_2(g) \rightleftharpoons 2SO_3(g)$$

(a) 83.7 kJ (b) –83.7 kJ (c) –20.6 kJ (d) 14.6 kJ (e) 20.6 kJ

Evaluation of Equilibrium Constants at Different Temperatures

b 109. The value of K_p at 390.°C for the reversible reaction given below is 7.95 x 10^{-3}. Calculate the value of K_p at 25°C. $\Delta H^0 = 162$ kJ/mol, R = 8.314 J/mol·K

$$A(g) + B(g) \rightleftharpoons C(g)$$

(a) 2.8 x 10^{-2} (b) 1.8 x 10^{-18} (c) 6.3 x 10^{-12}
(d) 3.0 x 10^{-14} (e) 3.4 x 10^{13}

b 110. The value of K_p at 390.°C for the reversible reaction given below is 7.95 x 10^{-3}. Calculate the value of K_p at 25°C. $\Delta H^0 = 162$ kJ/mol, R = 8.314 J/mol·K

$$A(g) + B(g) \rightleftharpoons C(g)$$

$$\ln\left(\frac{K_{T_2}}{K_{T_1}}\right) = \frac{\Delta H^0}{R}\left(\frac{1}{T_1} - \frac{1}{T_2}\right)$$

(a) 2.8 x 10^{-2} (b) 1.8 x 10^{-18} (c) 6.3 x 10^{-12}
(d) 3.0 x 10^{-14} (e) 3.4 x 10^{13}

d 111. If the K_p at 2000. K for the reaction given below is 2.1 x 10^{-3} and the K_p at 27°C is 5.0 x 10^{-30}, calculate ΔH^0 for this reaction.

$$A(g) + B(g) \rightleftharpoons C(g)$$

(a) 2.6 kJ   (b) 1.44 J   (c) –179 kJ   (d) 180 kJ   (e) 13.9 kJ

d 112. If the K_p at 2000. K for the reaction given below is 2.1 x 10^{-3} and the K_p at 27°C is 5.0 x 10^{-30}, calculate ΔH^0 for this reaction.

$$A(g) + B(g) \rightleftharpoons C(g)$$

$$\ln\left(\frac{K_{T_2}}{K_{T_1}}\right) = \frac{\Delta H^0}{R}\left(\frac{1}{T_1} - \frac{1}{T_2}\right)$$

(a) 2.6 kJ   (b) 1.44 J   (c) –179 kJ   (d) 180 kJ   (e) 13.9 kJ

18 Ionic Equilibria I: Acids and Bases

The following values will be useful for problems in this chapter.

Acid	K	Substance or Species	K
HF	$K_a = 7.2 \times 10^{-4}$	NH_3	$K_b = 1.8 \times 10^{-5}$
HNO_2	$K_a = 4.5 \times 10^{-4}$	$(CH_3)_3N$	$K_b = 7.4 \times 10^{-5}$
CH_3COOH	$K_a = 1.8 \times 10^{-5}$	$[Co(OH_2)_6]^{2+}$	$K_a = 5.0 \times 10^{-10}$
HOCl	$K_a = 3.5 \times 10^{-8}$	$[Fe(OH_2)_6]^{2+}$	$K_a = 3.0 \times 10^{-10}$
HOBr	$K_a = 2.5 \times 10^{-9}$	$[Fe(OH_2)_6]^{3+}$	$K_a = 4.0 \times 10^{-3}$
HOCN	$K_a = 3.5 \times 10^{-4}$	$[Be(OH_2)_4]^{2+}$	$K_a = 1.0 \times 10^{-5}$
HCN	$K_a = 4.0 \times 10^{-10}$	$[Cu(OH_2)_4]^{2+}$	$K_a = 1.0 \times 10^{-8}$
H_2SO_4	K_{a1} = very large	HBO_2	$K_a = 6.0 \times 10^{-10}$
	$K_{a2} = 1.2 \times 10^{-2}$	$(COOH)_2$	$K_{a1} = 5.9 \times 10^{-2}$
H_2CO_3	$K_{a1} = 4.2 \times 10^{-7}$		$K_{a2} = 6.4 \times 10^{-5}$
	$K_{a2} = 4.8 \times 10^{-11}$		

****

Review of Strong Electrolytes

d 1. Which one of the following is a **weak** acid?

(a) HNO_3 (b) HI (c) HBr (d) H_2SO_3 (e) $HClO_4$

d 2. Which of the numbered responses lists (all) the **strong acid(s)** below and no weak acids?

I. H_2CO_3 II. HOCl III. HNO_2 IV. $HClO_4$

(a) I and III (b) II and IV (c) II, III, and IV
(d) IV (e) another acid or another combination of acids

b 3. Of the following, which acids are **weak** acids?

I. HBr II. HF III. HNO_3 IV. HNO_2 V. H_2CO_3 VI. H_3AsO_4

(a) I, II, and III (b) II, IV, V, and VI (c) I, II, and V
(d) IV, V, and VI (e) another combination

a 4. Which of the following is a soluble, strong base?

(a) RbOH (b) NH_3 (c) $Al(OH)_3$ (d) $Co(OH)_3$ (e) $Co(OH)_2$

d 5. Which of the following is **not** a soluble base?

(a) Sr(OH)$_2$ (b) NaOH (c) KOH (d) Mn(OH)$_2$ (e) CsOH

d 6. Which one of the following is an **insoluble** base?

(a) Ca(OH)$_2$ (b) Sr(OH)$_2$ (c) Ba(OH)$_2$ (d) Cu(OH)$_2$ (e) CsOH

b 7. Which one of the following salts is soluble in water?

(a) AgCl (b) NaClO$_3$ (c) BaSO$_4$ (d) FeS (e) CaCO$_3$

c 8. Which one of the following salts is **insoluble** in water?

(a) NaCl (b) Cr(NO$_3$)$_3$ (c) BaSO$_4$
(d) Fe$_2$(SO$_4$)$_3$ (e) Cu(CH$_3$COO)$_2$

e 9. Which one of the following salts is **insoluble** in water?

(a) Ba(CH$_3$COO)$_2$ (b) BaCl$_2$ (c) CaCl$_2$
(d) Ca(NO$_3$)$_2$ (e) Ca$_3$(PO$_4$)$_2$

c 10. Which salt is produced from a strong acid and a strong base?

(a) NH$_4$Cl (b) NaClO (c) KClO$_3$ (d) Na$_2$CO$_3$ (e) LiNO$_2$

c 11. Which salt is **not** produced from a strong acid and a strong base?

(a) SrCl$_2$ (b) Ca(NO$_3$)$_2$ (c) Li$_2$SO$_3$ (d) CsBr$_2$ (e) NaI

d 12. Which one of the following substances is **not** a strong electrolyte?

(a) NH$_4$Cl (b) HClO$_4$ (c) HNO$_3$ (d) NH$_3$ (e) Mg(NO$_3$)$_2$

c 13. Which one of the following is a **soluble strong electrolyte**?

(a) HNO$_2$ (b) H$_2$CO$_3$ (c) Ca(OH)$_2$
(d) Mg(OH)$_2$ (e) BaCO$_3$

b 14. Which solution should be the best conductor of electrical current because it would contain the largest number of ions in aqueous solution?

(a) 0.10 M NH$_4$NO$_3$ (b) 0.10 M K$_2$SO$_4$ (c) 0.20 M BaSO$_4$
(d) 0.10 M NaCl (e) 0.10 M HClO$_4$

d 15. Each response gives a pair of solutions. Which pair of solutions should conduct electrical current equally well because they contain equal numbers of ions?

(a) 0.10 M NH$_3$ and 0.10 M NH$_4$Cl (b) 0.10 M HBr and 0.10 M HF
(c) 0.10 M NaCl and 0.10 M Na$_2$SO$_4$ (d) 0.10 M NaNO$_3$ and 0.10 M HNO$_3$
(e) 0.10 M HNO$_3$ and 0.10 M HNO$_2$

d 16. Consider a 0.020 M Fe$_2$(SO$_4$)$_3$ solution. Which one of the responses includes all of the following statements that are correct, and no incorrect statements?

    I.   The solution is 0.020 M in Fe^{3+}.
    II.   The solution is 0.030 M in SO$_4^{2-}$.
    III.   The total concentration of ions in the solution is 0.050 M.
    IV.   The solution is 0.040 M in Fe^{3+}.
    V.   The solution is 0.060 M in SO$_4^{2-}$.
    VI.   The total concentration of ions in the solution is 0.10 M.

(a) I and II   (b) I, II, and III   (c) IV and V
(d) IV, V, and VI   (e) none of these

The Autoionization of Water

e 17. Under what condition is the H$_3$O$^+$ concentration in water expected to be **zero**?

(a) in a solution of a very strong acid   (b) in a solution of a very strong base
(c) in a solution of a very weak acid   (d) in a solution of a very weak base
(e) never

e 18. In a sample of pure water, only one of the following statements is **always** true at all conditions of temperature and pressure. Which one is always true?

(a) [H$_3$O$^+$] = 1.0 x 10^{-7} M   (b) [OH$^-$] = 1.0 x 10^{-7} M
(c) pH = 7.0   (d) pOH = 7.0
(e) [H$_3$O$^+$] = [OH$^-$]

c 19. Calculate the concentrations of H$_3$O$^+$ and OH$^-$ ions in a 0.25 M HClO$_4$ solution.

(a) [H$_3$O$^+$] = 0.25 M, [OH$^-$] = 0.25 M
(b) [H$_3$O$^+$] = 0.25 M, [OH$^-$] = 4.0 M
(c) [H$_3$O$^+$] = 0.25 M, [OH$^-$] = 4.0 x 10^{-14} M
(d) [H$_3$O$^+$] = 0.50 M, [OH$^-$] = 2.0 x 10^{-14} M
(e) [H$_3$O$^+$] = 1.0 x 10^{-7} M, [OH$^-$] = 1.0 x 10^{-7} M

d 20. Calculate the concentrations of H$_3$O$^+$ and OH$^-$ ions in a 0.050 M Ba(OH)$_2$ solution.

(a) [H$_3$O$^+$] = 0.050 M, [OH$^-$] = 0.050 M
(b) [H$_3$O$^+$] = 1.0 x 10^{-7} M, [OH$^-$] = 0.10 M
(c) [H$_3$O$^+$] = 2.0 x 10^{-13} M, [OH$^-$] = 0.050 M
(d) [H$_3$O$^+$] = 1.0 x 10^{-13} M, [OH$^-$] = 0.10 M
(e) [H$_3$O$^+$] = 0.10 M, [OH$^-$] = 1.0 x 10^{-13} M

c 21. At 60°C, K$_w$ = 9.6 x 10^{-14}. What are the concentrations of the H$_3$O$^+$ and OH$^-$ ions in pure water that is neutral at 60°C?

(a) [H$_3$O$^+$] = [OH$^-$] = 4.8 x 10^{-14} (b) [H$_3$O$^+$] = [OH$^-$] = 4.8 x 10^{-7}
(c) [H$_3$O$^+$] = [OH$^-$] = 3.1 x 10^{-7} (d) [H$_3$O$^+$] = [OH$^-$] = 1.0 x 10^{-7}
(e) [H$_3$O$^+$] = 1.0 x 10^{-7}, [OH$^-$] = 9.6 x 10^{-7}

The pH and pOH Scales

e 22. A solution having a pH of 1.4 would be described as _____.

(a) distinctly basic　　(b) slightly basic　　(c) neutral
(d) slightly acidic　　(e) distinctly acidic

b 23. A solution in which the pH is 8.5 would be described as

(a) very basic　　(b) slightly basic　　(c) neutral
(d) slightly acidic　　(e) very acidic

a 24. A solution in which the pH is 12.5 would be described as _____.

(a) distinctly basic　　(b) slightly basic　　(c) neutral
(d) slightly acidic　　(e) distinctly acidic

c 25. Calculate the pH of a solution that has the H_3O^+ concentration of 0.50 M.

(a) −0.30　　(b) 13.70　　(c) 0.30　　(d) 7.30　　(e) 0.50

e 26. Calculate the pOH of a solution that has the OH^- concentration of 0.50 M.

(a) 0.50　　(b) 14.30　　(c) 6.70　　(d) 13.70　　(e) 0.30

d 27. Calculate the pH of a solution in which $[OH^-] = 2.50 \times 10^{-4}$ M.

(a) 0.40　　(b) 3.60　　(c) −3.60　　(d) 10.40　　(e) 13.60

d 28. What is the pOH of a solution in which $[H_3O^+] = 3.60 \times 10^{-10}$ M?

(a) 8.56　　(b) 5.44　　(c) 9.44　　(d) 4.56　　(e) 4.32

a 29. What is the concentration of H_3O^+ ions in a solution in which pH = 4.32?

(a) 4.8×10^{-5} M　　(b) 6.2×10^{-4} M　　(c) 5.1×10^{-4} M
(d) 8.6×10^{-5} M　　(e) 3.5×10^{-4} M

d 30. The pH of a solution is 4.80. What is the concentration of hydroxide ions in this solution?

(a) 4.2×10^{-9} M　　(b) 1.6×10^{-5} M　　(c) 3.6×10^{-12} M
(d) 6.3×10^{-10} M　　(e) 2.0×10^{-8} M

b 31. What is the hydrogen ion concentration in a solution having a pOH of 3.62?

(a) 2.6×10^{-11} M　　(b) 4.2×10^{-11} M　　(c) 3.8×10^{-4} M
(d) 3.8×10^{-4} M　　(e) 5.1×10^{-10} M

c 32. What is the pH of 0.250 M $HClO_4$ solution?

(a) 1.40　　(b) 2.05　　(c) 0.60　　(d) 4.12　　(e) 1.67

a 33. What is the pH of a 0.400 M HNO$_3$ solution?

(a) 0.40 (b) 2.05 (c) 0.60 (d) 4.12 (e) 1.67

d 34. Calculate the pH of 0.075 M KOH.

(a) 10.40 (b) 11.12 (c) 11.46 (d) 12.88 (e) 13.26

e 35. Calculate the pH of a solution that is 0.00030 M in Ba(OH)$_2$.

(a) 10.47 (b) 9.94 (c) 8.63 (d) 12.42 (e) 10.78

b 36. What is the pH of 400. mL of solution containing 0.0112 gram of HNO$_3$?

(a) 4.15 (b) 3.35 (c) 10.65 (d) 3.75 (e) 2.95

d 37. What is the pH of 500. mL of solution containing 0.0124 gram of Ca(OH)$_2$?

(a) 2.96 (b) 3.17 (c) 9.68 (d) 10.83 (e) 11.04

c 38. The pH of a solution of hydrochloric acid is 2.80. What is the molarity of the acid?

(a) 6.3 x 10^{-3} M (b) 4.2 x 10^{-3} M (c) 1.6 x 10^{-3} M
(d) 6.3 x 10^{-2} M (e) 4.2 x 10^{-2} M

a 39. The pH of a solution of Ba(OH)$_2$ is 9.40. What is the molarity of this solution of base?

(a) 1.3 x 10^{-5} M (b) 1.8 x 10^{-5} M (c) 6.0 x 10^{-4} M
(d) 8.3 x 10^{-4} M (e) 2.5 x 10^{-5} M

a 40. If enough base is added to a solution to cause the pH to increase from 7.5 to 8.5, the _____.

(a) [OH$^-$] increases by a factor of 10 (b) [H$_3$O$^+$] increases by a factor of 10
(c) [OH$^-$] increases by 1 M (d) [H$_3$O$^+$] increases by 1 M
(e) [OH$^-$] increases by a factor of 8.5/7.5

b 41. We add enough acid to a solution to cause the pH to decrease from 6.5 to 5.5. This means that

(a) [OH$^-$] increases by a factor of 10. (b) [H$_3$O$^+$] increases by a factor of 10.
(c) [OH$^-$] increases by 1 M. (d) [H$_3$O$^+$] increases by 1 M.
(e) [H$_3$O$^+$] increases by a factor of 6.5/5.5.

Ionization Constants for Weak Monoprotic Acids and Bases

e 42. For a given weak acid, HA, the value of K_a _____.

(a) will change with pH (b) cannot be less than 10^{-7} (c) cannot be greater than 10^{-7}
(d) does not change with temperature (e) is calculated from experimental data

c 43. Which response includes all of the following that are **weak** acids, and no strong acids?

  I. H_2SO_3 II. $HClO$ III. H_2SO_4 IV. $HClO_3$ V. $HClO_4$ VI. C_6H_5COOH

  (a) I, II, and III   (b) II, IV, V, and VI   (c) I, II, and VI
  (d) IV, V, and VI   (e) I and III

d 44. In a solution containing only a weak monoprotic acid HA, $[H_3O^+]$ is _____ $[A^-]$; if the solution is not very dilute, the concentration of nonionized HA is approximately equal to the _____ of the solution.

  (a) greater than, molarity   (b) less than, molarity   (c) equal to, pH
  (d) equal to, molarity   (e) less than, pH

e 45. Consider calculations of $[H_3O^+]$ in each of the following solutions. Do not go through the calculations. For which calculation is it **not** reasonable to assume that "x" is much less than the initial concentration? The x represents concentration ionized.

  (a) 0.20 M H_2O_2 $K_a = 2.4 \times 10^{-12}$
  (b) 0.010 M HCN $K_a = 4.0 \times 10^{-10}$
  (c) 0.010 M H_2S $K_{a1} = 1.0 \times 10^{-7}$, $K_{a2} = 1.0 \times 10^{-19}$
  (d) 1.00 M NH_3 $K_b = 1.8 \times 10^{-5}$
  (e) 0.010 M $(COOH)_2$ $K_{a1} = 5.9 \times 10^{-2}$, $K_{a2} = 6.4 \times 10^{-5}$

c 46. We make a 1.0 M solution of an unknown acid, HX. With a pH meter, we determine that the pH of the solution is 4.00. Which of the following statements about the HX is true?

  (a) HX is a strong acid.
  (b) HX is a weak acid with a K_a value of about 10^{-4}.
  (c) HX is a weak acid with a K_a value of about 10^{-8}.
  (d) HX is a weak acid with a K_a value of about 10^{-10}.
  (e) HX would probably be a good acid-base indicator.

b 47. Dichloroacetic acid is a weak monoprotic acid. A 0.100 M solution of this acid has $[H_3O^+] = 0.0070\ M$. What is the value of K_a for $Cl_2HCCOOH$? The reaction is
  $$Cl_2HCCOOH + H_2O \rightleftharpoons H_3O^+ + Cl_2HCCOO^-$$

  (a) 1.8×10^{-3} (b) 5.3×10^{-4} (c) 7.5×10^{-3}
  (d) 1.9×10^3 (e) 11.6

d 48. The $[H_3O^+] = 2.0 \times 10^{-4}\ M$ for a 0.020 M solution of a weak acid. Calculate the pK_a for this acid.

  (a) 1.70   (b) 3.70   (c) 2.00   (d) 5.70   (e) 4.69

e 49. A 0.00100 M solution of a weak acid, HX, is 3.0% ionized. Calculate the ionization constant for the acid.

  (a) 4.8×10^{-4} (b) 5.5×10^{-10} (c) 3.0×10^{-2}
  (d) 8.2×10^{-6} (e) 9.3×10^{-7}

e 50. Calculate the ionization constant for a weak acid, HA, that is 1.30% ionized in 0.100 M solution.

(a) 1.3×10^{-3}
(b) 9.9×10^{-2}
(c) 1.6×10^{-6}
(d) 1.8×10^{-5}
(e) 1.7×10^{-5}

a 51. Calculate the pK_a for a weak acid, HA, that is 2.3% ionized in 0.080 M solution?

(a) 4.37
(b) 4.71
(c) 1.66
(d) 2.33
(e) 3.09

c 52. The pH of a 0.10 M solution of a monoprotic acid is 2.85. What is the value of the ionization constant of the acid?

(a) 6.3×10^{-5}
(b) 3.8×10^{-6}
(c) 2.0×10^{-5}
(d) 4.0×10^{-8}
(e) 7.2×10^{-6}

c 53. The pH of a 0.20 M solution of a weak monoprotic acid is 3.70. What is the value of the ionization constant for the acid?

(a) 7.0×10^{-4}
(b) 4.0×10^{-6}
(c) 2.0×10^{-7}
(d) 1.8×10^{-5}
(e) 6.1×10^{-5}

c 54. The ionization constant for the hypothetical weak acid, HA, is 1.0×10^{-5}. What is the equilibrium concentration of $[H_3O^+]$ in 0.20 M HA solution?

(a) 4.3×10^{-3} M
(b) 8.1×10^{-4} M
(c) 1.4×10^{-3} M
(d) 1.0×10^{-5} M
(e) 5.0×10^{-4} M

a 55. Calculate the $[H_3O^+]$ in 0.010 M HOCl solution. $K_a = 3.5 \times 10^{-8}$

(a) 1.9×10^{-5} M
(b) 3.6×10^{-5} M
(c) 5.8×10^{-5} M
(d) 4.0×10^{-6} M
(e) 7.2×10^{-6} M

a 56. Calculate the value of $[H_3O^+]$ in a 0.010 M HOBr solution. $K_a = 2.5 \times 10^{-9}$

(a) 5.0×10^{-6} M
(b) 5.0×10^{-5} M
(c) 2.5×10^{-7} M
(d) 2.5×10^{-11} M
(e) 5.0×10^{-7} M

b 57. Calculate the pH of 0.10 M HCN solution. $K_a = 4.0 \times 10^{-10}$

(a) 6.75
(b) 5.20
(c) 8.42
(d) 9.52
(e) 10.4

d 58. Calculate the pH of 0.020 M HOBr solution. $K_a = 2.5 \times 10^{-9}$

(a) 4.80
(b) 4.91
(c) 5.02
(d) 5.15
(e) 5.38

a 59. What is the $[OCl^-]$ in 0.10 M hypochlorous acid, HOCl? $K_a = 3.5 \times 10^{-8}$

(a) 5.9×10^{-5} M
(b) 8.4×10^{-4} M
(c) 6.1×10^{-4} M
(d) 4.2×10^{-6} M
(e) 3.6×10^{-7} M

a 60. What is the value of [OH⁻] in a 0.015 M CH₃COOH solution? $K_a = 1.8 \times 10^{-5}$

(a) 1.9×10^{-11} M (b) 2.0×10^{-6} M (c) 1.0×10^{-9} M
(d) 5.0×10^{-8} M (e) 5.0×10^{-7} M

c 61. The pH of a 0.100 M solution of a weak acid, HA, is 3.50. Calculate the percent ionization of the acid in 0.100 M solution.

(a) 0.016% (b) 0.078% (c) 0.32% (d) 0.68% (e) 1.6%

c 62. What is the percent ionization of 0.20 M HNO₂? $K_a = 4.5 \times 10^{-4}$

(a) 1.0% (b) 2.8% (c) 4.6% (d) 5.3% (e) 5.9%

b 63. Calculate the original molarity of a solution of acetic acid that is 3.0% ionized. $K_a = 1.8 \times 10^{-5}$

(a) 0.076 M (b) 1.9×10^{-2} M (c) 0.038 M
(d) 3.4×10^{-4} M (e) 6.7×10^{-4} M

e*64. How many moles of nitrous acid, HNO₂, are required initially to prepare 2.5 liters of a solution of pH = 3.00? $K_a = 4.5 \times 10^{-4}$

(a) 1.8×10^{-4} mol (b) 6.2×10^{-2} mol (c) 1.7×10^{-4} mol
(d) 3.6×10^{-4} mol (e) 8.0×10^{-3} mol

e 65. Assume that five weak acids, identified only by numbers (I, II, III, IV, and V), have the following ionization constants.

Acid	Ionization Constant (K_a value)
I	1.0×10^{-3}
II	3.0×10^{-5}
III	2.6×10^{-7}
IV	4.0×10^{-9}
V	7.3×10^{-11}

A 0.10 M solution of which acid would have the highest pH?

(a) I (b) II (c) III (d) IV (e) V

e 66. Assume that five weak acids, identified only by numbers (I, II, III, IV, and V), have the following ionization constants. Which acid has the largest pK_a value?

Acid	Ionization Constant (K_a value)
I	1.0×10^{-3}
II	3.0×10^{-5}
III	2.6×10^{-7}
IV	4.0×10^{-9}
V	7.3×10^{-11}

(a) I (b) II (c) III (d) IV (e) V

Chapter 18   285

a 67. Assume that five weak acids, identified only by numbers (I, II, III, IV, and V), have the following ionization constants.

Acid	Ionization Constant (K_a value)
I	1.0×10^{-3}
II	3.0×10^{-5}
III	2.6×10^{-7}
IV	4.0×10^{-9}
V	7.3×10^{-11}

The anion of which acid is the weakest base?

(a) I   (b) II   (c) III   (d) IV   (e) V

e 68. The $[OH^-] = 1.3 \times 10^{-6}\ M$ for a 0.025 M solution of a weak base. Calculate the value of K_b for this weak base.

(a) 5.2×10^{-5} (b) 3.1×10^{-7} (c) 7.7×10^{-9}
(d) 4.0×10^{-8} (e) 6.8×10^{-11}

b 69. The pH of a 0.12 M solution of a weak base is 10.30. What is the value of pK_b for this weak base?

(a) 3.70   (b) 6.48   (c) 10.30   (d) 5.44   (e) 4.49

c 70. Calculate the pH of 0.10 M solution of aqueous ammonia. $K_b = 1.8 \times 10^{-5}$

(a) 9.36   (b) 10.89   (c) 11.11   (d) 12.00   (e) 2.89

b 71. Calculate the value of $[H_3O^+]$ in a 0.25 M solution of aqueous ammonia. $K_b = 1.8 \times 10^{-5}$

(a) $2.1 \times 10^{-3}\ M$ (b) $4.7 \times 10^{-12}\ M$ (c) $2.3 \times 10^{-9}\ M$
(d) $4.3 \times 10^{-10}\ M$ (e) $2.4 \times 10^{-11}\ M$

d 72. Trimethylamine ionizes as follows in water. What concentration of trimethylammonium ion, $(CH_3)_3NH^+$, is present in $9.0 \times 10^{-2}\ M\ (CH_3)_3N$? $K_b = 7.4 \times 10^{-5}$
$$(CH_3)_3N + H_2O \rightleftharpoons (CH_3)_3NH^+ + OH^-$$

(a) $1.6 \times 10^{-4}\ M$ (b) $5.2 \times 10^{-3}\ M$ (c) $3.8 \times 10^{-4}\ M$
(d) $2.6 \times 10^{-3}\ M$ (e) $2.7 \times 10^{-5}\ M$

a 73. What is the percent ionization for a $1.0 \times 10^{-3}\ M$ solution of pyridine? $K_b = 1.5 \times 10^{-9}$

(a) 0.12%   (b) 1.6%   (c) 2.8%   (d) 0.045%   (e) 0.67%

a 74. Trimethylamine ionizes as follows in water. What is the percent ionization for a $9.0 \times 10^{-2}\ M$ solution of $(CH_3)_3N$? $K_b = 7.4 \times 10^{-5}$
$$(CH_3)_3N + H_2O \rightleftharpoons (CH_3)_3NH^+ + OH^-$$

(a) 2.9%   (b) 0.030%   (c) 0.18%   (d) 0.42%   (e) 5.8%

Harcourt, Inc

Polyprotic Acids

a 75. Which answer includes all of the following **true** statements about 0.10 M H$_3$PO$_4$ solution?

    I.   The species present in highest concentration is nonionized H$_3$PO$_4$.
    II.   The species present in highest concentration is H$_2$PO$_4^{2-}$ ion.
    III.   The species present in lowest concentration is H$_2$PO$_4^{2-}$ ion.
    IV. $K_{a2} > K_{a3}$

(a) I and IV   (b) II and IV   (c) I and III
(d) II and III   (e) none of these

a 76. Consider a solution that is 0.10 M in a weak **triprotic** acid which is represented by the general formula H$_3$A with the following ionization constants.

For H$_3$A: $K_{a1} = 1.0 \times 10^{-3}$, $K_{a2} = 1.0 \times 10^{-8}$, $K_{a3} = 1.0 \times 10^{-12}$

What is the pH of the solution?

(a) 2.02   (b) 2.15   (c) 2.25   (d) 2.35   (e) 2.54

d 77. Consider a solution that is 0.10 M in a weak **triprotic** acid which is represented by the general formula H$_3$A with the following ionization constants.

For H$_3$A: $K_{a1} = 1.0 \times 10^{-3}$, $K_{a2} = 1.0 \times 10^{-8}$, $K_{a3} = 1.0 \times 10^{-12}$

What is the concentration of HA^{2-}?

(a) 4.5×10^{-3} M   (b) 1.0×10^{-4} M   (c) 3.2×10^{-5} M
(d) 1.0×10^{-8} M   (e) 6.2×10^{-7} M

c 78. Calculate the pH of a 0.10 M solution of a hypothetical triprotic acid H$_3$A, with $K_{a1} = 6.0 \times 10^{-3}$, $K_{a2} = 2.0 \times 10^{-8}$, and $K_{a3} = 1.0 \times 10^{-14}$.

(a) 1.48   (b) 1.61   (c) 1.66   (d) 1.84   (e) 2.06

a 79. What is the concentration of A^{3-} ions at equilibrium for a 0.10 M solution of a hypothetical triprotic acid H$_3$A, with $K_{a1} = 6.0 \times 10^{-3}$, $K_{a2} = 2.0 \times 10^{-8}$, and $K_{a3} = 1.0 \times 10^{-14}$?

(a) 9.1×10^{-21} M   (b) 6.2×10^{-18} M   (c) 3.1×10^{-15} M
(d) 1.0×10^{-14} M   (e) 4.8×10^{-19} M

e 80. Suppose that a sample of pure water is saturated with gaseous CO$_2$ to form a solution of carbonic acid. Which response has the following species arranged in the order of decreasing concentrations at equilibrium (highest concentration first, lowest concentration last)?

(a) H$_3$O$^+$>H$_2$CO$_3$>HCO$_3^-$>CO$_3^{2-}$ (b) H$_2$CO$_3$>HCO$_3^-$>H$_3$O$^+$>CO$_3^{2-}$
(c) CO$_3^{2-}$>H$_3$O$^+$>HCO$_3^-$>H$_2$CO$_3$ (d) HCO$_3^-$>H$_2$CO$_3$>CO$_3^{2-}$>H$_3$O$^+$
(e) H$_2$CO$_3$>H$_3$O$^+$>HCO$_3^-$>CO$_3^{2-}$

Harcourt, Inc

e 81. The hypothetical weak acid H_2A ionizes as shown below. Calculate the $[HA^-]$ in 0.20 M H_2A.

$$H_2A \rightleftharpoons H^+ + HA^- \quad K_{a1} = 1.0 \times 10^{-7}$$
$$HA^- \rightleftharpoons H^+ + A^{2-} \quad K_{a2} = 5.0 \times 10^{-11}$$

(a) $6.3 \times 10^{-5} M$ (b) $1.0 \times 10^{-7} M$ (c) $3.0 \times 10^{-4} M$
(d) $2.2 \times 10^{-6} M$ (e) $1.4 \times 10^{-4} M$

c 82. The hypothetical weak acid H_2A ionizes as shown below. Calculate the $[A^{2-}]$ in 0.20 M H_2A.

$$H_2A \rightleftharpoons H^+ + HA^- \quad K_{a1} = 1.0 \times 10^{-7}$$
$$HA^- \rightleftharpoons H^+ + A^{2-} \quad K_{a2} = 5.0 \times 10^{-11}$$

(a) $1.0 \times 10^{-7} M$ (b) $5.8 \times 10^{-14} M$ (c) $5.0 \times 10^{-11} M$
(d) $4.6 \times 10^{-13} M$ (e) $3.8 \times 10^{-18} M$

c 83. Calculate the pH of $3.2 \times 10^{-3} M$ H_2CO_3 solution.

(a) 4.09 (b) 4.30 (c) 4.44 (d) 4.94 (e) 5.56

c 84. What is the $[OH^-]$ in 0.20 M oxalic acid, $(COOH)_2$, solution?

(a) $4.1 \times 10^{-10} M$ (b) $7.4 \times 10^{-11} M$ (c) $1.2 \times 10^{-13} M$
(d) $3.2 \times 10^{-12} M$ (e) $3.8 \times 10^{-12} M$

c 85. Calculate the $[SO_4^{2-}]$ in 0.20 M H_2SO_4.

(a) $4.3 \times 10^{-2} M$ (b) $8.4 \times 10^{-2} M$ (c) $1.1 \times 10^{-2} M$
(d) $6.4 \times 10^{-1} M$ (e) $2.5 \times 10^{-1} M$

Solvolysis

d 86. Which of the following statements is **false**?

(a) Solvolysis is the reaction of a substance with the solvent in which it is dissolved.
(b) Hydrolysis is the reaction of a substance with water.
(c) The anion of a weak acid reacts with water to form nonionized acid and OH^- ions.
(d) According to Brønsted-Lowry the anions of strong acids are very strong bases.
(e) The conjugate acid of a strong base is a very weak acid.

d 87. Which of the following anions is the strongest base?

(a) Cl^- (b) NO_3^- (c) ClO_4^- (d) F^- (e) Br^-

Salts of Strong Bases and Strong Acids

d 88. Which one of the following salts produces neutral solutions when it is dissolved in water?

(a) NaCN (b) NaOCl (c) NaF (d) NaBr (e) $NaCH_3COO$

c 89. Which one of the following salts produces neutral solutions when it is dissolved in water?

(a) NH₄F (b) LiOCl (c) BaBr₂ (d) CaSO₃ (e) (NH₄)₂SO₄

e 90. Which response includes all of the following salts that give neutral aqueous solutions, and no other salts?

I. KNO₃   II. BaCl₂   III. NaCH₃COO   IV. NH₄Cl   V. NaBr

(a) I and II
(d) IV and V
(b) I, II, III, and V
(e) I, II, and V
(c) III and IV

d 91. Which response includes all the following salts that hydrolyze in dilute aqueous solution, and no other salts?

I. KCN   II. NaF   III. Na₂CO₃   IV. KCl   V. NH₄Br

(a) I, III, and IV
(d) I, II, III, and V
(b) II, III, and V
(e) I, IV, and V
(c) II, IV, and V

Salts of Strong Bases and Weak Acids

e 92. The value of K_a for benzoic acid, C_6H_5COOH, is 6.3 x 10⁻⁵. When sodium benzoate is dissolved in water, the reaction that occurs is _____, resulting in a solution that is _____.

(a) Na⁺ + C₆H₅COO⁻ ⇌ NaC₆H₅COO; neutral
(b) C₆H₅COO⁻ + H₂O ⇌ H₃O⁺ + C₆H₄COO²⁻; acidic
(c) C₆H₅COOH + H₂O ⇌ H₃O⁺ + C₆H₅COO⁻; acidic
(d) No reaction; neutral
(e) C₆H₅COO⁻ + H₂O ⇌ CH₆H₅COOH + OH⁻; basic

d 93. When solid NaCN is added to water, the pH _____.

(a) remains at 7
(b) becomes greater than 7 because of hydrolysis of Na⁺
(c) becomes less than 7 because of hydrolysis of Na⁺
(d) becomes greater than 7 because of hydrolysis of CN⁻
(e) becomes less than 7 because of hydrolysis of CN⁻

b 94. Which one of the following salts produces basic solutions when it is dissolved in water?

(a) NaNO₃
(d) NH₄I
(b) NH₄OCl
(e) KCl
(c) NH₄Br

c 95. Which one of the following salts produces basic solutions when it is dissolved in water?

(a) NH₄NO₃ (b) NaBr (c) NaF (d) NaNO₃ (e) NaI

b 96. Which response includes all the following salts whose aqueous solutions are basic, and no others?

I. KI   II. KBr   III. KNO$_3$ IV. KCN   V. KOCl

(a) I and V   (b) IV and V   (c) II, III, and IV
(d) III, IV, and V   (e) II, III, IV, and V

c 97. Which response includes all of the following salts that give basic aqueous solutions, and no other salts?

I. KCl   II. BaBr$_2$ III. KCH$_3$COO   IV. NH$_4$NO$_3$ V. NaF

(a) I, II, and V   (b) III and IV   (c) III and V
(d) IV and V   (e) III, IV, and V

e*98. A 0.10 M solution of which of the following salts is most basic?

(a) NaF   (b) NaNO$_2$ (c) NaCH$_3$COO   (d) NaOCl   (e) NaCN

d*99. A 0.10 M solution of which of the following would be most basic?

(a) RbI   (b) NH$_4$NO$_3$ (c) KCH$_3$COO   (d) NaCN   (e) BaCl$_2$

a*100. In 0.10 M solution, which of the following salts hydrolyzes to the least extent?

(a) NaF   (b) NaNO$_2$ (c) NaCH$_3$COO   (d) NaOCl   (e) NaCN

d 101. Calculate the (base) hydrolysis constant for the hypochlorite ion, OCl$^-$.

(a) 3.5×10^{-8} (b) 1.8×10^{-5} (c) 5.6×10^{-10}
(d) 2.9×10^{-7} (e) 3.5×10^{-6}

c 102. Evaluate the (base) hydrolysis constant for sodium cyanate, NaOCN.

(a) 3.7×10^{-12} (b) 6.4×10^{-9} (c) 2.9×10^{-11}
(d) 4.0×10^{-10} (e) 6.8×10^{-10}

b 103. What is the value of the (base) hydrolysis constant for NaBO$_2$, sodium metaborate?

(a) 3.2×10^{-6} (b) 1.7×10^{-5} (c) 1.8×10^{-6}
(d) 6.0×10^{4} (e) 2.2×10^{-7}

b*104. Calculate the acid ionization constant of an unknown monoprotic weak acid, HA, if its salt, NaA, has a (base) hydrolysis constant of 6.2×10^{-9}.

(a) 6.0×10^{-7} (b) 1.6×10^{-6} (c) 4.5×10^{-7}
(d) 5.6×10^{-8} (e) 4.3×10^{-6}

b*105. The (base) hydrolysis constant for the anion, A⁻, of a weak acid, HA, is 2.9×10^{-7}. What is the ionization constant for HA?

   (a) 9.1×10^{-20} (b) 3.4×10^{-8} (c) 1.1×10^{-2}
   (d) 7.1×10^{-6} (e) 1.4×10^{-4}

e 106. Calculate the pH of 0.15 M NaCN solution.

   (a) 2.72   (b) 7.00   (c) 8.08   (d) 10.23   (e) 11.29

c 107. Calculate the pH of 0.14 M NaF solution.

   (a) 8.09   (b) 8.12   (c) 8.14   (d) 8.18   (e) 8.21

d 108. What is the pH of 0.25 M KCN solution?

   (a) 1.17   (b) 2.60   (c) 10.42   (d) 11.40   (e) 12.83

d 109. Calculate the pH of 0.10 M solution of $NaBO_2$.

   (a) 9.84   (b) 12.89   (c) 10.48   (d) 11.11   (e) 2.89

d 110. Calculate the pH of 0.050 M $Ba(CN)_2$ solution. $Ba(CN)_2$ is a soluble ionic compound.

   (a) 2.80   (b) 2.96   (c) 11.04   (d) 11.20   (e) 12.40

e 111. Calculate the percent hydrolysis of the hypochlorite ion in 0.10 M NaOCl solution.

   (a) 0.0012%   (b) 0.024%   (c) 0.056%   (d) 0.10%   (e) 0.17%

a 112. What is the percent hydrolysis in 0.075 M sodium acetate, $NaCH_3COO$, solution?

   (a) 0.0087%   (b) 0.012%   (c) 0.0064%   (d) 0.0038%   (e) 0.043%

c 113. What is the percent hydrolysis in 0.20 M $NaNO_2$ solution?

   (a) 0.0064%   (b) 0.0047%   (c) 0.0010%   (d) 0.00061%   (e) 0.00092%

c 114. Calculate the [OH⁻] in 0.050 M potassium fluoride, KF.

   (a) 4.7×10^{-7} M   (b) 6.2×10^{-7} M   (c) 8.3×10^{-7} M
   (d) 1.4×10^{-6} M   (e) 2.2×10^{-6} M

e 115. Calculate the [OH⁻] in 0.20 M $NaNO_2$.

   (a) 4.8×10^{-9} M   (b) 1.4×10^{-8} M   (c) 1.2×10^{-7} M
   (d) 6.5×10^{-7} M   (e) 2.1×10^{-6} M

d 116. What is the [OH⁻] in 0.20 M sodium cyanate, NaOCN, solution?

   (a) 3.7×10^{-7} M   (b) 4.6×10^{-7} M   (c) 5.5×10^{-7} M
   (d) 2.4×10^{-6} M   (e) 8.7×10^{-7} M

e 117. What is the [H$_3$O$^+$] in 0.40 M NaCN solution?

(a) 6.4 x 10^{-10} M (b) 3.3 x 10^{-11} M (c) 1.4 x 10^{-2} M
(d) 4.8 x 10^{-3} M (e) 3.2 x 10^{-12} M

b 118. Calculate the [H$_3$O$^+$] in 0.030 M potassium fluoride, KF.

(a) 2.5 x 10^{-8} M (b) 1.5 x 10^{-8} M (c) 8.3 x 10^{-9} M
(d) 6.8 x 10^{-4} M (e) 5.5 x 10^{-9} M

d 119. Calculate the [H$_3$O$^+$] in a 0.20 M KCN solution.

(a) 2.2 x 10^{-3} M (b) 1.0 x 10^{-7} M (c) 2.4 x 10^{-11} M
(d) 4.5 x 10^{-12} M (e) 3.0 x 10^{-12} M

Salts of Weak Bases and Strong Acids

e 120. Which one of the following salts produces acidic solutions when it is dissolved in water?

(a) KCH$_3$COO (b) KF (c) KOCl (d) KBr (e) NH$_4$NO$_3$

c 121. Which one of the following salts produces acidic aqueous solutions?

(a) sodium chloride (b) sodium acetate (c) ammonium chloride
(d) calcium nitrate (e) rubidium perchlorate

e 122. Which response includes all of the following salts that give acidic aqueous solutions, and no other salts?

I. KCl II. BaBr$_2$ III. KCH$_3$COO IV. NH$_4$NO$_3$ V. NaF

(a) I, II, and V (b) III and IV (c) III and V
(d) IV and V (e) IV

e 123. Which response includes all the following salts whose aqueous solutions are acidic, and no others?

I. NH$_4$Cl II. NH$_4$NO$_3$ III. NH$_4$Br IV. NH$_4$I V. NH$_4$NO$_2$

(a) I and V (b) I, III, and V (c) II, III, and IV
(d) I, II, III, and IV (e) I, II, III, IV, and V

d 124. The reaction that occurs when NH$_4$Br dissolves in water is _____.

(a) NH$_4^+$ + OH$^-$ ⇌ NH$_4$OH
(b) Br$^-$ + H$_3$O$^+$ ⇌ HBr + H$_2$O
(c) NH$_3$ + H$_2$O ⇌ NH$_4^+$ + OH$^-$
(d) NH$_4^+$ + H$_2$O ⇌ NH$_3$ + H$_3$O$^+$
(e) Br$^-$ + H$_2$O ⇌ HBr + OH$^-$

b 125. The value of K_b for methylamine, $(CH_3)NH_2$, is 5.0 x 10^{-4}. When methylammonium chloride is dissolved in water, the reaction that occurs is _____, resulting in a solution that is _____.

(a) $(CH_3)NH_2 + H_2O \rightleftharpoons (CH_3)NH_3^+ + OH^-$; basic
(b) $(CH_3)NH_3^+ + H_2O \rightleftharpoons (CH_3)NH_2 + H_3O^+$; acidic
(c) $(CH_3)NH_3^+ + OH^- \rightleftharpoons (CH_3)NH_2 + H_2O$; acidic
(d) $(CH_3)NH_3^+ + OH^- \rightleftharpoons (CH_3)NH_2 + H_2O$; basic
(e) $(CH_3)NH_2 + H_3O^+ \rightleftharpoons (CH_3)NH_3^+ + H_2O$; acidic

c 126. Consider solutions of the five indicated salts dissolved in water. Which one could not possibly have the pH designated?

	Salt Solution	pH
(a)	NaCl	7.00
(b)	NaF	8.16
(c)	NH_4Cl	7.64
(d)	KCN	9.48
(e)	NH_4NO_3	5.90

c 127. The hydrolysis constant that would be used to calculate the pH in NH_4Cl solution is given by

(a) K_w/K_a (HCl)   (b) K_w/K_a (Cl$^-$)   (c) K_w/K_b (NH_3)
(d) K_w/K_b (NH_4^+)   (e) K_w/K_b (NH_4Cl)

e*128. Given the following hydrolysis constants, which of the salts has the weakest parent **acid**?

	salt	K_a or K_b for salt
(a)	$NaClO_2$	3.3 x 10^{-7}
(b)	NH_4NO_3	5.6 x 10^{-10}
(c)	KCH_3COO	5.6 x 10^{-10}
(d)	C_6H_5NHBr	5.0 x 10^{-6}
(e)	NaCN	2.5 x 10^{-5}

d 129. Calculate the hydrolysis constant for the ammonium ion, NH_4^+.

(a) 2.5 x 10^{-5} (b) 1.0 x 10^{-7} (c) 4.0 x 10^{-10}
(d) 5.6 x 10^{-10} (e) 5.5 x 10^{-4}

e 130. Calculate the pH of 0.030 M NH_4Cl.

(a) 4.78   (b) 4.90   (c) 5.12   (d) 5.28   (e) 5.39

e 131. Calculate the pH of 0.050 M NH_4Cl.

(a) 4.62   (b) 4.84   (c) 5.04   (d) 5.16   (e) 5.28

a 132. Calculate the pH of 0.15 M NH$_4$NO$_3$ solution.

    (a) 5.04   (b) 5.20   (c) 5.36   (d) 8.80   (e) 8.96

b 133. Calculate the pH of 0.020 M NH$_4$Br.

    (a) 5.56   (b) 5.48   (c) 5.21   (d) 5.15   (e) 5.06

e 134. What is the pH of 0.15 M solution of trimethylammonium nitrate, (CH$_3$)$_3$NHNO$_3$, a salt?

    (a) 8.66   (b) 9.20   (c) 8.88   (d) 5.12   (e) 5.34

d 135. What is the [H$_3$O$^+$] in 0.060 M NH$_4$Cl?

    (a) 8.7 x 10^{-6} M   (b) 7.6 x 10^{-6} M   (c) 6.6 x 10^{-6} M
    (d) 5.8 x 10^{-6} M   (e) 4.5 x 10^{-6} M

b 136. Calculate the [H$_3$O$^+$] in a 0.10 M solution of NH$_4$NO$_3$.

    (a) 1.7 x 10^{-5} M   (b) 7.5 x 10^{-6} M   (c) 2.2 x 10^{-6} M
    (d) 5.8 x 10^{-7} M   (e) 1.3 x 10^{-9} M

b 137. Calculate the [OH$^-$] in a 0.20 M NH$_4$Cl solution.

    (a) 1.3 x 10^{-9} M   (b) 9.5 x 10^{-10} M   (c) 1.0 x 10^{-7} M
    (d) 1.2 x 10^{-6} M   (e) 1.7 x 10^{-4} M

a 138. What is the percent hydrolysis of a 0.15 M solution of (CH$_3$)$_3$NHNO$_3$?

    (a) 0.0031%   (b) 0.0068%   (c) 0.0094%
    (d) 0.011%   (e) 0.022%

Salts of Weak Bases and Weak Acids

c 139. A 0.10 M solution of which of the following salts is neutral?

    (a) NH$_4$F   (b) NH$_4$NO$_2$ (c) NH$_4$CH$_3$COO
    (d) NH$_4$OCl   (e) NH$_4$CN

e*140. A 0.10 M solution of which of the following salts is most basic?

    (a) NH$_4$F   (b) NH$_4$NO$_2$ (c) NH$_4$CH$_3$COO
    (d) NH$_4$OCl   (e) NH$_4$CN

b*141. A 0.10 M solution of which of the following salts would be most basic?

    (a) NH$_4$OBr   (b) NaOBr   (c) KOCl
    (d) NaF   (e) Ca(CH$_3$COO)$_2$

a*142. Aqueous solutions of one of the following is acidic. Which one?

(a) NH$_4$NO$_2$
(b) NH$_4$CH$_3$COO
(c) NH$_4$OCl
(d) NH$_4$OBr
(e) NH$_4$CN

a*143. A 0.10 M solution of which of the following salts is most acidic?

(a) NH$_4$F
(b) NH$_4$NO$_2$
(c) NH$_4$CH$_3$COO
(d) NH$_4$OCl
(e) NH$_4$CN

a 144. Which response contains all the salts whose aqueous solutions are **neutral**, and no other salts?

I. NH$_4$NO$_3$
II. NaCN
III. KCl
IV. NH$_4$Br
V. LiCl
VI. CaCl$_2$
VII. CH$_3$NH$_3$Cl
VIII. KNO$_2$
IX. NH$_4$CH$_3$COO

(a) III, V, VI, and IX
(b) II, III, V, VI, and VII
(c) III, V, and VI
(d) I, IV, VII, and IX
(e) II, III, VI, and VIII

e 145. Which response contains all the salts whose aqueous solutions are **basic**, and no other salts?

I. NH$_4$NO$_3$
II. NaCN
III. KCl
IV. NH$_4$Br
V. LiCl
VI. CaCl$_2$
VII. CH$_3$NH$_3$Cl
VIII. KNO$_2$
IX. NH$_4$CH$_3$COO

(a) II, III, V, VI, and VIII
(b) I, IV, and VII
(c) II, VIII, and IX
(d) I, IV, VII, and IX
(e) II and VIII

b 146. Which response contains all the salts whose aqueous solutions are **acidic**, and no other salts?

I. NH$_4$NO$_3$
II. NaCN
III. KCl
IV. NH$_4$Br
V. LiCl
VI. CaCl$_2$
VII. CH$_3$NH$_3$Cl
VIII. KNO$_2$
IX. NH$_4$CH$_3$COO

(a) II, III, V, VI, and VIII
(b) I, IV, and VII
(c) II, VIII, and IX
(d) I, IV, VII, and IX
(e) II and VIII

a 147. If 1.0 mole of ammonium cyanide, NH$_4$CN, was dissolved in 1.0 liter of water, the pH of the solution would be _____ .

For NH$_3$: $K_b = 1.8 \times 10^{-5}$
For HCN: $K_a = 4.0 \times 10^{-10}$

(a) greater than 7
(b) impossible to predict
(c) equal to 7
(d) less than 7
(e) close to 1

d 148. If 1.0 mole of ammonium nitrite, NH$_4$NO$_2$, was dissolved in 1.0 liter of water, the pH of the solution would be _____ .

For NH$_3$: K_b = 1.8 x 10^{-5} For HNO$_2$: K_a = 4.5 x 10^{-4}

(a) greater than 7 (b) impossible to predict (c) equal to 7
(d) less than 7 (e) close to 14

Salts That Contain Small, Highly Charged Cations

d 149. Which response includes all the following salts that give acidic aqueous solutions, and no other salts?

I. AlCl$_3$ II. Cr(NO$_3$)$_3$ III. FeCl$_3$ IV. CaCl$_2$ V. BiCl$_3$

(a) I and II (b) I, II, III, and IV (c) IV
(d) I, II, III, and V (e) II and IV

a 150. What is the pH of 0.30 M FeCl$_2$ solution? The hydrated ferrous ion is [Fe(OH$_2$)$_6$]$^{2+}$.

(a) 5.02   (b) 4.41   (c) 3.76   (d) 3.22   (e) 2.86

a 151. What is the pH of 0.30 M FeCl$_2$ solution? The hydrated ferrous ion is [Fe(OH$_2$)$_6$]$^{2+}$.
For [Fe(OH$_2$)$_6$]$^{2+}$, K_a = 3.0 x 10^{-10}

(a) 5.02   (b) 4.41   (c) 3.76   (d) 3.22   (e) 2.86

b 152. Calculate the pH of a 0.10 M solution of BeCl$_2$ in water. The hydrated beryllium ion is [Be(OH$_2$)$_4$]$^{2+}$.

(a) 5.50   (b) 3.00   (c) 1.57   (d) 3.74   (e) 1.89

b 153. Calculate the pH of a 0.10 M solution of BeCl$_2$ in water. The hydrated beryllium ion is [Be(OH$_2$)$_4$]$^{2+}$.   For [Be(OH$_2$)$_4$]$^{2+}$, K_a = 1.0 x 10^{-5}

(a) 5.50   (b) 3.00   (c) 1.57   (d) 3.74   (e) 1.89

e 154. What is the pH of 0.025 M CuCl$_2$?

(a) 6.70   (b) 9.20   (c) 7.30   (d) 3.60   (e) 4.80

b 155. What is the pH of 0.10 M CuCl$_2$ solution?   For [Cu(OH$_2$)$_4$]$^{2+}$, K_a = 1.0 x 10^{-8}

(a) 4.40   (b) 4.49   (c) 4.58   (d) 4.63   (e) 4.68

d 156. Calculate the pH of 0.15 M Co(NO$_3$)$_2$.

(a) 4.52   (b) 4.74   (c) 4.88   (d) 5.06   (e) 5.28

d 157. Calculate the pH of 0.15 M Co(NO$_3$)$_2$.   For [Co(OH$_2$)$_6$]$^{2+}$, $K_a = 5.0 \times 10^{-10}$

(a) 4.52   (b) 4.74   (c) 4.88   (d) 5.06   (e) 5.28

b 158. Calculate the pH of 0.050 M Fe(NO$_3$)$_3$ solution. The hydrated Fe^{3+} ion is [Fe(OH$_2$)$_6$]$^{3+}$.

(a) 1.76   (b) 1.85   (c) 1.90   (d) 2.02   (e) 2.26

b 159. Calculate the pH of 0.050 M Fe(NO$_3$)$_3$ solution. The hydrated Fe^{3+} ion is [Fe(OH$_2$)$_6$]$^{3+}$. For [Fe(OH$_2$)$_6$]$^{3+}$, $K_a = 4.0 \times 10^{-3}$

(a) 1.76   (b) 1.85   (c) 1.90   (d) 2.02   (e) 2.26

e 160. What is the percent hydrolysis of Fe^{2+} in a 0.30 M FeCl$_2$ solution?

(a) 0.20%   (b) 5.2 x 10^{-4}%   (c) 0.067%
(d) 0.044%   (e) 0.0032%

c 161. What is the percent hydrolysis in a 0.10 M aqueous solution of BeCl$_2$?

(a) 16%   (b) 8.2%   (c) 1.0%   (d) 12%   (e) 27%

19 Ionic Equilibria II: Buffers and Titration Curves

The following equilibrium constants will be useful for some of the problems.

Substance	Constant	Substance	Constant
HCO_2H	$K_a = 1.8 \times 10^{-4}$	H_2CO_3	$K_1 = 4.2 \times 10^{-7}$
HNO_2	$K_a = 4.5 \times 10^{-4}$		$K_2 = 4.8 \times 10^{-11}$
$HOCl$	$K_a = 3.5 \times 10^{-8}$	$(COOH)_2$	$K_1 = 5.9 \times 10^{-2}$
HF	$K_a = 7.2 \times 10^{-4}$		$K_2 = 6.4 \times 10^{-5}$
HCN	$K_a = 4.0 \times 10^{-10}$	CH_3COOH	$K_a = 1.8 \times 10^{-5}$
H_2SO_4	$K_1 = $ very large	$C_6H_5NH_2$	$K_b = 4.2 \times 10^{-10}$
	$K_2 = 1.2 \times 10^{-2}$	NH_3	$K_b = 1.8 \times 10^{-5}$

****

The Common Ion Effect and Buffer Solutions

e 1. What is the $[H_3O^+]$ of a solution that is 0.0100 M in HOCl and 0.0300 M in NaOCl?

(a) $2.14 \times 10^{-7} M$ (b) $1.45 \times 10^{-7} M$ (c) $7.41 \times 10^{-8} M$
(d) $2.29 \times 10^{-8} M$ (e) $1.17 \times 10^{-8} M$

c 2. Calculate the $[H_3O^+]$ of a solution that is 0.20 M in HF and 0.10 M in NaF.

(a) $3.2 \times 10^{-4} M$ (b) $4.0 \times 10^{-6} M$ (c) $1.4 \times 10^{-3} M$
(d) $6.3 \times 10^{-5} M$ (e) $5.0 \times 10^{-3} M$

d 3. Calculate the pH of a solution that is 0.20 M in $NaCH_3COO$ and 0.10 M in CH_3COOH.

(a) 4.63   (b) 4.74   (c) 4.95   (d) 5.05   (e) 5.22

e 4. Calculate the pH of a solution that is 0.10 M in acetic acid and 0.30 M in sodium acetate.

(a) 4.74   (b) 4.87   (c) 4.92   (d) 5.06   (e) 5.22

d 5. Calculate the pH of a solution that is 0.12 M in HOCl and 0.20 M in NaOCl.

(a) 6.76   (b) 6.32   (c) 7.24   (d) 7.68   (e) 7.76

a 6. What is the pH of a solution which is 0.0400 M in formic acid, HCO_2H, and 0.0600 M in sodium formate, NaHCOO?

(a) 3.92   (b) 3.96   (c) 4.00   (d) 9.52   (e) 4.08

d 7. What is the pH of a solution that is 0.20 M in HCN and 0.15 M in KCN?

(a) 7.60   (b) 8.40   (c) 8.80   (d) 9.27   (e) 10.10

b 8. Calculate the pH of 500. mL of 0.100 M acetic acid, CH_3COOH, which also contains 0.100 mole of sodium acetate, $NaCH_3COO$.

(a) 4.77   (b) 5.05   (c) 5.32   (d) 5.68   (e) 6.42

c 9. Calculate the pH of 500. mL of 0.200 M acetic acid, CH_3COOH, to which 0.0750 mol of sodium acetate, $NaCH_3COO$, has been added.

(a) 3.87   (b) 3.94   (c) 4.62   (d) 4.09   (e) 4.16

e*10. Calculate the pH of a solution that is 0.050 M in HCN and 0.025 M in $Ba(CN)_2$.

(a) 7.60   (b) 8.80   (c) 9.10   (d) 10.20   (e) 9.40

e*11. Calculate the pH of a solution which is 0.400 M in HNO_2 and 1.875 M in $Ca(NO_2)_2$.

(a) 3.64   (b) 4.06   (c) 4.87   (d) 5.68   (e) 4.32

b 12. What is the pH of an aqueous solution containing CH_3COOH and $NaCH_3COO$ in a molar ratio of 4.0 to 1.0?

(a) 4.37   (b) 4.14   (c) 9.25   (d) 5.35   (e) 4.74

b 13. A solution that is 0.20 M in NH_3 is also 0.30 M in NH_4Cl. What is the $[OH^-]$ in this solution?

(a) 4.5 x 10^{-6} M   (b) 1.2 x 10^{-5} M   (c) 7.4 x 10^{-5} M
(d) 6.4 x 10^{-4} M   (e) 2.4 x 10^{-7} M

c 14. What is the $[H_3O^+]$ of a solution that is 0.010 M in aqueous NH_3 and 0.030 M in NH_4NO_3?

(a) 7.2 x 10^{-6} M   (b) 6.0 x 10^{-6} M   (c) 1.7 x 10^{-9} M
(d) 1.4 x 10^{-9} M   (e) 1.2 x 10^{-9} M

d 15. What is the pH of a solution that is 0.080 M in aqueous ammonia and 0.040 M in NH_4Cl?

(a) 2.92   (b) 4.44   (c) 7.00   (d) 9.56   (e) 11.08

b 16. What is the pH of a solution that is 0.30 M in aniline, $C_6H_5NH_2$, and 0.15 M in anilinium chloride, $C_6H_5NH_3^+Cl^-$? Aniline ionizes as follows.
$C_6H_5NH_2 + H_2O \rightleftharpoons C_6H_5NH_3^+ + OH^-$

(a) 4.32   (b) 4.92   (c) 5.62   (d) 5.74   (e) 5.95

c 17. What is the pH of a solution containing aqueous ammonia and ammonium chloride in a molar ratio of 5.0 to 3.0?

(a) 9.40   (b) 9.44   (c) 9.48   (d) 9.52   (e) 9.56

Buffering Action

c 18. Which one of the following combinations **cannot** produce a buffer solution?

(a) HNO_2 and $NaNO_2$ (b) HCN and NaCN   (c) $HClO_4$ and $NaClO_4$
(d) NH_3 and $(NH_4)_2SO_4$ (e) NH_3 and NH_4Br

b 19 Which one of the following combinations is **not** a buffer solution?

(a) NH_3 - $(NH_4)_2SO_4$ (b) HBr - KBr   (c) HCN - NaCN
(d) NH_3 - NH_4Br (e) CH_3COOH - $NaCH_3COO$

b 20. Choose the response that includes all the following combinations that **are** buffer solutions and none that are not buffer solutions. All components are present in 0.10 M concentrations.

I.   HCN and NaCN   II. NH_3 and NH_4Cl
III. HNO_3 and NH_4NO_3 IV. H_2SO_4 and Na_2SO_4

(a) I, III, and IV   (b) I and II   (c) II, III, and IV
(d) III and IV   (e) I and III

d 21. Which response contains all the statements **true** of buffer solutions, and no false statements?

I.   A buffer solution could consist of equal concentrations of ammonia and ammonium bromide.
II. A buffer solution could consist of equal concentrations of perchloric acid, $HClO_4$, and sodium perchlorate.
III. A buffer solution will change only slightly in pH upon addition of acid or base.
IV. In a buffer solution containing benzoic acid, C_6H_5COOH, and sodium benzoate, NaC_6H_5COO, the species that reacts with added $[OH^-]$ is the benzoate ion.

(a) II, III, and IV   (b) I and IV   (c) II and III
(d) I and III   (e) another combination

b 22. Consider a solution which is 0.15 M in HF and 0.10 M in KF. Which response contains all the **true** statements, and no others?

I.   If NaOH is added, potassium ion reacts with hydroxide ion.
II. If a small amount of NaOH is added, the pH increases very slightly.
III. If HNO_3 is added, hydrogen ion reacts with fluoride ion.
IV. If more KF is added, the pH decreases.

(a) I, III, and IV   (b) II and III   (c) II and IV
(d) III   (e) another combination

d 23. A solution is initially 0.100 M in HOCl and 0.300 M in NaOCl. What is the pH if 0.030 mol of solid NaOH is added to 1.00 L of this solution? Assume no volume change.

(a) 5.24   (b) 5.38   (c) 8.02   (d) 8.13   (e) 9.06

d 24. What is the pH of a solution that is 0.20 M in HOCl and 0.15 M NaOCl after 0.050 mol HCl/L has been bubbled into the solution?

(a) 7.85   (b) 6.15   (c) 6.95   (d) 7.06   (e) 7.45

e 25. If 100. mL of 0.040 M NaOH solution is added to 100. mL of solution which is 0.10 M in CH_3COOH and 0.10 M in $NaCH_3COO$, what be will the pH of the new solution?

(a) 4.74   (b) 4.81   (c) 4.89   (d) 5.00   (e) 5.11

e 26. If 0.040 moles of solid NaOH is added to 1.0 liter of a solution that is 0.10 M in NH_3 and 0.20 M in NH_4Cl, what will be the pH of the resulting solution? Assume no volume change due to the addition of the NaOH.

(a) 4.80   (b) 8.95   (c) 5.05   (d) 8.65   (e) 9.20

b 27. If 0.050 moles of gaseous HCl is bubbled in 1.0 liter of a solution that is 0.15 M in NH_3 and 0.10 M in NH_4Cl, what will be the pH of the resulting solution?

(a) 7.05   (b) 9.08   (c) 4.57   (d) 9.43   (e) 4.92

d 28. If 100. mL of 0.030 M HCl solution is added to 100. mL of buffer solution which is 0.10 M in NH_3 and 0.10 M in NH_4Cl, what will be the pH of the new solution?

(a) 5.01   (b) 4.48   (c) 7.88   (d) 8.99   (e) 9.52

a 29. If 0.40 g of solid NaOH is added to 1.0 liter of a buffer solution that is 0.10 M in CH_3COOH and 0.10 M in $NaCH_3COO$, how will the pH of the solution change?

(a)   The pH increases from 4.74 to 4.83.
(b)   The pH decreases from 7.00 to 4.83.
(c)   The pH does not change.
(d)   The pH decreases from 4.74 to 4.65.
(e)   The pH increases from 4.74 to 7.00.

Preparation of Buffer Solutions

d 30. Calculate the pH for a buffer solution prepared by mixing 100. mL of 0.10 M HF and 200. mL of 0.10 M KF.

(a) 2.82   (b) 2.96   (c) 3.32   (d) 3.44   (e) 3.53

d 31. If 400. mL of 0.100 M CH_3COOH and 200. mL of 0.100 M $NaCH_3COO$ solutions are mixed, what is the pH of the resulting solution?

(a) 3.09   (b) 3.33   (c) 3.78   (d) 4.44   (e) 4.60

b 32. A buffer solution is prepared by mixing 250. mL of 1.00 M CH_3COOH with 500. mL of 0.500 M calcium acetate, $Ca(CH_3COO)_2$. Calculate the pH.

(a) 5.36   (b) 5.05   (c) 4.74   (d) 4.58   (e) 4.40

d 33. A buffer solution is prepared by mixing 250. mL of 1.00 M HNO$_2$ with 500. mL of 0.500 M calcium nitrite, Ca(NO$_2$)$_2$. Calculate the pH.

(a) 4.43   (b) 3.35   (c) 3.05   (d) 3.65   (e) 4.88

e 34. What is the [H$_3$O$^+$] of a solution resulting from the mixture of 10.0 mL of 0.200 M KF with 20.0 mL of 0.100 M hydrofluoric acid, HF?

(a) 5.6 x 10^{-5} M   (b) 3.4 x 10^{-2} M   (c) 1.7 x 10^{-3} M
(d) 4.1 x 10^{-3} M   (e) 7.2 x 10^{-4} M

d*35. It is desired to buffer a solution at pH = 4.30. What molar ratio of CH$_3$COOH to NaCH$_3$COO should be used?

(a) 1.2/1   (b) 0.8/1   (c) 0.12/1   (d) 2.8/1   (e) 6.2/1

d*36. Calculate the ratio [CH$_3$COOH]/[NaCH$_3$COO] that gives a solution with pH = 5.00.
[CH$_3$COOH]/[NaCH$_3$COO] _____.

(a) 0.28   (b) 0.36   (c) 0.44   (d) 0.56   (e) 0.63

a 37. What relative number of moles of sodium acetate and acetic acid in water must be used to prepare a buffer with pH = 5.08?

(a) 2.2 mol salt to 1.0 mol acid   (b) 1.6 mol salt to 1.0 mol acid
(c) 1.0 mol salt to 1.0 mol acid   (d) 0.45 mol salt to 1.0 mol acid
(e) 0.75 mol salt to 1.0 mol acid

d 38. Calculate the mass of solid NaCH$_3$COO that must be added to 1.0 liter of 0.20 M CH$_3$COOH solution so that the pH of the resulting solution will be 5.00. Assume no volume change due to addition of solid NaCH$_3$COO.

(a) 15 g   (b) 20 g   (c) 25 g   (d) 30 g   (e) 35 g

b 39. If 0.090 mole of solid NaOH is added to 1.0 liter of 0.180 M CH$_3$COOH, what will be the pH of the resulting solution? Assume no volume change due to addition of NaOH.

(a) 4.51   (b) 4.74   (c) 5.08   (d) 5.70   (e) 5.94

b 40. One of the following buffer solutions has pOH = 5.05. Which one? Hint: Solve the general problem rather than 5 specific problems.

(a) 0.10 M NH$_3$ and 0.10 M NH$_4$Cl   (b) 0.10 M NH$_3$ and 0.20 M NH$_4$Cl
(c) 0.20 M NH$_3$ and 0.10 M NH$_4$Cl   (d) 0.050 M NH$_3$ and 0.20 M NH$_4$Cl
(e) 0.20 M NH$_3$ and 0.050 M NH$_4$Cl

b 41. Calculate the pH for a buffer solution prepared by mixing 100. mL of 0.60 M NH$_3$ and 200. mL of 0.45 M NH$_4$Cl.

(a) 8.65   (b) 9.08   (c) 9.87   (d) 4.90   (e) 6.62

c*42. Calculate the ratio [NH₃]/[NH₄Cl] that gives a solution with pH = 9.48.

(a) 0.33 (b) 3.0 (c) 1.7 (d) 0.61 (e) 4.5

d 43. How many moles of NH₄Cl must be added to 1.00 liter of 0.010 M aqueous NH₃ to adjust the pH to 10.00?

(a) 3.6 x 10⁻⁵ mol (b) 3.6 x 10⁻³ mol (c) 2.7 x 10⁻³ mol
(d) 1.8 x 10⁻³ mol (e) 1.8 x 10⁻² mol

a 44. How much NH₄Cl must be added to 2.00 liters of 0.200 M aqueous ammonia to give a solution with pH = 8.20? Assume no volume change due to the addition of NH₄Cl.

(a) 246 g (b) 166 g (c) 2.25 g (d) 123 g (e) 14.6 g

e 45. If 0.10 mole of HCl is bubbled into 1.0 liter of 0.25 M NH₃ solution, what will be the pH of the resulting solution?

(a) 9.08 (b) 4.56 (c) 9.68 (d) 4.92 (e) 9.44

Acid-Base Indicators

e 46. Which response is **false**? An acid-base indicator _____.

(a) might be an acid
(b) might have only one highly colored form
(c) might be a base
(d) might have two highly colored forms
(e) has a color-change range that is independent of its own ionization constant value

b 47. The nonionized form of an acid indicator is yellow, and its anion is blue. The K_a of this indicator is 10^{-5}. What will be the approximate pH range over which this indicator changes color?

(a) 3-5 (b) 4-6 (c) 5-7 (d) 8-10 (e) 9-11

c 48. The nonionized form of an acid indicator is yellow, and its anion is blue. The K_a of this indicator is 10^{-5}. What will be the color of the indicator in a solution of pH 3?

(a) red (b) orange (c) yellow (d) green (e) blue

d 49. The equivalence point of the titration of an unknown base with HCl is at pH = 5.90. What would be the K_a for the **best** choice of an indicator for this titration?

(a) 10⁻² (b) 10⁻⁴ (c) 10⁻⁸ (d) 10⁻⁶ (e) 10⁻⁷

c 50. The equivalence point of the titration of an unknown acid with NaOH is at pH = 8.10. What would be the K_a for the **best** choice of an indicator for this titration?

(a) 10⁻² (b) 10⁻⁴ (c) 10⁻⁸ (d) 10⁻⁶ (e) 10⁻⁷

d 51. Consider an indicator that ionizes as shown below for which $K_{HIn} = 1.0 \times 10^{-4}$.

$$HIn + H_2O \rightleftharpoons H_3O^+ + In^-$$
yellow   red

Which of the responses contains all the **true** statements and no others?

I. The predominant color in its acid range is yellow.
II. In the middle of the pH range of its color change, a solution containing the indicator probably will be orange.
III. At pH = 7.00 a solution containing this indicator (and no other colored species) will be red. (Hint: Write the equilibrium constant expression for the indicator.)
IV. At pH = 7.00 most of the indicator is in the nonionized form.

(a) I and III (b) II and IV (c) III and IV (d) I, II, and III (e) another combination

Strong Acid/Strong Base Titration Curves

c 52. Which is the best indicator for the titration of HCl solution with NaOH solution, i.e. the most nearly ideal one?

	Acid Range Color	pH Range	Basic Range Color
(a)	pink	1.2-2.8	yellow
(b)	blue	3.4-4.6	yellow
(c)	yellow	6.5-7.8	purple
(d)	colorless	9.9-11.1	red
(e)	none of these indicators		

b 53. What is the pH at the point in a titration at which 20.00 mL of 1.000 M KOH has been added to 25.00 mL of 1.000 M HBr?

(a) 1.67 (b) 0.95 (c) 3.84 (d) 2.71 (e) 1.22

a 54. Calculate the pH of a solution prepared by adding 80.0 mL of 0.100 M NaOH solution to 100. mL of 0.100 M HNO$_3$ solution.

(a) 1.95 (b) 2.02 (c) 2.08 (d) 2.16 (e) 2.24

d 55. Calculate the pH of a solution prepared by adding 115 mL of 0.100 M NaOH to 100. mL of 0.100 M HNO$_3$ solution.

(a) 11.60 (b) 11.68 (c) 11.76 (d) 11.84 (e) 11.92

b 56. If 50.00 mL of 0.1000 M NaOH is titrated with 0.1000 M HCl, what is the pH of the solution after 30.00 mL of HCl solution has been added?

(a) 13.00 (b) 12.40 (c) 12.60 (d) 12.00 (e) 12.80

d 57. Calculate the pH of the solution resulting from the addition of 30.0 mL of 0.200 M HClO$_4$ to 60.0 mL of 0.150 M NaOH.

(a) 1.47 (b) 7.00 (c) 11.88 (d) 12.52 (e) 13.06

d 58. Calculate the pH of the solution resulting from the addition of 40.0 mL of 0.200 M HClO$_4$ to 60.0 mL of 0.150 M NaOH.

(a) 1.47   (b) 2.00   (c) 11.88   (d) 12.00   (e) 13.06

b 59. What is the pH of the solution resulting from the addition of 30.0 mL of 0.200 M HClO$_4$ to 20.0 mL of 0.150 M NaOH?

(a) 1.00   (b) 1.22   (c) 1.36   (d) 1.48   (e) 1.67

a 60. What is the pH of the solution resulting from the addition of 40.0 mL of 0.200 M HClO$_4$ to 20.0 mL of 0.150 M NaOH?

(a) 1.08   (b) 1.25   (c) 1.36   (d) 1.48   (e) 1.67

b 61. Which titration curve could describe the titration of a solution of HCl by addition of a solution of KOH?

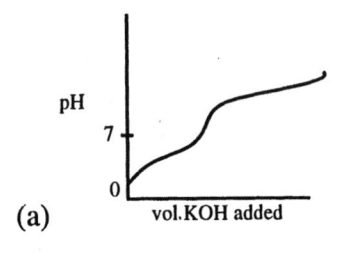

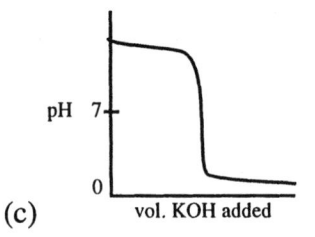

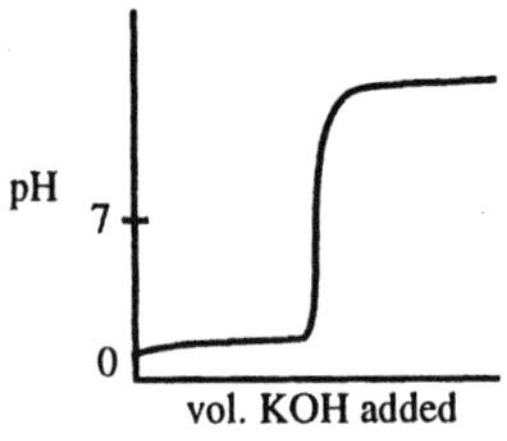

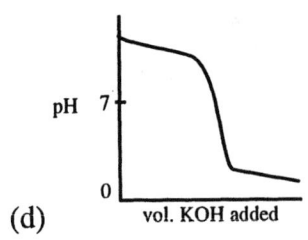

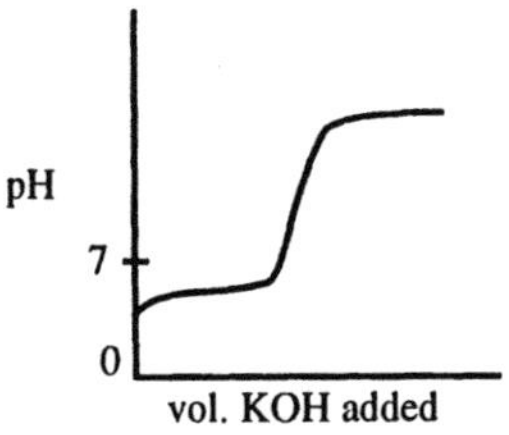

c 62. Which titration curve could describe the titration of a solution of NaOH by addition of a solution of HCl?

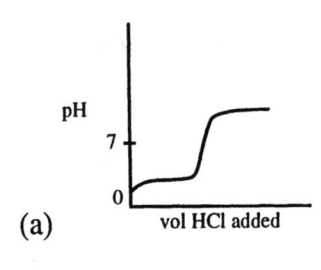

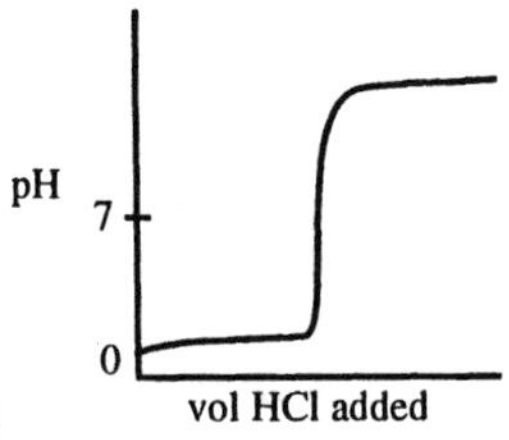

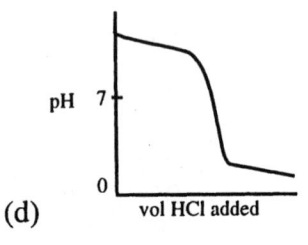

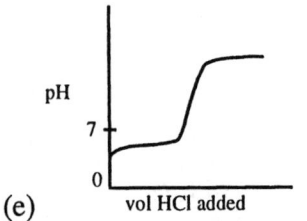

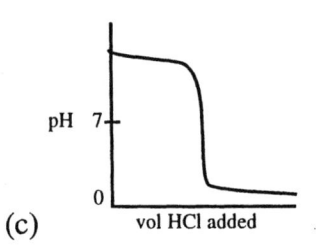

Weak Acid/Strong Base and Weak Base/Strong Acid Titration Curves

c 63. When a weak acid is titrated with a strong base, the pH at the equivalence point is **always** _____.

(a) 7
(b) less than 7
(c) greater than 7
(d) is less than 1
(e) greater than 4

b 64. When a weak base is titrated with a strong acid, the pH at the equivalence point is **always** _____.

(a) 7
(b) less than 7
(c) greater than 7
(d) less than 1
(e) greater than 14

d 65. Which indicator could be used to titrate CH_3COOH with NaOH solution?

	Acid Range Color	pH Range	Basic Range Color
(a)	pink	1.2-2.8	yellow
(b)	blue	3.4-4.6	yellow
(c)	yellow	6.5-7.8	purple
(d)	colorless	8.3-9.9	red
(e)	none of these indicators		

b 66. Which indicator could be used to titrate aqueous NH_3 with HCl solution?

	Acid Range Color	pH Range	Basic Range Color
(a)	pink	1.2-2.8	yellow
(b)	blue	3.4-4.6	yellow
(c)	yellow	6.5-7.8	purple
(d)	colorless	8.3-9.9	red
(e)	none of these indicators		

b 67. Consider the titration of a solution of a weak acid by adding a solution of a strong base. Which response includes all the following statements that are true, and no others?

I. The end-point cannot be detected using an indicator.
II. The pH at the equivalence point is 7.00.
III. The solution is buffered before the equivalence point.
IV. The solution is buffered both before and after the equivalence point.
V. Methyl orange (pH color range 3.1-4.4) would be a better indicator for this titration than would thymol blue (pH range 8.0-9.6).

(a) II and IV
(b) III
(c) IV and V
(d) I and III
(e) IV and V

c 68. What is the $[H_3O^+]$ in a solution resulting from mixing of 100. mL of 0.100 M HCN and 10.0 mL of 0.100 M KOH?

(a) $9.1 \times 10^{-3} M$
(b) $5.7 \times 10^{-6} M$
(c) $3.6 \times 10^{-9} M$
(d) $8.2 \times 10^{-5} M$
(e) $2.0 \times 10^{-8} M$

b 69. Calculate the pH of a solution prepared by adding 60.0 mL of 0.100 M NaOH to 100. mL of 0.100 M CH_3COOH solution.

(a) 4.56 (b) 4.92 (c) 5.00 (d) 5.08 (e) 5.16

e 70. What is the pH of the solution resulting from the addition of 20.0 mL of 0.0100 M NaOH solution to 30.0 mL of 0.0100 M acetic acid, CH_3COOH?

(a) 4.56 (b) 4.73 (c) 4.88 (d) 4.96 (e) 5.05

c 71. What is the pH at the equivalence point in the titration of 50.0 mL of 0.100 M hydrofluoric acid, HF, with 0.100 M NaOH?

(a) 5.88 (b) 6.08 (c) 7.92 (d) 8.12 (e) 8.56

c 72. Calculate the pH of a solution resulting from the addition of 20.0 mL of 0.100 M HCl to 50.0 mL of 0.100 M NH_3.

(a) 9.26 (b) 4.91 (c) 9.44 (d) 4.60 (e) 9.08

d 73. Calculate the pH of a solution resulting from the addition of 50.0 mL of 0.30 M HNO_3 to 50.0 mL of 0.30 M NH_3.

(a) 4.92 (b) 7.46 (c) 6.18 (d) 5.04 (e) 8.30

d 74. Which of the following titrations could the following curve describe?

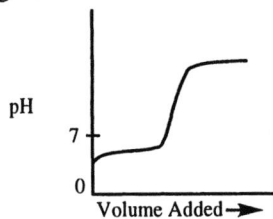

(a) KOH added to HNO_3
(b) HCl added to aqueous NH_3
(c) HNO_3 added to KOH
(d) NaOH added to HF
(e) CH_3COOH added to aqueous NH_3

b 75. Which of the following titrations could the following curve describe?

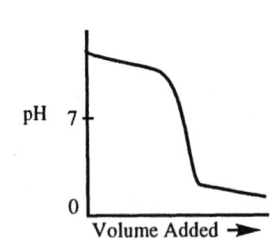

(a) NaOH added to HF
(b) HCl added to aqueous NH_3
(c) CH_3COOH added to aqueous NH_3
(d) KOH added to $HClO_4$
(e) HNO_3 added to NaOH

e 76. Which titration curve could describe the titration of a solution of CH_3COOH by addition of a solution of KOH?

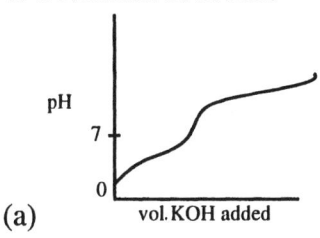

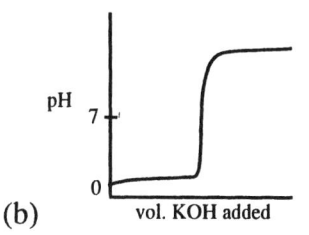

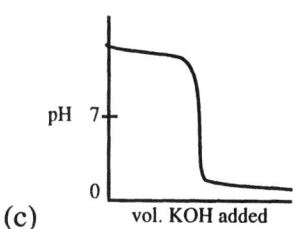

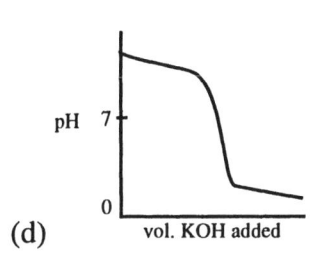

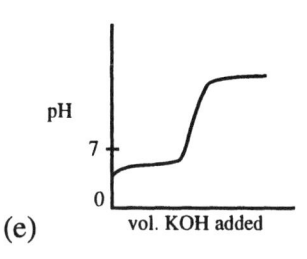

d 77. Which titration curve could describe the titration of a solution of NH_3 by addition of a solution of HCl?

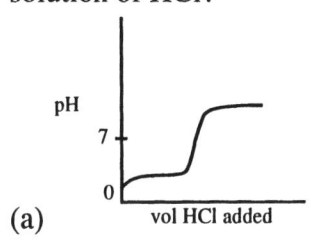

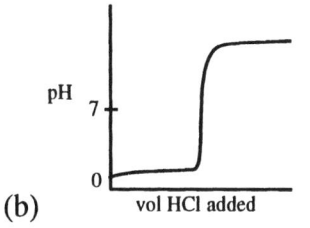

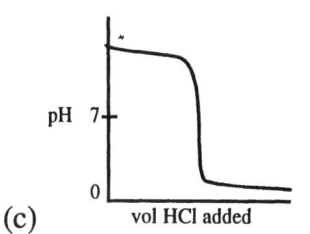

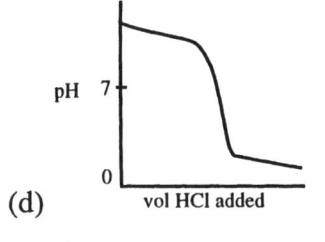

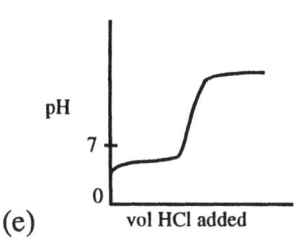

d 78. A 50.0 mL solution of a 0.10 M weak monoprotic acid was mixed with 20.0 mL of 0.10 M KOH, and the resulting solution was diluted to 100. mL. The pH of the solution was found to be 5.25. What is the ionization constant for the acid?

(a) 1.2×10^{-4} (b) 4.6×10^{-4} (c) 1.4×10^{-6}
(d) 3.8×10^{-6} (e) 6.4×10^{-7}

Weak Acid/Weak Base Titration Curves

d 79. When a weak base is titrated with a weak acid, the pH at the equivalence point is always _____.

(a) 7 (b) less than 7 (c) greater than 7
(d) depends on the value of K_b for the base and K_a for the acid.
(e) zero

e 80. Which indicator could be used to titrate aqueous ammonia with CH_3COOH?

	Acid Range Color	pH Range	Basic Range Color
(a)	pink	1.2-2.8	yellow
(b)	blue	3.4-4.6	yellow
(c)	yellow	6.5-7.8	purple
(d)	colorless	8.3-9.9	red
(e)	none of these indicators		

e 81. Consider the titration of 30.0 mL of 0.0200 M nitrous acid, HNO_2, by adding 0.0500 M aqueous ammonia to it. Which statement is true? The solution is buffered _____.

(a) before the equivalence point only
(b) at the equivalence point only
(c) after the equivalence point only
(d) before and at the equivalence point
(e) before and after the equivalence point

c*82. Consider the titration of 30.0 mL of 0.0200 M nitrous acid, HNO_2, by adding 0.0500 M aqueous ammonia to it. Which statement is true? The pH at the equivalence point _____.

(a) is greater than 7
(b) is equal to 7
(c) is less than 7
(d) is less than 0.02
(e) zero

a 83. Which titration curve could describe the titration of a solution of CH_3COOH by addition of a solution of NH_3?

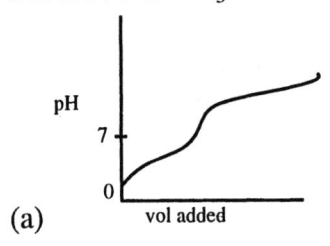

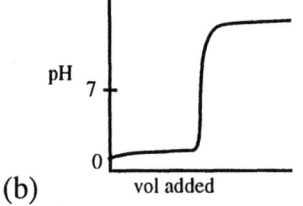

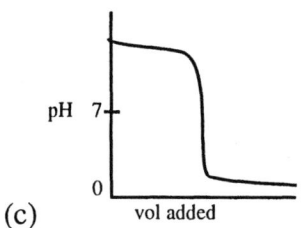

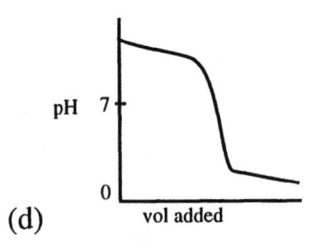

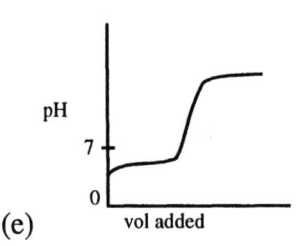

a 84. Consider the titrations of the pairs of aqueous acids and bases listed at the left. Which pair is **incorrectly** described (at the right) in terms of the acidic, basic, or neutral character of the solutions at the equivalence point?

	Acid-Base Pair	Character at Equivalence Point
(a)	$HClO + NaOH$	acidic
(b)	$HNO_3 + Ca(OH)_2$	neutral
(c)	$CH_3COOH + KOH$	basic
(d)	$HCl + NH_3$	acidic
(e)	$CH_3COOH + NH_3$	neutral

20 Ionic Equilibria III: The Solubility Product Principle

Solubility Product Constants

c 1. Which of the following solubility product expressions is **incorrect**?

(a) $K_{sp\,(Ag_2S)} = [Ag^+]^2[S^{2-}]$
(b) $K_{sp\,(CaF_2)} = [Ca^{2+}][F^-]^2$
(c) $K_{sp\,(Sb_2S_3)} = [Sb^{2+}]^3[S^{3-}]^2$
(d) $K_{sp\,(CuS)} = [Cu^{2+}][S^{2-}]$
(e) $K_{sp\,(Ag_3PO_4)} = [Ag^+]^3[PO_4^{3-}]$

e 2. Which of the following solubility product expressions is **incorrect**?

(a) $K_{sp\,(PbI_2)} = [Pb^{2+}][I^-]^2$
(b) $K_{sp\,(AgCl)} = [Ag^+][Cl^-]$
(c) $K_{sp\,(Hg_2Cl_2)} = [Hg_2^{2+}][Cl^-]^2$
(d) $K_{sp\,(MnCO_3)} = [Mn^{2+}][CO_3^{2-}]$
(e) $K_{sp\,(SrCrO_4)} = [Sr^{2+}][Cr^{6+}][O^{2-}]^4$

d 3. Which of the following solubility product expressions is **incorrect**?

(a) $K_{sp\,(HgI_2)} = [Hg^{2+}][I^-]^2$
(b) $K_{sp\,(CuCN)} = [Cu^+][CN^-]$
(c) $K_{sp\,(BaSO_4)} = [Ba^{2+}][SO_4^{2-}]$
(d) $K_{sp\,(BaF_2)} = [Ba^{2+}][F_2^{2-}]$
(e) $K_{sp\,(Ag_3PO_4)} = [Ag^+]^3[PO_4^{3-}]$

b*4. Which of the following solubility product expressions is **correct**?

(a) $K_{sp\,(Ag_3PO_4)} = [Ag^+][PO_4^{3-}]^3$
(b) $K_{sp\,(Hg_2Cl_2)} = [Hg_2^{2+}][Cl^-]^2$
(c) $K_{sp\,(BaF_2)} = [Ba^{2+}][F_2^{2-}]$
(d) $K_{sp\,(Sb_2S_3)} = [Sb^{2+}]^2[S^{3-}]^3$
(e) $K_{sp\,(MnCO_3)} = [Mn^{2+}][C^{4+}][O^{2-}]^3$

Determination of Solubility Product Constants

e 5. How is the K_{sp} of $Ca_3(PO_4)_2$ related to s, the molar solubility of $Ca_3(PO_4)_2$?

(a) $K_{sp} = 4s^5$ (b) $K_{sp} = 27s^3$ (c) $K_{sp} = 18s^5$ (d) $K_{sp} = 54s^4$ (e) $K_{sp} = 108s^5$

b 6. The molar solubility of $BaCO_3$ is 9.0 x 10^{-5} M at 25°C. What is the solubility product constant for $BaCO_3$?

(a) 1.2 x 10^{-8} (b) 8.1 x 10^{-9} (c) 5.3 x 10^{-12}
(d) 4.0 x 10^{-15} (e) 6.7 x 10^{-11}

e 7. Calculate the solubility product constant for lead fluoride, PbF_2. Its molar solubility is 2.1 x 10^{-3} mole per liter at 25°C.

(a) 4.4 x 10^{-6} (b) 4.2 x 10^{-7} (c) 9.3 x 10^{-7}
(d) 9.3 x 10^{-9} (e) 3.7 x 10^{-8}

c 8. At 25°C, 1.4 x 10⁻⁵ mole of $Cd(OH)_2$ dissolves to give 1.0 liter of saturated aqueous solution. What is the solubility product for $Cd(OH)_2$?

(a) 1.7×10^{-5} (b) 2.9×10^{-10} (c) 1.1×10^{-14}
(d) 5.8×10^{-15} (e) 4.1×10^{-12}

c 9. One liter of saturated zinc hydroxide solution contains 0.000222 g of dissolved $Zn(OH)_2$. Calculate K_{sp} for $Zn(OH)_2$.

(a) 4.8×10^{-12} (b) 2.6×10^{-16} (c) 4.5×10^{-17}
(d) 7.1×10^{-14} (e) 1.1×10^{-17}

b 10. The solubility of $Ce(OH)_3$ is 9.93 x 10⁻⁴ g per liter at 25°C. Calculate the solubility product for $Ce(OH)_3$.

(a) 5.2×10^{-6} (b) 2.0×10^{-20} (c) 7.3×10^{-20}
(d) 4.0×10^{-14} (e) 1.6×10^{-18}

d 11. The solubility of $Fe(OH)_2$ is 3.00 x 10⁻³ g in 2.00 liters at 18°C. What is its K_{sp} at 18°C?

(a) 4.92×10^{-12} (b) 6.03×10^{-9} (c) 2.44×10^{-10}
(d) 1.86×10^{-14} (e) 2.11×10^{-11}

b 12. The solubility of cerium periodate, $Ce(IO_4)_4$, is 0.015 g/mL of water at 25°C. Calculate its solubility product constant. Its formula weight is 904 g/mol.

(a) 8.1×10^{-14} (b) 3.2×10^{-7} (c) 9.4×10^{-24}
(d) 4.1×10^{-12} (e) 6.9×10^{-22}

c 13. The solubility of bismuth sulfide is 1.8 x 10⁻⁵ g/100. mL of water at 18°C. Calculate the K_{sp} for Bi_2S_3 at 18°C.

(a) 6.0×10^{-36} (b) 3.4×10^{-23} (c) 5.7×10^{-31}
(d) 5.8×10^{-12} (e) 3.2×10^{-8}

e 14. Magnesium hydroxide is a slightly soluble substance. If the pH of a saturated solution of $Mg(OH)_2$ is 10.49 at 25°C, calculate K_{sp} for $Mg(OH)_2$.

(a) 8.8×10^{-16} (b) 4.2×10^{-15} (c) 6.0×10^{-10}
(d) 4.4×10^{-14} (e) 1.5×10^{-11}

Uses of Solubility Product Constants

e 15. How is the molar solubility, s, related to the K_{sp} for $Fe(OH)_3$; i.e., s = _____?

(a) $(K_{sp})^{1/4}$ (b) $(K_{sp}/4)^{1/3}$ (c) $(K_{sp})^{1/2}$ (d) $(K_{sp}/16)^{1/4}$ (e) $(K_{sp}/27)^{1/4}$

a 16. Calculate the molar solubility of AgCl at 25°C. $K_{sp} = 1.8 \times 10^{-10}$.

(a) $1.3 \times 10^{-5}\ M$ (b) $1.7 \times 10^{-5}\ M$ (c) $4.2 \times 10^{-5}\ M$
(d) $4.6 \times 10^{-5}\ M$ (e) $1.8 \times 10^{-10}\ M$

c 17. The value of K_{sp} for SrSO$_4$ is 2.8 x 10^{-7}. What is the molar solubility of SrSO$_4$?

(a) 7.6 x 10^{-7} M (b) 5.8 x 10^{-13} M (c) 5.3 x 10^{-4} M
(d) 5.7 x 10^{-3} M (e) 1.3 x 10^{-8} M

c 18. The K_{sp} for magnesium arsenate is 2.1 x 10^{-20} at 25°C. What is the molar solubility of Mg$_3$(AsO$_4$)$_2$ at 25°C?

(a) 6.7 x 10^{-3} M (b) 3.6 x 10^{-4} M (c) 4.5 x 10^{-5} M
(d) 7.0 x 10^{-5} M (e) 1.4 x 10^{-6} M

b 19. K_{sp} for silver arsenate is 1.1 x 10^{-20}. What is the molar solubility of Ag$_3$AsO$_4$?

(a) 2.2 x 10^{-7} M (b) 4.5 x 10^{-6} M (c) 6.2 x 10^{-8} M
(d) 4.1 x 10^{-7} M (e) 1.0 x 10^{-10} M

c 20. Calculate the molar solubility of aluminum hydroxide at 25°C. Its K_{sp} is 1.9 x 10^{-33}.

(a) 1.2 x 10^{-9} M (b) 4.2 x 10^{-15} M (c) 2.9 x 10^{-9} M
(d) 4.4 x 10^{-14} M (e) 9.3 x 10^{-9} M

e 21. Which of the following compounds has the lowest molar solubility in water at 25°C? As Ag$_4$Fe(CN)$_6$ dissolves, it dissociates as follows.

$$Ag_4[Fe(CN)_6](s) \rightleftharpoons 4Ag^+ + [Fe(CN)_6]^{4-}$$

	Compound		K_{sp}
(a)	Ag$_2$CO$_3$		8.1 x 10^{-12}
(b)	Ag$_4$Fe(CN)$_6$	(see above)	1.6 x 10^{-41}
(c)	Ag$_3$PO$_4$		1.3 x 10^{-20}
(d)	Ag$_2$SO$_4$		1.7 x 10^{-5}
(e)	Bi(OH)$_3$		3.2 x 10^{-40}

e 22. Calculate the concentration of OH$^-$ ions in a saturated Mn(OH)$_2$ solution. The solubility product for Mn(OH)$_2$ is 4.6 x 10^{-14}.

(a) 1.0 x 10^{-5} M (b) 2.0 x 10^{-5} M (c) 1.6 x 10^{-5} M
(d) 3.2 x 10^{-5} M (e) 4.5 x 10^{-5} M

b 23. Calculate the concentration of F$^-$ ions in saturated CaF$_2$ solution at 25°C. K_{sp} = 3.9 x 10^{-11}.

(a) 2.1 x 10^{-4} M (b) 4.3 x 10^{-4} M (c) 0.016 M
(d) 0.032 M (e) 0.10 M

b 24. Calculate [Al^{3+}] in a saturated Al(OH)$_3$ solution at 25°C. K_{sp} = 1.9 x 10^{-33}

(a) 6.6 x 10^{-9} M (b) 2.9 x 10^{-9} M (c) 1.5 x 10^{-17} M
(d) 8.4 x 10^{-18} M (e) 1.5 x 10^{-7} M

b*25. Calculate the pH of a saturated aqueous solution of Co(OH)$_2$. K_{sp} is 2.5 x 10^{-16}.

(a) 8.60 (b) 8.90 (c) 9.10 (d) 9.20 (e) 9.40

b 26. What mass of Zn(OH)$_2$ is contained in 1.0 liter of saturated solution? $K_{sp} = 4.5 \times 10^{-17}$.

(a) 0.00011 g (b) 0.00022 g (c) 0.00044 g
(d) 0.010 g (e) 0.016 g

e 27. What mass of CaF$_2$ is contained in 1.0 liter of saturated solution? $K_{sp} = 3.9 \times 10^{-11}$.

(a) 0.0010 g (b) 0.0028 g (c) 0.0034 g (d) 0.010 g (e) 0.017 g

a*28. Calculate the solubility of Mg(OH)$_2$ in g per 100. mL of water. $K_{sp} = 1.5 \times 10^{-11}$.

(a) 9.1×10^{-4} g/100. mL (b) 3.6×10^{-3} g/100. mL (c) 6.6×10^{-5} g/100. mL
(d) 1.8×10^{-3} g/100. mL (e) 5.0×10^{-6} g/100. mL

a*29. The solubility product constant for MgF$_2$ is 6.4×10^{-9}. How many grams of MgF$_2$ will dissolve in 150 mL of H$_2$O at 25°C?

(a) 1.1×10^{-2} g (b) 5.8×10^{-3} g (c) 4.7×10^{-1} g
(d) 6.0×10^{-4} g (e) 2.2×10^{-3} g

b 30. Given the following values of K_{sp} for four slightly soluble sulfides at 25°C.

Sulfide	K_{sp}	Sulfide	K_{sp}
CdS	3.6×10^{-29}	PbS	8.4×10^{-28}
CuS	8.7×10^{-36}	MnS	5.1×10^{-15}

Which one of the following ions exists in the lowest concentration in a solution in which the sulfide ion concentration had been fixed at some (fairly large) constant value at 25°C?

(a) Cd^{2+} (b) Cu^{2+} (c) Pb^{2+} (d) Mn^{2+}
(e) Cannot be answered without knowing the sulfide ion concentration.

b 31. AgCl would be least soluble at 25°C in _____.

(a) pure water (b) 0.1 M CaCl$_2$ (c) 0.1 M HCl
(d) 0.1 M HNO$_3$ (e) It is equally soluble in all of the preceding substances.

d 32. K_{sp} for lead fluoride is 3.7×10^{-8}. What is the molar solubility of PbF$_2$ in 0.10 M NaF?

(a) 4.6×10^{-5} M (b) 1.9×10^{-4} M (c) 3.7×10^{-7} M
(d) 3.7×10^{-6} M (e) 8.5×10^{-8} M

b 33. How many **moles** of PbSO$_4$ will dissolve in 1.0 liter of 0.010 M K$_2$SO$_4$? The solubility product for PbSO$_4$ is 1.8×10^{-8}.

(a) 5.0×10^7 (b) 1.8×10^{-6} (c) 1.8×10^{-4} (d) 1.8×10^{-2} (e) 5.0×10^{-3}

a 34. How many **grams** of AgCl will dissolve in 1.0 L of 0.25 M KCl? $K_{sp\ (AgCl)} = 1.8 \times 10^{-10}$

(a) 1.0×10^{-7} (b) 2.6×10^{-8} (c) 6.4×10^{-9} (d) 7.8×10^{-8} (e) 2.5×10^{-9}

d 35. How many grams of MgF$_2$ will dissolve in 150. mL of 0.100 M NaF solution? K_{sp} for MgF$_2$ = 6.4 x 10^{-9}

(a) 6.2 x 10^{-7} g   (b) 4.1 x 10^{-6} g   (c) 1.0 x 10^{-5} g
(d) 6.0 x 10^{-6} g   (e) 9.3 x 10^{-7} g

e 36. If aqueous solutions of the following pairs of compounds are mixed, which combination might result in the formation of a precipitate?

(a) MgCl$_2$ + Na$_2$S   (b) NH$_4$ClO$_4$ + K$_2$S   (c) BaCl$_2$ + Ca(CH$_3$COO)$_2$
(d) CrCl$_3$ + Ni(NO$_3$)$_2$ (e) Na$_3$PO$_4$ + Cr(NO$_3$)$_3$

d 37. We mix each of the following pairs of solutions together. In which case would we **not** expect a precipitate (solid) to be formed?

(a) KCl solution and AgNO$_3$ solution   (b) Pb(NO$_3$)$_2$ solution and H$_2$SO$_4$ solution
(c) CuCl$_2$ solution and Na$_2$S solution   (d) CaCl$_2$ solution and HNO$_3$ solution
(e) A precipitate would be likely to form in every case.

a 38. When we mix together, from separate sources, the ions of a slightly soluble ionic salt, the salt will begin to precipitate if Q_{sp} _____ K_{sp}, and will continue to precipitate until Q_{sp} _____ K_{sp}.

(a) is greater than; equals   (b) is less than; is greater than
(c) is less than; equals   (d) equals; is less than   (e) equals; is greater than

d 39. The K_{sp} for Fe(IO$_3$)$_3$ is 10^{-14}. We mix two solutions, one containing Fe^{3+} and one containing IO$_3^-$ ions at 25°C. At the instant of mixing, [Fe^{3+}] = 10^{-4} M and [IO$_3^-$] = 10^{-5} M. Which one of the following statements is true?

(a) A precipitate forms, because $Q_{sp} > K_{sp}$.
(b) A precipitate forms, because $Q_{sp} < K_{sp}$.
(c) No precipitate forms, because $Q_{sp} > K_{sp}$.
(d) No precipitate forms, because $Q_{sp} < K_{sp}$.
(e) None of the preceding statements is true.

e 40. K_{sp} for CaF$_2$ is 3.9 x 10^{-11} and K_{sp} for PbF$_2$ is 3.7 x 10^{-8}. If 200. mL **each** of 5.0 x 10^{-3} M NaF, 2.0 x 10^{-5} M Ca(NO$_3$)$_2$, and 3.0 x 10^{-3} M Pb(NO$_3$)$_2$ solutions are mixed,

(a) only CaF$_2$ will precipitate.   (b) only PbF$_2$ will precipitate.
(c) both CaF$_2$ and PbF$_2$ will precipitate, and the precipitate should be visible.
(d) both CaF$_2$ and PbF$_2$ will precipitate, but the precipitate should **not** be visible.
(e) neither CaF$_2$ nor PbF$_2$ will precipitate.

c 41. If NaCl is added to a 0.010 M solution of AgNO$_3$ in water at 25°C, at what [Cl$^-$] does precipitation of AgCl begin? K_{sp} for AgCl = 1.8 x 10^{-10}.

(a) 1.0 x 10^{-10} M (b) 1.3 x 10^{-6} M (c) 1.8 x 10^{-8} M
(d) 1.8 x 10^{-12} M (e) 1.3 x 10^{-10} M

d 42. Calculate the [Ca^{2+}] required to start the precipitation of calcium fluoride, CaF$_2$, from a solution containing 0.0025 M F$^-$ at 25°C. K_{sp} for CaF$_2$ = 3.9 x 10^{-11}.

(a) 6.4 x 10^{-7} M (b) 5.2 x 10^{-10} M (c) 4.8 x 10^{-3} M
(d) 6.2 x 10^{-6} M (e) 1.6 x 10^{-8} M

b 43. What concentration of Al^{3+} must be exceeded in a solution buffered at pH = 4.80 in order to initiate precipitation of Al(OH)$_3$ at 25°C? K_{sp} for Al(OH)$_3$ = 1.9 x 10^{-33}.

(a) 6.3 x 10^{-10} M (b) 7.6 x 10^{-6} M (c) 4.1 x 10^{-4} M
(d) 5.3 x 10^{-7} M (e) 2.2 x 10^{-8} M

c 44. Calculate the concentration of sulfate ion in a saturated solution of barium sulfate to which barium chloride has been added until [Ba^{+2}] = 0.1 M at 25°C. K_{sp} for BaSO$_4$ = 1 x 10^{-10}.

(a) 1 x 10^{-5} M (b) 1 x 10^{-8} M (c) 1 x 10^{-9} M
(d) 1 x 10^{-7} M (e) 1 x 10^{-6} M

e 45. If solid KOH is added to 0.10 M Mg(NO$_3$)$_2$ solution until the pH reaches 12.00, what concentration of Mg^{2+} remains in solution? Assume no volume change due to addition of solid KOH. K_{sp} for Mg(OH)$_2$ is 1.5 x 10^{-11}.

(a) 3.8 x 10^{-8} M (b) 1.0 x 10^{-2} M (c) 5.0 x 10^{-3} M
(d) 1.5 x 10^{-9} M (e) 1.5 x 10^{-7} M

b 46. What is the minimum concentration of bromide ion required to limit the concentration of silver ion in solution to a maximum of 1.0 x 10^{-9} M? K_{sp} for AgBr is 3.3 x 10^{-13}.

(a) 1.0 x 10^{-9} M (b) 3.3 x 10^{-4} M (c) 1.1 x 10^{-7} M
(d) 5.0 x 10^{-6} M (e) 3.3 x 10^{-22} M

e 47. What is the minimum concentration of iodide ion required to limit the concentration of lead ion in solution to a maximum of 1.0 x 10^{-6} M? K_{sp} for PbI$_2$ is 8.7 x 10^{-9}.

(a) 1.7 x 10^{-2} M (b) 1.8 x 10^{-1} M (c) 7.6 x 10^{-5} M
(d) 8.7 x 10^{-3} M (e) 9.3 x 10^{-2} M

Fractional Precipitation

e 48. Which of the following responses contains all of the **true** statements?

I. An ion may never be **completely** precipitated from solution, even when the precipitating reagent is used in large excess.
II. If solid NaCl is added slowly to a **saturated solution** of NiS in contact with solid NiS, nothing significant will happen to the NiS.
III. Consider two relatively insoluble 1:1 salts (i.e., one cation and one anion per formula unit) containing the same cation, whose solubility products differ by 10^4. More than 90% of one anion can be precipitated before the other precipitates when the common cation is added to a solution of the two anions.

(a) I and III   (b) I and II   (c) II and III   (d) III   (e) I, II, and III d 49. A solution is 0.012 M in $Pb(NO_3)_2$ and 0.20 M in $Sr(NO_3)_2$. Solid Na_2SO_4 is added until a precipitate just begins to form. The precipitate is _____ and the concentration of sulfate ion at this point is _____. K_{sp} for $PbSO_4$ is 1.8 x 10^{-8} and for $SrSO_4$ is 2.8 x 10^{-7}.

(a) $SrSO_4$; 2.6 x 10^{-7} M
(b) $SrSO_4$; 8.3 x 10^{-7} M
(c) $PbSO_4$; 6.3 x 10^{-6} M
(d) $PbSO_4$; 1.5 x 10^{-6} M
(e) $SrSO_4$; 1.4 x 10^{-6} M

c 50. Suppose we have a solution that is 0.10 M in Ca^{2+} and also 0.10 M in Pb^{2+} ions. We wish to separate these ions by fractional precipitation of one of them as the carbonate. We do this by adding a source of carbonate ions, such as solid Na_2CO_3. Use the K_{sp} values, 4.8 x 10^{-9} for $CaCO_3$ and 1.5 x 10^{-13} for $PbCO_3$. Neglect any change in volume due to addition of solid sodium carbonate. Which will begin to precipitate from solution first — $CaCO_3$ or $PbCO_3$ — and what will be the $[CO_3^{2-}]$ when this begins (whether or not we can see it!)?

(a) $CaCO_3$, 4.8 x 10^{-8} M
(b) $CaCO_3$, 4.8 x 10^{-9} M
(c) $PbCO_3$, 1.5 x 10^{-12} M
(d) $PbCO_3$, 1.5 x 10^{-13} M
(e) $PbCO_3$, 7.2 x 10^{-21} M

d 51. A solution is 0.10 M in $Pb(NO_3)_2$ and 0.10 M in $AgNO_3$. Solid NaI is added until the second solid compound just begins to precipitate. Which compound precipitated first — PbI_2 or AgI — and what was the $[I^-]$ when the second compound began to precipitate?
K_{sp} = 8.7 x 10^{-9} for PbI_2 K_{sp} = 1.5 x 10^{-16} for AgI

(a) PbI_2, 5.9 x 10^{-4} M
(b) AgI, 9.3 x 10^{-5} M
(c) PbI_2, 4.4 x 10^{-3} M
(d) AgI, 2.9 x 10^{-4} M
(e) AgI, 8.7 x 10^{-8} M

d 52. Solid Na_2SO_4 is added to a solution that is 0.30 M in both Sr^{2+} and Pb^{2+}. Assuming no volume change, what will be the $[Pb^{2+}]$ at the point at which $SrSO_4$ just begins to precipitate at 25°C? K_{sp} for $SrSO_4$ = 2.8 x 10^{-7} and for $PbSO_4$ = 1.8 x 10^{-8}.

(a) 0.24 M (b) 0.16 M (c) 0.30 M (d) 0.019 M (e) 0.040 M

d 53. The solubility product for barium chromate, $BaCrO_4$, is 2.0 x 10^{-10} and the solubility product for lead chromate, $PbCrO_4$, is 1.8 x 10^{-14}. If solid Na_2CrO_4 is added slowly to a solution which is 0.010 M in $Ba(NO_3)_2$ and 0.010 M in $Pb(NO_3)_2$, what will the concentration of Pb^{2+} be **just before** $BaCrO_4$ begins to precipitate? Assume no volume change due to the addition of solid Na_2CrO_4.

(a) 4.0 x 10^{-7} M
(b) 2.0 x 10^{-8} M
(c) 1.8 x 10^{-14} M
(d) 9.0 x 10^{-7} M
(e) 1.4 x 10^{-5} M

e 54. A solution is 0.0010 M in both Ag^+ and Au^+. Some solid NaCl is added slowly until the second solid compound just begins to precipitate. What is the concentration of Au^+ ions at this point? K_{sp} for AgCl = 1.8 x 10^{-10} and for AuCl is 2.0 x 10^{-13}.

(a) 2.0 x 10^{-10} M
(b) 4.5 x 10^{-7} M
(c) 1.8 x 10^{-7} M
(d) 3.0 x 10^{-4} M
(e) 1.1 x 10^{-6} M

d 55. If solid AgNO$_3$ is slowly added to a solution that is 0.050 M in NaI and 0.12 M in NaCl, what is the concentration of I$^-$ when AgCl just begins to precipitate? K_{sp} for AgCl = 1.8 x 10^{-10}, K_{sp} for AgI = 1.5 x 10^{-16}

(a) 1.0 x 10^{-9} M (b) 9.2 x 10^{-9} M (c) 8.5 x 10^{-8} M
(d) 1.0 x 10^{-7} M (e) 6.7 x 10^{-7} M

a 56. At 25°C the K_{sp} for MgCO$_3$ is 4.0 x 10^{-5} and that for BaCO$_3$ is 8.1 x 10^{-9}. Assuming both magnesium ions and barium ions are present in 1.0 x 10^{-3} M concentrations, what is the equilibrium concentration of carbonate ions that will precipitate the maximum concentration of Ba^{2+} without precipitating any Mg^{2+} at 25°C?

(a) 4.0 x 10^{-2} M (b) 4.2 x 10^{-3} M (c) 3.5 x 10^{-3} M
(d) 5.8 x 10^{-2} M (e) 3.2 x 10^{-2} M

d 57. Solid lead nitrate is added slowly to a solution that is 0.010 M in sodium sulfate and 0.010 M in sodium chromate at 25°C. What percent of the chromate ions remain in solution, i.e., unprecipitated, just before lead sulfate begins to precipitate? K_{sp} for PbSO$_4$ = 1.8 x 10^{-8}, K_{sp} for PbCrO$_4$ = 1.8 x 10^{-14}

(a) 0.18%   (b) 0.018%   (c) 0.0018%   (d) 0.00010%   (e) 0.00018%

a 58. Solid silver nitrate is added slowly to a solution that is 0.0010 M in sodium chloride and 0.0010 M in sodium bromide. What % of the bromide ions remain in solution, i.e., unprecipitated, just before silver chloride begins to precipitate? K_{sp} for AgCl = 1.8 x 10^{-10}, K_{sp} for AgBr = 3.3 x 10^{-13}

(a) 0.18%   (b) 0.018%   (c) 0.0010%   (d) 0.00010%   (e) 0.0018%

b 59. A solution is 1.0 x 10^{-3} M in Mg(NO$_3$)$_2$ and 1.0 x 10^{-3} M in Cd(NO$_3$)$_2$. Some solid NaOH is added slowly. What percentage of the Cd^{2+} originally in solution has precipitated as Cd(OH)$_2$ at the point at which Mg(OH)$_2$ **just begins** to precipitate? K_{sp} for Cd(OH)$_2$ is 1.2 x 10^{-14} and K_{sp} for Mg(OH)$_2$ is 1.5 x 10^{-11}.

(a) 98.12%   (b) 99.92%   (c) 99.63%   (d) 98.73%   (e) 99.52%

d*60. A solution is 1.0 x 10^{-4} M in sodium sulfate, Na$_2$SO$_4$, and 1.0 x 10^{-4} M in sodium selenate, Na$_2$SeO$_4$. Solid barium chloride, BaCl$_2$, is added slowly and one of the two salts, BaSO$_4$ **or** BaSeO$_4$, begins to precipitate. What percentage of the anion that precipitates first has precipitated at the point at which the second anion begins to precipitate? K_{sp} for BaSO$_4$ is 1.1 x 10^{-10} and for BaSeO$_4$ is 2.8 x 10^{-11}.

(a) 10%   (b) 25%   (c) 90%   (d) 75%   (e) 50%

d 61. Solid AgNO$_3$ is added slowly to a solution which is 0.0010 M in NaCl and 0.0010 M in NaBr. The K_{sp} for AgCl is 1.8 x 10^{-10} and the K_{sp} for AgBr is 3.3 x 10^{-13}. What concentration of Br$^-$ remains in solution when 50.% of the Cl$^-$ has precipitated?

(a) 3.6 x 10^{-7} M (b) 1.2 x 10^{-8} M (c) 3.3 x 10^{-10} M
(d) 9.2 x 10^{-7} M (e) 4.2 x 10^{-9} M

Simultaneous Equilibria Involving Slightly Soluble Compounds

d 62. Consider an aqueous solution containing 0.10 M magnesium nitrate and some ammonia. Which response contains all of the following statements that are correct, and no others? K_{sp} for $Mg(OH)_2 = 1.5 \times 10^{-11}$.

    I.   The solution is saturated with respect to magnesium nitrate.
    II.   If the concentration of ammonia is high enough, magnesium hydroxide will precipitate.
    III.   If ammonium chloride is added, the hydroxide ion concentration in solution will be reduced.
    IV.   Nitrate ions, from any ammonium nitrate added, increase the hydroxide ion concentration by reacting with water to form nonionized nitric acid molecules and hydroxide ions. This could cause precipitation of magnesium hydroxide.

(a) I, II, and III   (b) I and III   (c) II and IV
(d) II and III   (e) another combination

d 63. What concentration of aqueous NH_3 is necessary to just start precipitation of $Mn(OH)_2$ from a 0.020 M solution $MnSO_4$? K_b for NH_3 is 1.8×10^{-5} and K_{sp} for $Mn(OH)_2$ is 4.6×10^{-14}.

(a) $1.4 \times 10^{-5} M$ (b) $3.7 \times 10^{-7} M$ (c) $1.6 \times 10^{-6} M$
(d) $1.3 \times 10^{-7} M$ (e) $8.4 \times 10^{-2} M$

c 64. If a solution is to be made 0.010 M in $Mg(NO_3)_2$ **and** 0.20 M in aqueous NH_3, how many mol/L of NH_4Cl are required to prevent the precipitation of $Mg(OH)_2$ at 25°C? The K_b for aqueous NH_3 is 1.8×10^{-5} and the K_{sp} for $Mg(OH)_2$ is 1.5×10^{-11}.

(a) 0.054   (b) 0.56   (c) 0.092   (d) 1.8   (e) 0.86

b 65. What minimum concentration of ammonium ion (from ammonium chloride) is necessary to prevent the precipitation of $Mn(OH)_2$ from a solution which is to be made 0.030 M in $Mn(NO_3)_2$ and 0.030 M in aqueous ammonia? The solubility product for $Mn(OH)_2$ is 4.6×10^{-14}, and the ionization constant for aqueous ammonia is 1.8×10^{-5}.

(a) 0.18 M (b) 0.44 M (c) 0.54 M (d) 0.72 M (e) 0.88 M

e 66. If 1.0 liter of solution is to be made 0.010 M in $Mg(NO_3)_2$ and 0.10 M in aqueous ammonia, how many moles of NH_4Cl are necessary to prevent the precipitation of magnesium hydroxide? The solubility product for $Mg(OH)_2$ is 1.5×10^{-11}, and the ionization constant for aqueous ammonia is 1.8×10^{-5}.

(a) 0.018 mol   (b) 0.016 mol   (c) 0.020 mol
(d) 0.040 mol   (e) 0.046 mol

b 67. A 2.0-liter volume of a solution is $4.2 \times 10^{-5} M$ in $Mn(NO_3)_2$ and 0.10 M in ammonia. How many moles of NH_4NO_3 are required to prevent the precipitation of $Mn(OH)_2$? K_{sp} for $Mn(OH)_2 = 4.6 \times 10^{-14}$ and K_b for $NH_3 = 1.8 \times 10^{-5}$.

(a) 0.62 mol   (b) 0.11 mol   (c) 0.048 mol   (d) 0.24 mol   (e) 0.033 mol

e*68. A 50.0 mL-sample of 1.8 M aqueous NH$_3$ is mixed with an equal volume of a solution containing 0.010 mol of MgCl$_2$. What mass of NH$_4$Cl must be added to the resulting solution to prevent precipitation of Mg(OH)$_2$? K_{sp} for Mg(OH)$_2$ = 1.5 x 10^{-11} and K_b for NH$_3$ = 1.8 x 10^{-5}.

(a) 4.4 g (b) 5.7 g (c) 6.2 g (d) 6.6 g (e) 7.1 g

Dissolving Precipitates

a 69. The solubility of magnesium carbonate, MgCO$_3$, can be increased by acidifying the solution in which MgCO$_3$ is suspended. This is an example of dissolution via _____.

(a) formation of a weak electrolyte
(b) reduction of the anion
(c) oxidation of the anion
(d) formation of a complex ion
(e) changing an ion to another species which is not a weak electrolyte

d 70. Which of the following salts would **not** be expected to be more soluble in acidic solution than in pure water?

(a) Fe(OH)$_3$ (b) FeS (c) ZnCO$_3$ (d) AgI (e) Hg$_2$(CH$_3$COO)$_2$

c 71. Which of the following salts would be expected to be more soluble in acidic solution than in pure water?

(a) AgBr (b) Hg$_2$I$_2$ (c) CaF$_2$ (d) PbCl$_2$ (e) BaSO$_4$

e 72. How might one test if PbS has been completely precipitated from a saturated H$_2$S solution?

(a) heat the solution   (b) add a strong acid   (c) add a weak acid
(d) add more H$_2$S   (e) add a weak base

e 73. Consider the following system: Some solid AgCl is in contact with its saturated aqueous solution. Gaseous NH$_3$ is then bubbled through the solution slowly with swirling until all the AgCl dissolves. Choose the response which includes all the **true** statements, and no others.

I.   At the point at which all the AgCl has dissolved [Ag$^+$] = [Cl$^-$].
II. The AgCl dissolves due to formation of the complex ion, [Ag(NH$_3$)$_2$]$^+$.
III. As the NH$_3$ is added to the solution, the [Cl$^-$] increases.
IV. The solution remains saturated with silver chloride as long as any solid AgCl remains.
V.   The solubility of silver chloride in aqueous ammonia solution is greater than in pure water.

(a) I, II, and IV   (b) II, III, and IV   (c) I, IV, and V
(d) II and V   (e) II, III, IV, and V

c 74. Consider the following two complex ions of Cd^{2+} and their dissociation constants, K_d.

	K_d
$[Cd(NH_3)_4]^{2+}$	1.0×10^{-7}
$[Cd(CN)_4]^{2-}$	7.8×10^{-18}

Which response contains all the following statements that are true, and no others?

I. $[Cd(CN)_4]^{2-}$ is more dissociated in 0.010 M solution than $[Cd(NH_3)_4]^{2+}$ is in 0.010 M solution.
II. Adding strong acid to a solution that is 0.010 M in $[Cd(NH_3)_4]^{2+}$ would tend to dissociate the complex ion.
III. Adding strong acid to a solution that is 0.010 M in $[Cd(CN)_4]^{2-}$ would tend to dissociate the complex ion.
IV. If one wished to dissolve a certain weight of insoluble CdI_2 by formation of $[Cd(CN)_4]^{2-}$ ions by adding either NaCN or HCN, fewer moles of NaCN would be required.

(a) I, III, and IV   (b) I and II   (c) II, III, and IV
(d) III   (e) another combination

b*75. Calculate the number of moles of ammonia needed to dissolve 0.0010 mole of silver chloride in 4000. mL of solution at 25°C. K_{sp} for AgCl is 1.8×10^{-10} and the dissociation constant for $[Ag(NH_3)_2]^+$ is 6.3×10^{-8}.

(a) 0.16 mole   (b) 0.021 mole   (c) 14 moles
(d) 0.28 mole   (e) 1.5 moles

c 76. Calculate the minimum concentration of NH_3 necessary in 1.0 liter of solution to just dissolve 3.0×10^{-3} mole of AgBr at 25°C. K_{sp} for AgBr = 3.3×10^{-13}, K_d for $[Ag(NH_3)_2]^+$ = 6.3×10^{-8}.

(a) 6.2 M   (b) 8.6 M   (c) 1.3 M   (d) 0.18 M   (e) 0.067 M

c*77. What concentration of NH_3 must be present in 4.0×10^{-3} M Ag^+ to prevent the precipitation of AgCl when $[Cl^-]$ reaches 1.0×10^{-3} M? K_{sp} for AgCl = 1.8×10^{-10} and K_d for $[Ag(NH_3)_2]^+$ = 6.3×10^{-8}.

(a) 7.4×10^{-4} M   (b) 6.1×10^{-3} M   (c) 4.5×10^{-2} M
(d) 1.6×10^{-2} M   (e) 8.5×10^{-1} M

b 78. Calculate the concentration of Cu^{2+} ions in a 0.010 M $[Cu(NH_3)_4]^{2+}$ solution at 25°C. K_d for $[Cu(NH_3)_4]^{2+}$ = 8.5×10^{-13}

(a) 1.3×10^{-3} M   (b) 5.1×10^{-4} M   (c) 6.3×10^{-5} M
(d) 8.7×10^5 M   (e) 1.3×10^{-6} M

a 79. Calculate the $[Zn^{2+}]$ in a solution which is initially 0.30 M in $[Zn(CN)_4]^{2-}$ ions at 25°C. K_d for $[Zn(CN)_4]^{2-}$ = 1×10^{-19}.

(a) 4×10^{-5} M   (b) 6×10^{-8} M   (c) 5×10^{-6} M
(d) 1×10^{-3} M   (e) 3×10^{-4} M

b 80. Calculate the concentration of free Ni^{2+} in 1.0 M $[Ni(NH_3)_6](NO_3)_2$ at 25°C. K_d for $[Ni(NH_3)_6]^{2+}$ is 1.8×10^{-9}.

(a) 1.6×10^{-5} M (b) 0.012 M (c) 3.2×10^{-4} M
(d) 0.084 M (e) 4.4×10^{-5} M

a 81. What is the concentration of free Cd^{2+} ions in a solution that is 0.30 M in $Na_2[Cd(CN)_4]$? K_d for $[Cd(CN)_4]^{2-}$ is 7.8×10^{-18}.

(a) 9.8×10^{-5} M (b) 6.2×10^{-4} M (c) 8.6×10^{-5} M
(d) 1.6×10^{-4} M (e) 2.8×10^{-4} M

c 82. Calculate the concentration of cadmium ions in a 0.10 M $[Cd(CN)_4]^{2-}$ solution. The complex ion dissociates: $[Cd(CN)_4]^{2-} \rightleftharpoons Cd^{2+} + 4CN^-$, and its dissociation constant is 7.8×10^{-18}.

(a) 3.0×10^{-5} M (b) 5.7×10^{-5} M (c) 7.9×10^{-5} M
(d) 9.8×10^{-5} M (e) 2.2×10^{-4} M

d 83. Calculate the concentration of cyanide ions in a 0.10 M $[Cd(CN)_4]^{2-}$ solution. The complex ion dissociates: $[Cd(CN)_4]^{2-} \rightleftharpoons Cd^{2+} + 4CN^-$, and its dissociation constant is 7.8×10^{-18}.

(a) 1.1×10^{-4} M (b) 7.9×10^{-5} M (c) 9.4×10^{-6} M
(d) 3.2×10^{-4} M (e) 8.8×10^{-4} M

21 Electrochemistry

Electrodes

c 1. In **any** electrochemical cell, the cathode is **always** _____.

(a) the positive electrode (b) the negative electrode
(c) the electrode at which some species gains electrons
(d) the electrode at which some species loses electrons
(e) the electrode at which oxidation occurs

d 2. In **any** electrochemical cell, the anode is **always** _____.

(a) the positive electrode (b) the negative electrode
(c) the electrode at which some species gains electrons
(d) the electrode at which some species loses electrons
(e) the electrode at which reduction occurs

The Electrolysis of Molten Salts

d 3. During the electrolysis of molten sodium bromide, sodium ions move _____.

(a) to the anode, which is positively charged
(b) to the anode, which is negatively charged
(c) to the cathode, which is positively charged
(d) to the cathode, which is negatively charged
(e) through the wire to the battery

c 4. The half-reaction that occurs at the cathode during the electrolysis of molten sodium bromide is _____.

(a) $2Br^- \rightarrow Br_2 + 2e^-$ (b) $Br_2 + 2e^- \rightarrow 2Br^-$
(c) $Na^+ + e^- \rightarrow Na$ (d) $Na \rightarrow Na^+ + e^-$
(e) $2H_2O + 2e^- \rightarrow 2OH^- + H_2$

a 5. The half-reaction that occurs at the anode during the electrolysis of molten sodium bromide is _____.

(a) $2Br^- \rightarrow Br_2 + 2e^-$ (b) $Br_2 + 2e^- \rightarrow 2Br^-$
(c) $Na^+ + e^- \rightarrow Na$ (d) $Na \rightarrow Na^+ + e^-$
(e) $2H_2O + 2e^- \rightarrow 2OH^- + H_2$

b 6. What product is formed at the anode when molten sodium chloride, NaCl, is electrolyzed using a Downs cell?

(a) O_2 (b) Cl_2 (c) NaOH (d) H_2 (e) Na metal

e 7. Which of the following statements concerning use of the Downs cell for the electrolysis of molten sodium chloride is **incorrect**?

    (a) The pale green gas, Cl₂, is liberated at the anode.
    (b) The anode is the positive electrode and the cathode is the negative electrode for this cell.
    (c) The molten sodium and gaseous chlorine products must be kept separated to prevent an explosive reaction.
    (d) The Downs cell is expensive to run because of the cost of construction, of electricity and of heat to melt the NaCl.
    (e) Although production of sodium metal by the electrolysis of aqueous NaCl is much cheaper, the electrolysis of molten NaCl is used because it gives a purer metal.

e 8. The electrolysis of **molten** lithium hydride, LiH, using inert electrodes produces metallic lithium and gaseous hydrogen. Lithium is produced at the ___(I)___, which is the ___(II)___ electrode. Electrons flow through the wire from the ___(III)___ to the other electrode.

	(I)	(II)	(III)
(a)	anode	negative	anode
(b)	cathode	positive	anode
(c)	cathode	positive	cathode
(d)	anode	positive	cathode
(e)	cathode	negative	anode

b 9. The electrolysis of **molten** lithium hydride, LiH, using inert electrodes produces metallic lithium and gaseous hydrogen. The hydrogen is produced by the ___(I)___ half-reaction, ___(II)___.

	(I)	(II)
(a)	reduction	2H⁺ + 2e⁻ → H₂(g)
(b)	oxidation	2H⁻ → H₂(g) + 2e⁻
(c)	reduction	2e⁻ + H₂O → H₂(g) + O²⁻
(d)	oxidation	2e⁻ + 2OH⁻ → H₂(g) + 2O²⁻
(e)	auto-oxidation-reduction	H⁺ + H⁻ → H₂(g)

The Electrolysis of Aqueous Salt Solutions

d 10. In an electrolytic cell, the electrode that acts as a source of electrons **to the solution** is called the _____; the chemical change that occurs at this electrode is called _____.

    (a) anode, oxidation   (b) anode, reduction
    (c) cathode, oxidation   (d) cathode, reduction
    (e) Cannot answer unless we know the species being oxidized and reduced.

a 11. Which one of the following is **not** obtained from the electrolysis of an **aqueous** solution of NaCl?

    (a) Na   (b) NaOH   (c) Cl₂   (d) H₂
    (e) All of the preceding are obtained from this electrolysis.

b 12. The electrolysis of an aqueous sodium chloride solution using inert electrodes produces gaseous chlorine at one electrode. At the other electrode gaseous hydrogen is produced, and the solution becomes basic around the electrode. Which of the following is the equation for the cathode half-reaction in this electrolytic cell?

(a) $2Cl^- \rightarrow Cl_2 + 2e^-$
(b) $Cl_2 + 2e^- \rightarrow 2Cl^-$
(b) $2H_2O + 2e^- \rightarrow H_2 + 2OH^-$
(d) $H_2 + 2OH^- \rightarrow 2H_2O + 2e^-$
(e) none of these

e 13. What reaction occurs at the anode during the electrolysis of aqueous Na_2SO_4?

(a) $Na^+ + e^- \rightarrow Na$
(c) $SO_4^{2-} \rightarrow SO_4 + 2e^-$
(b) $2H^+ + 2e^- \rightarrow H_2$
(d) $SO_4^{2-} \rightarrow SO_2 + O_2 + 2e^-$
(e) $2H_2O \rightarrow O_2 + 4H^+ + 4e^-$

b 14. Consider the electrolysis of an aqueous solution of aluminum fluoride. Which one of the following statements describes what will be observed?

(a) Al metal is produced at one electrode, and O_2 and H^+ produced at the other.
(b) O_2 and H^+ are produced at one electrode, and H_2 and OH^- produced at the other.
(c) Al metal is produced at one electrode, and F_2 produced at the other.
(d) Al metal is produced at one electrode, and O_2 and H^+ produced at the other.
(e) H_2 and OH^- are produced at one electrode, and F_2 produced at the other.

a 15. For the electrolysis of aqueous KCl solution using inert electrodes, chlorine gas is evolved at one electrode and hydrogen gas is evolved at the other electrode. The solution around the electrode at which hydrogen gas is evolved becomes basic as the electrolysis proceeds. Which reaction occurs at the anode?

(a) $2Cl^- \rightarrow Cl_2 + 2e^-$
(c) $Cl_2 + 2e^- \rightarrow 2Cl^-$
(b) $2H_2O \rightarrow O_2 + 4H^+ + 4e^-$
(d) $2H_2O + 2e^- \rightarrow H_2 + 2OH^-$
(e) None of the first four responses is correct.

e 16. For the electrolysis of aqueous KCl solution using inert electrodes, chlorine gas is evolved at one electrode and hydrogen gas is evolved at the other electrode. The solution around the electrode at which hydrogen gas is evolved becomes basic as the electrolysis proceeds. Which of the following describe, or are applicable to, the cathode and the reaction that occurs at the cathode?

I.	the positive electrode	II.	$2H_2O \rightarrow O_2 + 4H^+ + 4e^-$
III.	the negative electrode	IV.	$2H_2O + 2e^- \rightarrow H_2 + 2OH^-$
V.	$2Cl^- \rightarrow Cl_2 + 2e^-$	VI.	electrons enter external circuit (wire)
VII.	$Cl_2 + 2e^- \rightarrow 2Cl^-$	VIII.	electrons leave external circuit (wire)
IX.	oxidation	X.	reduction

(a) III, IV, VIII, and IX
(b) I, II, VI, and IX
(c) II, III, VI, and X
(d) I, IV, VIII, and X
(e) None of the first four responses is correct.

d 17. For the electrolysis of aqueous KCl solution using inert electrodes, chlorine gas is evolved at one electrode and hydrogen gas is evolved at the other electrode. The solution around the electrode at which hydrogen gas is evolved becomes basic as the electrolysis proceeds. Which of the following describe, or are applicable to, the electrode at which chlorine is produced and its reaction?

I.	the positive electrode	II.	$2H_2O \rightarrow O_2 + 4H^+ + 4e^-$
III.	the negative electrode	IV.	$2H_2O + 2e^- \rightarrow H_2 + 2OH^-$
V.	$2Cl^- \rightarrow Cl_2 + 2e^-$	VI.	electrons enter external circuit (wire)
VII.	$Cl_2 + 2e^- \rightarrow 2Cl^-$	VIII.	electrons leave external circuit (wire)
IX.	oxidation	X.	reduction

(a) III, V, VIII, and X (b) III, V, VI, and X (c) I, VII, VIII, and X
(d) I, V, VI, and IX (e) None of the first four responses is correct.

Faraday's Law of Electrolysis

e 18. If 1 faraday of electricity is passed through an electrolytic cell containing a solution of a metal salt, the metal may be reduced and deposited at the cathode. Below are given the masses of metals that would be deposited from different salt solutions. Which one is **incorrect**?

(a) 107.9 g Ag from AgCl (b) 31.8 g Cu from $CuSO_4$ (c) 63.6 g Cu from CuCl
(d) 65.7 g Au from $AuCl_3$ (e) 26.0 g Cr from $Cr_2(SO_4)_3$

c 19. How many coulombs of charge pass through a cell if 2.40 amperes of current are passed through the cell for 85.0 minutes?

(a) 2.04×10^2 C (b) 1.33×10^{-1} C (c) 1.22×10^4 C
(d) 2.12×10^3 C (e) 3.40 C

d 20. How many faradays (F) are passed through a cell if 4.50 amperes of current are passed through the cell for 2.50 hours?

(a) 6.82 F (b) 0.0207 F (c) 302 F (d) 0.420 F (e) 6.99×10^{-3} F

a 21. How many moles of chromium would be electroplated by passing a current of 5.2 amperes through a solution of $Cr_2(SO_4)_3$ for 45.0 minutes?

(a) 0.048 mol (b) 2.9 mol (c) 0.15 mol (d) 6.9 mol (e) 0.073 mol

c 22. An aqueous copper(II) sulfate solution is electrolyzed for 45 minutes. A 3.2 ampere current is used. What mass of copper is produced?

(a) 0.95 g (b) 1.9 g (c) 2.8 g (d) 4.6 g (e) 5.5 g

b 23. How many grams of metallic nickel can be produced by the electrolysis of aqueous nickel(II) chloride, $NiCl_2$, with a 0.350 ampere current for 5.00 hours?

(a) 1.19 g (b) 1.92 g (c) 7.66 g (d) 2.76 g (e) 3.83 g

b 24. What mass of chromium could be deposited by electrolysis of an aqueous solution of $Cr_2(SO_4)_3$ for 60.0 minutes using a steady current of 10.0 amperes?

(a) 3.25 g (b) 6.47 g (c) 17.3 g (d) 0.187 g (e) 0.373 g

e 25. Molten $AlCl_3$ is electrolyzed for 5.0 hours with a current of 0.40 ampere. Metallic aluminum is produced at one electrode and chlorine gas, Cl_2, is produced at the other. How many grams of aluminum are produced?

(a) 1.1 g (b) 0.22 g (c) 0.44 g (d) 0.58 g (e) 0.67 g

d 26. How many moles of oxygen, O_2, will be produced by the electrolysis of water with a current of 1.5 amperes for 6.0 hours?

(a) 0.014 mol (b) 0.34 mol (c) 0.74 mol
(d) 0.084 mol (e) 0.0014 mol

c 27. Molten $AlCl_3$ is electrolyzed for 5.0 hours with a current of 0.40 ampere. Metallic aluminum is produced at one electrode and chlorine gas, Cl_2, is produced at the other. How many liters of Cl_2 measured at STP are produced at the other electrode?

(a) 0.56 L (b) 0.63 L (c) 0.84 L (d) 0.98 L (e) 1.02 L

b 28. How many liters of F_2 at STP could be liberated from the electrolysis of molten NaF under a 2.16 ampere current for 60.0 minutes?

(a) 0.774 L (b) 0.902 L (c) 1.55 L (d) 1.80 L (e) 2.16 L

b 29. How many mL of chlorine gas can be obtained at STP by the electrolysis of molten NaCl using a 0.350 ampere current for 20.0 minutes?

(a) 195 mL (b) 48.7 mL (c) 133 mL (d) 97.5 mL (e) 582 mL

a 30. Calculate the quantity of charge necessary to produce 10 liters of $H_2(g)$ at STP from the electrolysis of water.

(a) 8.6 x 10^4 coulombs (b) 3.7 x 10^4 coulombs (c) 1.7 x 10^4 coulombs
(d) 5.3 x 10^4 coulombs (e) 4.8 x 10^4 coulombs

b 31. How long would a constant current of 18.0 amperes be required to flow in order for 9000. coulombs of charge to pass through a cell?

(a) 200. s (b) 500. s (c) 0.002 s (d) 50. s (e) 1.6 x 10^5 s

c 32. How many minutes would a 5.00 ampere current have to be applied to plate out 8.00 grams of copper metal from aqueous copper(II) sulfate solution?

(a) 14.6 minutes (b) 33.3 minutes (c) 81.0 minutes
(d) 124 minutes (e) 188 minutes

c*33. Elementary fluorine is prepared by electrolysis of molten KHF_2. How long would it take to produce 11.0 liters of F_2 at STP using a current of 15.0 amperes? The half-reaction is $HF_2^- \rightarrow HF + \frac{1}{2}F_2 + e^-$

(a) 3.64 hours
(b) 8.78 hours
(c) 1.76 hours
(d) 2.26 hours
(e) 2.74 hours

e*34. Current is passed through a cell where the half-reaction that occurs at the cathode is $5e^- + MnO_4^- + 8H^+ \rightarrow Mn^{2+} + 4H_2O$. All the MnO_4^- ions present in 25.0 mL of solution have been reduced after a current of 0.600 ampere has passed for 844 seconds. What was the original concentration of MnO_4^- ions?

(a) $7.10 \times 10^{-3}\ M$
(b) $1.02 \times 10^{-1}\ M$
(c) $0.21\ M$
(d) $1.47\ M$
(e) $4.20 \times 10^{-2}\ M$

Voltaic or Galvanic Cells

c 35. In voltaic cells, such as those diagrammed in your text, the salt bridge _____.

(a) is not necessary in order for the cell to work
(b) acts as a mechanism to allow mechanical mixing of the solutions
(c) allows charge balance to be maintained in the cell
(d) is tightly plugged with a firm agar gel through which ions cannot move
(e) drives electrons from one half-cell to the other

a 36. Oxidation occurs at the _____ in a voltaic cell and oxidation occurs at the _____ in an electrolytic cell.

(a) anode, anode
(b) cathode, cathode
(c) anode, cathode
(d) cathode, anode
(e) anode, salt bridge

e 37. Which response contains all the following statements that are **correct**, and no others?

I. In voltaic cells the flow of electrons is spontaneous.
II. In electrolytic cells electrons flow in the external circuit (through the wire) from the anode to the cathode.
III. In voltaic cells the cathode is the positive electrode.

(a) I
(b) III
(c) I and II
(d) II and III
(e) I, II, and III

c 38. A cell is constructed by immersing a strip of iron in a solution that is $1.0\ M$ in Fe^{2+} ions and connecting this electrode to a standard hydrogen electrode using a wire and a salt bridge. The following observations can be made. The strip of iron loses mass, and the concentration of Fe^{2+} ions increases. In the standard hydrogen electrode, the concentration of hydrogen ions decreases. Which of the following is the strongest reducing agent?

(a) Fe^{2+}
(b) H^+
(c) Fe
(d) H_2
(e) Fe^{3+}

c 39. A cell is constructed by immersing a strip of lead in a 1.0 M Pb(NO$_3$)$_2$ solution and a strip of silver in a 1.0 M AgNO$_3$ solution. The circuit is completed by a wire and a salt bridge. As the cell operates, the strip of silver gains mass (only silver), and the concentration of silver ions in the solution around the silver strip decreases, while the strip of lead loses mass, and the concentration of lead ions increases in the solution around the lead strip. Which of the following represents the reaction that occurs at the cathode in this cell?

(a) Pb^{2+} + 2e$^-$ → Pb
(b) Pb → Pb^{2+} + 2e$^-$
(c) Ag$^+$ + e$^-$ → Ag
(d) Ag → Ag$^+$ + e$^-$
(e) none of the above

b 40. A cell is constructed by immersing a strip of lead in a 1.0 M Pb(NO$_3$)$_2$ solution and a strip of silver in a 1.0 M AgNO$_3$ solution. The circuit is completed by a wire and a salt bridge. As the cell operates, the strip of silver gains mass (only silver), and the concentration of silver ions in the solution around the silver strip decreases, while the strip of lead loses mass, and the concentration of lead ions increases in the solution around the lead strip. Which equation represents the reaction that occurs at the negative electrode in the above cell?

(a) Pb^{2+} + 2e$^-$ → Pb
(b) Pb → Pb^{2+} + 2e$^-$
(c) Ag$^+$ + e$^-$ → Ag
(d) Ag → Ag$^+$ + e$^-$
(e) none of the above

d 41. A voltaic cell is constructed by immersing a strip of copper metal in 1.0 M CuSO$_4$ solution and a strip of aluminum in 0.50 M Al$_2$(SO$_4$)$_3$ solution. A wire and a salt bridge complete the circuit. The aluminum strip loses mass, and the concentration of aluminum ions in the solution increases. The copper electrode gains mass, and the concentration of copper ions decreases. Which of the following are applicable to the copper electrode?

I. The anode
II. The cathode
III. The positive electrode
IV. The electrode at which electrons are produced
V. The negative electrode
VI. The electrode at which electrons are used up

(a) I, III, and V (b) I, IV, and V (c) II, IV, and V (d) II, III, and VI
(e) None of the first four responses contains all the correct choices and no others.

b 42. A voltaic cell is constructed by immersing a strip of copper metal in 1.0 M CuSO$_4$ solution and a strip of aluminum in 0.50 M Al$_2$(SO$_4$)$_3$ solution. A wire and a salt bridge complete the circuit. The aluminum strip loses mass, and the concentration of aluminum ions in the solution increases. The copper electrode gains mass, and the concentration of copper ions decreases. Which of the following are applicable to the electrode at which oxidation occurs?

I. The anode
II. The cathode
III. The positive electrode
IV. The electrode at which electrons are produced
V. The negative electrode
VI. The electrode at which electrons are used up

(a) I, III, and V (b) I, IV, and V (c) II, V, and VI (d) II, III, and IV
(e) None of the first four responses contains all the correct choices and no others.

Standard Electrode Potentials

e 43. Which statement concerning a table of standard reduction potentials is **incorrect**?

    (a) On the left side of the table are cations, hydrogen ion and elemental nonmetals.
    (b) On the right side of the table are elemental metals, hydrogen and nonmetal anions.
    (c) Fluorine, F_2, is the strongest oxidizing agent.
    (d) The alkali metals (Group IA) are among the strongest reducing agents.
    (e) The more positive the reduction potential, the stronger the species on the left acts as a reducing agent.

c 44. Which of the following is the strongest oxidizing agent?
    (a) Al^{3+} (b) Al   (c) F_2 (d) F^- (e) H_2

c 45. Which of the following species is the strongest oxidizing agent?
    (a) Zn^{2+} (b) Zn   (c) Hg^{2+} (d) Hg   (e) Na

b 46. Which of the following species is the strongest reducing agent?
    (a) Zn^{2+} (b) Zn   (c) Hg^{2+} (d) Hg   (e) Ag

a 47. Which of the following is the strongest reducing agent?
    (a) Na   (b) Au   (c) Cu^{2+} (d) F_2 (e) Cd

d 48. Which of the following is the weakest reducing agent?
    (a) Al^{3+} (b) Al   (c) F_2 (d) F^- (e) K^+

e 49. Which of the following is the weakest oxidizing agent?
    (a) Cu^{2+} (b) Zn^{2+} (c) F^- (d) Na   (e) K^+

e 50. Which of the following metals is most easily oxidized?
    (a) Cd   (b) Cu   (c) Fe   (d) Ni   (e) Zn

a 51. Which one of the following statements about the oxidizing strength of the group IB metals (Cu, Ag, and Au) is **true**?

    (a) Cu is easier to oxidize than Au.
    (b) Au is easier to oxidize than Ag.
    (c) Ag is easier to oxidize than Cu.
    (d) Au is easier to oxidize than Cu.
    (e) Nothing can be decided about this from a table of electrode potentials.

d*52. Determine the order in which the following metal ions are reduced as increasing voltage is applied to electrolytic cells in which molten salts of these cations are present. List the first one reduced first, and the last one reduced last.

Ag^+, Mg^{2+}, Na^+, Fe^{3+}

(a) Mg^{2+}, Fe^{3+}, Ag^+, Na^+ (b) Na^+, Mg^{2+}, Fe^{3+}, Ag^+ (c) Fe^{3+}, Mg^{2+}, Na^+, Ag^+
(d) Ag^+, Fe^{3+}, Mg^{2+}, Na^+ (e) Na^+, Fe^{3+}, Ag^+, Mg^{2+}

b 53. At standard conditions, which of the following can oxidize Hg to Hg^{2+}?

(a) Cu^{2+} (b) Cl_2 (c) Ag   (d) Zn^{2+} (e) Zn

d 54. At standard conditions, which one of the following will reduce Ni^{2+} to Ni but will not reduce Mg^{2+} to Mg?

(a) Cd^{2+} (b) Sn   (c) Ca   (d) Cr   (e) K

Uses of Standard Electrode Potentials

e 55. Given the following standard electrode potentials:

Half-Reaction	E^0
$Al^{3+} + 3e^- \rightarrow Al$	−1.66 V
$Hg^{2+} + 2e^- \rightarrow Hg$	0.86 V

Suppose the standard Al^{3+}/Al and Hg^{2+}/Hg electrodes are hooked up to deliver energy. Which response contains all the true statements regarding the cell and no others?

I.   The anode is Al.   II.   The negative terminal is Hg.
III.   If the standard hydrogen electrode were substituted for the Hg^{2+}/Hg half-cell, the cell voltage would be 1.66 volts.
IV.   Anions in the salt bridge will flow toward the Hg^{2+}/Hg compartment.
V.   The concentration of Al^{3+} will increase.

(a) II, IV, and V   (b) I, II, and IV   (c) I and IV
(d) II and V   (e) I, III, and V

b 56. Which one of the following statements about the half-cell processes is true for the cell, $Cd|Cd^{2+}(1\ M)||Cu^{2+}(1\ M)|Cu$?

(a) Cu^{2+} is reduced at the anode.   (b) Cu^{2+} is reduced at the cathode.
(c) Cd^{2+} is reduced at the anode.   (d) Cd^{2+} is reduced at the cathode.
(e)   The spontaneous reaction that occurs in this cell is not a redox reaction.

d 57. A voltaic cell consists of a standard hydrogen electrode connected by a salt bridge and a wire to an electrode consisting of a strip of Cd metal dipping into a 1 M solution of $Cd(NO_3)_2$. When the cell produces current, the electrons flow through the wire from the _____ electrode to the _____ electrode. In this cell the _____ electrode acts as the cathode.

(a) Cd, H_2, Cd   (b) H_2, Cd, Cd   (c) H_2, Cd, H_2
(d) Cd, H_2, H_2 (e) None of the first four responses are correct.

d 58. Which of the following describes the net reaction that occurs in the cell, Cd|Cd²⁺(1 M)||Cu²⁺(1 M)|Cu?

(a) Cu + Cd²⁺ → Cu²⁺ + Cd
(b) Cu + Cd → Cu²⁺ + Cd²⁺
(c) Cu²⁺ + Cd²⁺ → Cu + Cd
(d) Cu²⁺ + Cd → Cu + Cd²⁺
(e) 2Cu + Cd²⁺ → 2Cu⁺ + Cd

b 59. Which one of the following reactions is spontaneous (in the direction given) under standard electrochemical conditions?

(a) Pb²⁺ + 2I⁻ → Pb + I₂
(b) Cu²⁺ + Fe → Cu + Fe²⁺
(c) 2Au + Pt²⁺ → 2Au⁺ + Pt
(d) Mg²⁺ + 2Br⁻ → Mg + Br₂
(e) 2Hg + 2Cl⁻ + 2H⁺ → Hg₂Cl₂ + H₂

d 60. What is the cell potential for a cell constructed by immersing a strip of copper in a 1.0 M CuSO₄ solution and a strip of tin in a 1.0 M SnSO₄ solution and completing the circuit by a wire and a salt bridge?

(a) –0.19 V   (b) +0.19 V   (c) –0.48 V   (d) +0.48 V   (e) +0.54 V

b 61. A voltaic cell is constructed by immersing a strip of copper metal in 1.0 M CuSO₄ solution and a strip of aluminum in 0.50 M Al₂(SO₄)₃ solution. A wire and a salt bridge complete the circuit. The aluminum strip loses mass, and the concentration of aluminum ions in the solution increases. The copper electrode gains mass, and the concentration of copper ions decreases. What is the cell potential?

(a) +1.28 V
(b) +2.00 V
(c) +2.34 V
(d) +2.50 V
(e) +3.66 V

a 62. What is the numerical value for the standard cell potential for the following reaction?
    2Cr³⁺(aq) + 3Cu(s) → 2Cr(s) + 3Cu²⁺(aq)

(a) –1.08 V   (b) –0.40 V   (c) 0.40 V   (d) 1.08 V   (e) 2.52 V

a 63. Calculate the standard cell potential for the cell, Cd|Cd²⁺(1 M)||Cu²⁺(1 M)|Cu.

(a) +0.74 V   (b) –0.74 V   (c) +0.06 V   (d) –0.06 V   (e) 0.00 V

Standard Electrode Potentials for Other Half-Reactions

a 64. Which of the following is the strongest oxidizing agent in 1 M solutions?

(a) MnO₄⁻ (in aqueous acidic solution)
(b) Ag⁺(aq)
(c) Cu²⁺(aq)
(d) Ca²⁺(aq)
(e) Li⁺(aq)

d 65. A voltaic cell is constructed by immersing a strip of cobalt metal into a 1.0 M solution of $Co(NO_3)_2$ and connecting this electrode by a wire and salt bridge to a strip of platinum in contact with a solution that is 0.50 M in $Fe_2(SO_4)_3$ and 1.0 M in $FeSO_4$. Consult the table of standard electrode potentials to determine the reaction that occurs. Which of the following is the anode half-reaction?

(a) $Fe^{3+} + e^- \rightarrow Fe^{2+}$
(b) $Co^{2+} + 2e^- \rightarrow Co$
(c) $2H^+ + 2e^- \rightarrow H_2$
(d) $Co \rightarrow Co^{2+} + 2e^-$
(e) $Fe^{2+} \rightarrow Fe^{3+} + e^-$

c 66. Given the following standard electrode potentials:

Half-Reaction	E^0
$O_2(g) + 4H^+ + 4e^- \rightarrow 2H_2O$	+1.23 V
$2CO_2(g) + 2H^+ + 2e^- \rightarrow (COOH)_2$	−0.49 V

Which response contains all the true statements and no others? (Assume all species are present under standard electrochemical conditions.)

I. H_2O will spontaneously oxidize $(COOH)_2$ to form CO_2.
II. $O_2(g)$ will spontaneously oxidize $(COOH)_2$ to form CO_2.
III. $(COOH)_2$ will spontaneously reduce $O_2(g)$ to form H_2O.
IV. H^+ will spontaneously reduce $(COOH)_2$ to form CO_2.
V. CO_2 will spontaneously oxidize H_2O to form $O_2(g)$.

(a) II, IV, and V
(b) I, III, and IV
(c) II and III
(d) I and IV
(e) III and V

b 67. Which response includes all of the following reactions (and no others) that can be used (in the direction given) for a galvanic cell under standard electrochemical conditions?

I. $Sn^{2+} + Mg^{2+} \rightarrow Sn^{4+} + Mg$
II. $Cu^{2+} + Fe \rightarrow Cu + Fe^{2+}$
III. $H_2O_2 + Sn^{4+} \rightarrow O_2 + Sn^{2+} + 2H^+$
IV. $2Hg + 2Cl^- + 2H^+ \rightarrow Hg_2Cl_2 + H_2$

(a) I and II
(b) II
(c) I, III, and IV
(d) II and IV
(e) another combination

e 68. Which response lists all of the following reactions that are spontaneous?

I. $F_2(g) + SO_2 + 2H_2O \rightarrow 2F^- + SO_4^{2-} + 4H^+$
II. $Cl_2(g) + Sn^{2+} \rightarrow 2Cl^- + Sn^{4+}$
III. $2NO(g) + 4H_2O + 3Br_2(l) \rightarrow 6Br^- + 2NO_3^- + 8H^+$
IV. $H_2O_2 + 2H^+ + 2I^- \rightarrow 2H_2O(l) + I_2(s)$

(a) I and III
(b) II and IV
(c) I, II, and III
(d) II, III, and IV
(e) I, II, III, and IV

e 69. Which response lists all of the following reactions that are spontaneous under standard electrochemical conditions?

I. $Cr(s) + 3Fe^{3+} \rightarrow 3Fe^{2+} + Cr^{3+}$
II. $2Fe^{2+} + Cu^{2+} \rightarrow 2Fe^{3+} + Cu$
III. $Li(s) + H^+ \rightarrow Li^+ + \frac{1}{2}H_2(g)$

(a) I (b) II (c) III
(d) II and III (e) I and III

d 70. Which one(s) of the following reactions is/are spontaneous under standard electrochemical conditions?

I. $2Al^{3+} + 3Cu \rightarrow 2Al + 3Cu^{2+}$
II. $Sn^{2+} + H_2 \rightarrow Sn + 2H^+$
III. $Hg^{2+} + Ni \rightarrow Hg + Ni^{2+}$
IV. $O_2 + 4H^+ + 2Cd \rightarrow 2H_2O + 2Cd^{2+}$

(a) I and II (b) III (c) II and IV
(d) III and IV (e) II, III, and IV

Corrosion and Corrosion Protection

d 71. Which statement concerning protecting metals against corrosion is **false**?

(a) A metal protected by a thin layer of a less easily oxidized metal will corrode even more rapidly if the layer is breached because an adverse electrochemical cell is created.
(b) A metal may be directly connected to a "sacrificial anode" (another metal that is more active and therefore preferentially oxidized).
(c) Some metals naturally form a protective film on their surface, often the metal oxide.
(d) Galvanizing (coating iron with the more active metal, zinc) is only protective as long as the coating layer remains intact. If the zinc coating is broken, there is no more protection.
(e) Nonmetal coatings such as paint may be used to prevent contact with oxygen.

The Nernst Equation

a 72. Calculate the reduction potential of the Cu^{2+}/Cu electrode when $[Cu^{2+}] = 1.0 \times 10^{-8}$ M.

(a) +0.10 V (b) +0.33 V (c) +0.34 V
(d) +0.35 V (e) +0.37 V

e 73. Calculate the reduction potential of the Zn^{2+}/Zn electrode when $[Zn^{2+}] = 1.0 \times 10^{-8}$ M.

(a) –0.73 V (b) –0.75 V (c) –0.76 V
(d) –0.77 V (e) –1.00 V

a 74. A cell is constructed by immersing a strip of silver in 0.10 M AgNO$_3$ solution and a strip of lead in 1.0 M Pb(NO$_3$)$_2$ solution. A wire and salt bridge complete the cell. What is the potential of the silver electrode in the cell?

(a) 0.74 V (b) 0.80 V (c) 0.83 V
(d) 0.86 V (e) 0.88 V

b 75. A cell is constructed by immersing a strip of silver in 0.10 M AgNO$_3$ solution and a strip of lead in 1.0 M Pb(NO$_3$)$_2$ solution. A wire and salt bridge complete the cell. What is the potential for the cell?

(a) 0.90 V (b) 0.87 V (c) 0.95 V
(d) 0.96 V (e) 0.98 V

a 76. Calculate E_{cell} for the reaction below when [Zn^{2+}] = 1.00 M, [H$^+$] = 1.00 x 10^{-6} M, and P_{H_2} = 1.00 atm.

$$Zn(s) + 2H^+ \rightarrow Zn^{2+} + H_2(g)$$

(a) +0.41 V (b) +0.053 V (c) 0.64 V
(d) +1.12 V (e) +0.76 V

b 77. Consider the following reaction: If a galvanic cell utilizes this reaction and the initial concentrations are [Hg$_2^{2+}$] = 0.10 M, [Sn^{2+}] = 0.30 M, and [Sn^{4+}] = 0.20 M, what will be E_{cell} under these conditions?

$$Hg_2^{2+}(aq) + Sn^{2+}(aq) \rightarrow Sn^{4+}(aq) + 2Hg(s)$$

(a) 0.57 V (b) 0.61 V (c) 0.74 V (d) 0.66 V (e) 0.68 V

c 78. What is the cell potential for the reaction below if [Cr$_2$O$_7^{2-}$] = 0.0010 M, [Cr^{3+}] = 0.150 M, [H$^+$] = 1.00 M, and [Br$^-$] = 0.450 M, and some bromine is also present initially?

$$Cr_2O_7^{2-} + 14H^+ + 6Br^- \rightarrow 3Br_2(l) + 2Cr^{3+} + 7H_2O(l)$$

(a) +0.28 V (b) +0.31 V (c) 0.22 V (d) +0.24 V (e) +0.29 V

e 79. Calculate the cell potential of the following voltaic cell.
Zn|Zn^{2+}(1.0 x 10^{-8} M)||Cu^{2+}(1.0 x 10^{-6} M)|Cu

(a) +1.06 V (b) +1.10 V (c) +1.03 V (d) +1.14 V (e) +1.16 V

c 80. Calculate the cell potential of the following voltaic cell at 25°C.
Mg|Mg^{2+}(1.0 x 10^{-6} M)||Ag$^+$(1.0 x 10^{-2} M)|Ag

(a) +3.17 V (b) +3.29 V (c) +3.23 V (d) +3.11 V (e) +1.63 V

d 81. Calculate E_{cell} for the following voltaic cell.
Ag|Ag$^+$(1.0 x 10^{-5} M)||Au^{3+}(1.0 x 10^{-1} M)|Au

(a) +0.78 V (b) +0.46 V (c) +0.88 V (d) +0.98 V (e) +2.58 V

a 82. Calculate the cell potential for the following voltaic cell.
$Cr|Cr^{3+}(1.0 \times 10^{-2} M)||Co^{2+}(1.0 \times 10^{-5} M)|Co$

(a) +0.35 V   (b) +0.91 V   (c) +0.57 V   (d) +0.28 V   (e) –1.13 V

Using Electrochemical Cells to Determine Concentrations

d 83. Calculate the value of the reaction quotient, Q, if the observed cell voltage is 1.53 V for the following voltaic cell.
$Al|Al^{3+}(x\ M)||Sn^{2+}(y\ M)|Sn$

(a) 27   (b) 10   (c) 13   (d) 0.1   (e) 7×10^{-27}

a 84. Calculate the Cd^{2+} concentration in the following cell if E_{cell} – 0.23 V.
$Cd(s)|Cd^{2+}(x\ M)||Ni^{2+}(1.00\ M)|Ni$

(a) 0.0019 M (b) $1.4 \times 10^{-5}\ M$ (c) 0.0036 M
(d) 0.015 M (e) 0.0086 M

e 85. Calculate the [H+] for the hydrogen half-cell of the following voltaic cell if the observed cell voltage is 2.02 V.
$Mg|Mg^{2+}(1.00\ M)||H^{+}(?\ M), H_2(1.00\ atm)|Pt$

(a) $8.0 \times 10^5\ M$ (b) $1.1 \times 10^{-3}\ M$ (c) $9.8 \times 10^{-1}\ M$
(d) $8.0 \times 10^{-3}\ M$ (e) $1.2 \times 10^{-6}\ M$

a 86. Calculate the pH for the hydrogen half-cell of the following voltaic cell if the observed cell voltage is 1.00 V.
$Mn|Mn^{2+}(1.00\ M)||H^{+}(pH = ?), H_2(1.00\ atm)|Pt$

(a) 3.04   (b) 4.31   (c) 1.52   (d) 3.62   (e) 2.78

e 87. A concentration cell is constructed by placing identical iron electrodes in two Fe^{2+} solutions. The potential of this cell is observed to be 0.047 V. If the more concentrated Fe^{2+} solution is 0.10 M, what is the concentration of the other Fe^{2+} solution?

(a) $1.5 \times 10^{-2}\ M$ (b) $2.8 \times 10^{-5}\ M$ (c) $3.5 \times 10^{-4}\ M$
(d) $9.2 \times 10^{-2}\ M$ (e) $2.6 \times 10^{-3}\ M$

The Relationship of E^0_{cell} to ΔG^0 and K

b 88. What is ΔG^0 at 25°C for the reaction below? ($F = 96,500$ J/V·mol e−)
$Cu^{2+} + Cd \rightarrow Cu + Cd^{2+}$

(a) –71.1 kJ   (b) –143 kJ   (c) 597 kJ
(d) 193 kJ   (e) +71.1 kJ

b 89. Calculate the Gibbs free energy change for the reaction below when initial concentrations of Cr^{3+} and Cu^{2+} are 1.00 M. (F = 96,500 J/V·mol e⁻)

$$2Cr^{3+}(aq) + 3Cu(s) \rightarrow 2Cr(s) + 3Cu^{2+}(aq)$$

(a) –232 kJ (b) 623 kJ (c) 313 kJ (d) 232 kJ (e) –523 kJ

b 90. Calculate ΔG^0 for the following reaction from its E^0_{cell} value. (F = 96,500 J/V·mol e⁻)

$$3Hg_2Cl_2 + 2Cr \rightarrow 2Cr^{3+} + 6Hg + 6Cl^-$$

(a) –1.12 x 10³ kJ (b) –585 kJ (c) –361 kJ
(d) 1.62 x 10³ kJ (e) –1.78 x 10³ kJ

b 91. What is ΔG^0 per mole of dichromate ions for the reduction of dichromate ions, $Cr_2O_7^{2-}$, to Cr^{3+} by bromide ions, Br^-, in acidic solution? (F = 96,500 J/V·mol e⁻)

$$Cr_2O_7^{2-} + 14H^+ + 6Br^- \rightarrow 3Br_2(\ell) + 2Cr^{3+} + 7H_2O(\ell)$$

(a) +26.3 kJ (b) –145 kJ (c) +145 kJ (d) –26.3 kJ (e) –53.6 kJ

d 92. Calculate ΔG for the reaction of the cell below under the stated conditions. (F = 96,500 J/V·mol e⁻)

$$Zn|Zn^{2+}(1.0 \times 10^{-8}\ M)||Cu^{2+}(1.0 \times 10^{-6}\ M)|Cu$$

(a) –163 kJ (b) –192 kJ (c) 201 kJ (d) –212 kJ (e) –268 kJ

e 93. A voltaic cell is constructed by immersing a strip of copper metal in 1.0 M $CuSO_4$ solution and a strip of aluminum in 0.50 M $Al_2(SO_4)_3$ solution. A wire and a salt bridge complete the circuit. The aluminum strip loses mass, and the concentration of aluminum ions in the solution increases. The copper electrode gains mass, and the concentration of copper ions decreases. Calculate the equilibrium constant at 25°C for the cell reaction. (F = 96,500 J/V·mol e⁻ and R = 8.314 J/mol·K) (Round off errors are large!!!)

(a) 4 x 10⁴¹ (b) 2 x 10⁸³ (c) 6 x 10⁹⁵
(d) 3 x 10¹⁰¹ (e) 9 x 10²⁰²

d 94. Calculate the equilibrium constant at 25°C for the reaction below. Answer is only **approximate**. (F = 96,500 J/V·mol e⁻ and R = 8.314 J/mol·K)

$$2Cr^{3+}(aq) + 3Cu(s) \rightarrow 2Cr(s) + 3Cu^{2+}(aq)$$

(a) 10⁻¹⁶² (b) 10⁺⁵⁶ (c) 10⁺⁸⁸ (d) 10⁻¹⁰⁹ (e) 10⁻³⁸

e 95. The standard reduction potentials for the half-reactions $Ag^+(aq) + e^- \rightarrow Ag(s)$ and $AgCl(s) + e^- \rightarrow Ag(s) + Cl^-(aq)$ are 0.7994 and 0.222 volt, respectively. Calculate K_{sp} for AgCl at 25°C. (F = 96,500 J/V·mol e⁻ and R = 8.314 J/mol·K)

(a) 4.7 x 10⁻¹² (b) 3.2 x 10⁻⁸ (c) 6.2 x 10⁻⁹
(d) 4.3 x 10⁻⁸ (e) 1.7 x 10⁻¹⁰

d 96. Given the standard electrode potentials below, calculate K_c at 25°C for the following reaction. ($F = 96,500$ J/V·mol e⁻ and $R = 8.314$ J/mol·K)

$$2Fe^{3+}(aq) + 2I^-(aq) \rightarrow 2Fe^{2+}(aq) + I_2(s)$$

	E^0
$Fe^{3+}(aq) + e^- \rightarrow Fe^{2+}(aq)$	+0.771 V
$I_2(s) + 2e^- \rightarrow 2I^-(aq)$	+0.535 V

(a) 1.6×10^{12} (b) 1.1×10^{-8} (c) 1.0×10^{-4}
(d) 9.6×10^7 (e) 9.7×10^3

a 97. The equilibrium constant, at 25°C, for the reaction below is 1.34×10^{77}. What is E^0 for this reaction?

$$2Tl + BrO_3^- + 6H^+ \rightarrow 2Tl^{3+} + Br^- + 3H_2O$$

(a) 0.76 V   (b) 0.90 V   (c) 0.31 V   (d) 0.22 V   (e) 1.03 V

b 98. The equilibrium constant, at 25°C, for the reaction below is 1.99×10^{20}. What is E^0 for this reaction?

$$NO_3^- + 3H^+ + Cu \rightarrow Cu^{2+} + HNO_2 + H_2O$$

(a) 0.090 V   (b) 0.60 V   (c) 0.88 V   (d) 1.05 V   (e) 0.21 V

Commercial Voltaic Cells

c 99. Which of the following voltaic cells is **incorrectly** labelled?

(a) Leclanche dry cell   primary
(b) alkaline dry cell   primary
(c) mercury battery   secondary
(d) lead storage battery   secondary
(e) nickel-cadmium cell   secondary

d 100. Which of the following statements concerning the hydrogen-oxygen fuel cell is **incorrect**?

(a) Fuel cells are voltaic cells in which the reactants are continuously supplied to the cell and the products are continuously removed.
(b) Most of the chemical energy from the formation of H–O bonds is converted directly into electrical energy.
(c) The efficiency of energy conversion of the fuel cell is 60-70% of the theoretical maximum, about twice the efficiency of burning hydrogen in a heat engine with a generator.
(d) The cathode reaction is: $H_2 + 2OH^- \rightarrow 2H_2O + 2e^-$.
(e) The H_2/O_2 cell is nonpolluting, only H_2O is released.

22 Metals I: Metallurgy

Occurrence of the Metals

c 1. Which of the following metallic elements is **not** present in substantial quantities in a healthy human body?

    (a) sodium   (b) potassium   (c) lead
    (d) calcium   (e) magnesium

e 2. Four of the following occur as native ores. One does not. Which one?

    (a) Ag   (b) Au   (c) Pt   (d) Cu   (e) Al

e 3. Which response contains all the metals listed below that occur as native ores, and no other metals?

    K, Mg, Li, Cu, Ag, Ba, Cr, Zn, Pt

    (a) Mg, Ag, Zn   (b) K, Li, Ba, Cr, Zn   (c) Ba, Cr, Pt
    (d) Cu, Ba   (e) Cu, Ag, Pt

c 4. Which one of the following ores is a sulfide ore?

    (a) kaolinite   (b) magnesite   (c) galena
    (d) magnetite   (e) bauxite

d 5. Which one of the following ores is **not** a sulfide ore?

    (a) chalcopyrite   (b) chalcocite   (c) galena
    (d) dolomite   (e) cinnabar

c 6. Which ore of the following pairs of minerals and ores do not match?

	Mineral	Ore
(a)	Fe_2O_3	hematite
(b)	$CaCO_3$	limestone
(c)	PbS	gypsum
(d)	$BaSO_4$	barite
(e)	FeS_2	iron pyrite

e 7. Which of the following statements about natural sources of the elements is **incorrect**?

    (a) Helium is obtained from wells in the United States and Russia.
    (b) Halide salts of many of group IA elements are found in oceans, salt lakes, and brine wells.
    (c) Most noble gases (other than helium) are obtained from air.
    (d) Gold, platinum, and palladium are found as uncombined native ores.
    (e) Nitrogen and oxygen are obtained from water.

338

e 8. Extraction of metals from one of the following classes of ores is difficult and expensive. Which class?

(a) oxide ores (b) sulfide ores (c) halide ores
(d) carbonate ores (e) silicate ores

c 9. Metals are seldom obtained from silicate minerals. Why?

(a) The amount of silicate minerals on earth is very small.
(b) Almost no desirable metals exist as silicate mineral ores.
(c) The extraction of metals from silicates is very difficult and costly.
(d) Those few metals that exist as silicate ores are more cheaply available as native ores.
(e) Extraction of metals from silicates produces environmentally dangerous wastes.

Metallurgy

e 10. Which of the following is not one of the five possible steps of metallurgy?

(a) mining the ore (b) pretreatment of the ore
(c) reduction of ore to free metal (d) refining or purifying the metal
(e) disposal of metal after it has been used

Pretreatment of Ores

d 11. Which one of the following is a metallurgical pretreatment process?

(a) alloying (b) distillation (c) reduction
(d) flotation (e) refining

e 12. The cyclone separator is used in _____.

(a) a blast furnace (b) roasting (c) smelting
(d) electrolysis (e) the pretreatment of ores

d 13. Which one of the following is **not** converted to a metal oxide upon heating in air?

(a) ZnS (b) Mg(OH)$_2$ (c) CaCO$_3$ (d) HgS (e) NiS

a 14. What mass of CO_2 would be produced by heating a 500.-kg sample of limestone that is 75.0% CaCO$_3$? Assume complete reaction and no other source of carbon dioxide.

(a) 165 kg (b) 220. kg (c) 180. kg (d) 225 kg (e) 188 kg

b 15. What volume of CO_2 at STP would be produced by heating a 100.-g sample of limestone that is 82.0% CaCO$_3$? Assume complete reaction, and no other source of carbon dioxide.

(a) 0.183 L (b) 18.4 L (c) 26.7 L (d) 22.4 L (e) 12.3 L

c 16. What mass of MgO would be produced by heating a 150.-g sample that is 65.0% Mg(OH)$_2$? Assume complete reaction and no other source of magnesium.

(a) 141 g (b) 97.5 g (c) 67.4 g (d) 104 g (e) 95.4 g

e 17. What mass of NiO would be produced by roasting a 180.-g sample that is 55.0% NiS? Assume complete reaction and no other source of nickel.

(a) 148 g (b) 66.6 g (c) 219 g (d) 8.15 g (e) 81.5 g

b 18. Consider the roasting of 10.0 tons of sphalerite which is 60.0% ZnS. Assume that all the ZnS is converted to ZnO, and no other sulfur sources are present. How many tons of SO$_2$ are produced?

(a) 2.36 tons   (b) 3.94 tons   (c) 5.34 tons
(d) 6.00 tons   (e) 6.57 tons

a 19. A 210.g-sample of sphalerite is roasted and produces 128.0 g of ZnO. Assuming complete reaction and no other source of zinc, calculate the percentage of ZnS in the sphalerite.

(a) 72.9% (b) 50.9% (c) 61.0% (d) 83.5% (e) 49.0%

Reduction to the Free Metals

e 20. Which of the following metals can be obtained from its ores **only** by electrolysis of a molten salt?

(a) Zn (b) Cr (c) Cu (d) Ag (e) Ca

b 21. Which of the following metals can be obtained from its ores **without** using electrolysis?

(a) K (b) Fe (c) Al (d) Na (e) Li

a 22. Which of the following species **cannot** function as a reducing agent in metallurgy?

(a) F$_2$ (b) H$_2$ (c) C (d) CO (e) Al

d 23. Which of the following arranges the four metals in order of decreasing activity?

	most active						least active
(a)	Au	>	Cu	>	Al	>	Na
(b)	Cu	>	Al	>	Au	>	Na
(c)	Cu	>	Na	>	Al	>	Au
(d)	Na	>	Al	>	Cu	>	Au
(e)	none of these						

a 24. Titanium metal is produced by converting the oxide (TiO$_2$) to the halide salt (TiCl$_4$) and then reducing the halide salt with an active metal. TiCl$_4$ + 2Mg $\xrightarrow{heat}$ Ti + 2MgCl$_2$. What other metal is produced using the same reduction process?

(a) U (b) Al (c) Na (d) Fe (e) Ca

d 25. What mass of Hg would be produced by heating a 120.g-sample of cinnabar that is 82.0% HgS? Assume complete reaction and no other source of mercury.

(a) 103 g (b) 91.1 g (c) 107 g (d) 84.8 g (e) 114 g

d 26. What mass of Cl_2 would be produced by the electrolysis of 219 grams of molten NaCl?

(a) 66.4 g (b) 200. g (c) 100. g (d) 133 g (e) 266 g

e 27. What mass of chlorine would be produced by the complete electrolysis of a sample of molten $CaCl_2$ if the process produced 56.9 g of Ca?

(a) 39.7 g (b) 167 g (c) 49.8 g (d) 25.8 g (e) 101 g

b 28. What volume of chlorine at STP would be produced by the complete electrolysis of 1.00 kilogram of molten KCl? (Assume chlorine behaves as an ideal gas.)

(a) 100. L (b) 150. L (c) 200. L (d) 250. L (e) 300. L

Refining of Metals

c 29. Which of the following involves the use of an induction heater?

(a) Hall process (b) blast furnace (c) zone refining
(d) electrolysis (e) reverberatory furnace

e 30. Which of the following statements concerning zone refining is **incorrect**?

(a) Zone refining is used to produce extremely pure metals.
(b) The process consists of an induction heater, which surrounds an impure solid bar of metal, passing from one end of the bar to the other end.
(c) As the heater moves along the bar, it melts portions of the bar.
(d) As the heating element moves away, the melted portions of the bar recrystallize.
(e) The bar is purified because the impurities are burned out of the bar.

d 31. Metals obtained from reduction processes are almost always impure. Several methods given below may be used for further refining (purification). Which response lists all of these methods and no others?

I. zone refining   II. roasting   III. electrolysis
IV. alloying   V. distillation   VI. flotation

(a) I, II and III   (b) II, IV and VI   (c) I, III and VI
(d) I, III and V   (e) I and IV

Magnesium

a 32. What is a major natural source of magnesium?

(a) sea water   (b) native ore   (c) river water
(d) impurity in coal   (e) a by-product of purification of Cu ore

b 33. Which of the following is the reaction for the precipitation of Mg²⁺ ions from sea water?

(a) $Mg^{2+}(aq) + 2HCl(g) \rightarrow MgCl_2(s) + 2H^+(aq)$
(b) $Mg^{2+}(aq) + Ca(OH)_2(s) \rightarrow Mg(OH)_2(s) + Ca^{2+}(aq)$
(c) $Mg(s) + 2NaOH(aq) \rightarrow Mg(OH)_2(s) + 2Na(s)$
(d) $Mg^{2+}(aq) + O_2(g) + H_2(g) \rightarrow Mg(OH)_2(s)$
(e) $Mg^{2+}(aq) + H_2S(g) \rightarrow MgS(s) + 2H^+(aq)$

c 34. Sea water is 0.13% Mg by mass. What mass of sea water would be required to produce 50. g of Mg metal if the yield was 85%?

(a) 450 g
(b) 330 g
(c) 4.5 x 10⁴ g
(d) 3.3 x 10⁴ g
(e) 1.9 x 10³ g

Aluminum

d 35. Four of the following statements are applicable to aluminum or to the metallurgy of aluminum. One is **not**. Which one?

(a) Aluminum is obtained from oxide ores.
(b) The amphoteric nature of aluminum hydroxide is the basis for an important step in the metallurgy of aluminum.
(c) Graphite anodes are used in the electrolytic cell in which aluminum is produced, and these anodes are gradually converted to oxides.
(d) Aluminum is widely used as a structural material because metallic aluminum is resistant to oxidation by air.
(e) Recycled aluminum accounts for more than half of the production of this metal.

d 36. The Hall-Heroult process electrolyzes a molten mixture of Al_2O_3 and cryolite ($Na_3[AlF_6]$) to reduce the aluminum cations to aluminum metal. Why is cryolite used in the mixture?

(a) It lowers the E^0 for the reduction of the aluminum cations.
(b) It raises the E^0 for the reduction of the aluminum cations.
(c) It increases the conductivity of the mixture.
(d) It lowers the melting point of the mixture, thereby lowering the cost of the energy needed to melt the mixture.
(e) It increases the yield by putting more aluminum in the mixture.

c 37. A new economical process for the production of aluminum has been developed on a commercial scale. It uses only about 30% as much electrical energy as the Hall-Heroult process. What is the name of this new process?

(a) Reynolds process
(b) zone process
(c) Alcoa chlorine process
(d) basic oxygen process
(e) blast process

Iron

e 38. Four of the following statements are applicable to the metallurgy of iron. Which one is **not**?

(a) The ore is roasted to remove water, decompose carbonates, and oxidize sulfides.
(b) Large amounts of coke are used in producing iron.
(c) Carbon monoxide is the reducing agent.
(d) Limestone is added to form slag with silicate impurities in the ore.
(e) Mercury compounds are added to the blast furnace to form the very insoluble HgS ($K_{sp} = 3 \times 10^{-53}$) to reduce pollution of the air by release of volatile sulfur-containing compounds.

e 39. The calcium silicate formed in the blast furnace reduction of iron ore is called _____.

(a) charge (b) gangue (c) pig iron (d) flux (e) slag

e*40. Which of the following is **not** a metal added to iron to form a steel?

(a) Mn (b) Cr (c) Ni (d) V (e) C

b 41. The two most desirable iron ores are Fe_2O_3 and Fe_3O_4. Which contains the highest percentage of iron, and what is that percentage?

(a) Fe_2O_3, 69.9% (b) Fe_3O_4, 72.4% (c) Fe_2O_3, 35.0%
(d) Fe_2O_3, 40.0% (e) Fe_3O_4, 77.7%

d 42. What mass of Fe would be contained in a 80.0-g sample of hematite ore that is 76.0% Fe_2O_3? Assume no other form of Fe is present.

(a) 55.9 g (b) 47.2 g (c) 24.3 g (d) 42.5 g (e) 60.8 g

b 43. What mass of Fe would be contained in a 80.0-g sample of magnetite ore that is 76.0% Fe_3O_4? Assume no other form of Fe is present.

(a) 24.3 g (b) 44.0 g (c) 56.0 g (d) 57.2 g (e) 26.1 g

Copper

d 44. One of the following occurs as a native ore to a significant extent in a few locations on the earth's surface, but it is obtained commercially from low-grade ores. Which one?

(a) Hg (b) Al (c) Fe (d) Cu (e) Ba

c 45. Which of the following is **not** a copper containing mineral?

(a) chalcopyrite (b) chalcocite (c) limestone
(d) malachite (e) azurite

c*46. Which equation best represents the reaction that occurs during the roasting of chalcocite, which contains copper(I) sulfide?

(a) $2Cu_2SO_3 + O_2 \xrightarrow{heat} 2Cu_2SO_4$
(b) $2CuS + 3O_2 \rightarrow 2CuO + 2SO_2$
(c) $Cu_2S + O_2 \xrightarrow{heat} 2Cu + SO_2$
(d) $Cu_2S + O_2 \rightarrow 2CuO + S$
(e) none of these

b 47. Four of the following statements are applicable to the metallurgy of copper. Which one is **not** applicable?

(a) Copper is obtained primarily from low-grade sulfide ores.
(b) Coke is used as a reducing agent in the production of copper.
(c) The two main classes of copper ores are sulfides and basic carbonates.
(d) Copper is refined electrolytically by dissolving large bars of impure copper and redepositing copper onto thin plates of very pure copper in a copper(II) sulfate/sulfuric acid bath.
(e) Metals that are less active than copper are obtained from the anode mud when copper is refined, and the value of these less active metals may be great enough to pay for the refining of copper.

d 48. Which one of the following elements would **not** be found in the anode mud of an electrolytic cell used to refine copper?

(a) Au   (b) Ag   (c) Pt   (d) K   (e) Se

c 49. A sample of azurite is 70.0% $Cu_3(CO_3)_2(OH)_2$. Calculate the percent copper in the ore.

(a) 18.4%   (b) 27.6%   (c) 38.7%   (d) 55.3%   (e) 61.2%

a 50. What mass of Cu would be present in 36.0 g of pure malachite, $CuCO_3 \cdot Cu(OH)_2$?

(a) 20.7 g   (b) 18.1 g   (c) 12.2 g   (d) 23.2 g   (e) 13.9 g

b 51. The only compound containing copper in an ore is chalcopyrite, $CuFeS_2$. If a 120.g-sample of this ore contains 25.6 g of copper, what is the percentage of $CuFeS_2$ in the sample?

(a) 73.9%   (b) 61.6%   (c) 50.8%   (d) 42.8%   (e) 21.3%

Gold

d 52. Gold has been used and valued since ancient times. Which of the following is **not** a reason?

(a) It was available in the uncombined free state as a native ore.
(b) Its color became a symbol of wealth and beauty.
(c) It is an inactive metal that does not corrode or discolor.
(d) Its hardness allowed it to accept and hold a sharp edge.
(e) It could be easily worked by artisans to create objects of great beauty.

d 53. Gold is often recovered by alloying it with mercury. How is the gold and mercury then separated?

- (a) The mercury is oxidized and dissolved in water.
- (b) The gold is reduced to the free metal which sinks to the bottom of the mercury and the liquid mercury is poured off.
- (c) The gold is reacted with NaCN to form a complex ion that is soluble in water.
- (d) The mercury is distilled away.
- (e) Electrolytic reduction is used to separate the gold and mercury.

d 54. Sodium cyanide is an important reagent in the metallurgy of _____.

(a) Mg   (b) Fe   (c) Cu   (d) Au   (e) K

c 55. Why is the cyanide process for gold purification preferred over the use of mercury?

- (a) Mercury is much, much more poisonous than cyanide.
- (b) Costly electrolysis is required to separate the gold from mercury.
- (c) Mercury persists in the environment for a long time, and its poisoning is cumulative.
- (d) Gold is too reactive a metal and may react dangerously with mercury.
- (e) Mercury is a liquid and is much more difficult to handle safely compared to cyanide.

23 Metals II: Properties and Reactions

Group IA Metals: Properties and Occurrence

c 1. Which one of the following properties of the alkali metals does **not** increase as the group is descended?

(a) density   (b) atomic radius   (c) first ionization energy
(d) ionic radius   (e) all of these increase

e 2. All the isotopes of _____ are radioactive.

(a) Li   (b) K   (c) Rb   (d) Cs   (e) Fr

a 3. Which one of the following cations has the greatest ability to polarize anions?

(a) Li+   (b) Na+   (c) K+   (d) Rb+   (e) Cs+

d 4. Which statement concerning alkali metals is **incorrect**?

(a) Alkali metals are not found free in nature because they are so easily oxidized.
(b) The free metals, except lithium, are soft, silvery metals that can be cut with a knife.
(c) The free metals are very corrosive.
(d) All of the free metals have melting points above 100°C.
(e) These metals are excellent electrical and thermal conductors.

e 5. Which statement concerning Li+ ions is **incorrect**?

(a) Li+ ions have a higher charge density than other IA ions.
(b) Li+ ions can polarize large anions.
(c) Li+ ions exert a stronger attraction for H_2O molecules than do other IA cations.
(d) The magnitude of the standard reduction potential of Li is unexpectedly large.
(e) Li compounds have more ionic character than other corresponding IA metal compounds.

e*6. What is **not a contributing factor** to the fact that the magnitude of the standard reduction potential of Li is unexpectedly large compared to other IA metals?

(a) The Li+ ion is very small.
(b) The charge density (ratio of charge to size) is very high for Li+ ions.
(c) The Li+ ions exert a stronger attraction for H_2O molecules than do other IA ions.
(d) During reduction the H_2O molecules must be stripped off the Li+ ion, which is a very endothermic process.
(e) Lithium has the smallest atomic weight of the IA metals.

Reactions of the Group IA Metals

b 7. Which of the following compounds is a peroxide?

(a) Li_2O (b) Na_2O_2 (c) KO_2 (d) MnO_2 (e) TiO_2

b 8. Li, as is often true for elements of the second period, differs in many ways from the other members of its family. Li compounds often resemble the compounds of what element?

(a) Be (b) Mg (c) Al (d) Na (e) B

a 9. Li compounds resemble Mg compounds in some ways. This is known as _____.

(a) diagonal similarities (b) inert pair effect (c) transition properties
(d) allotropic properties (e) isotopic properties

c 10. Which of the following is **not** an example of the diagonal similarities of Li and Mg compounds?

(a) Li, unlike other IA metals, combines with N_2 to form a nitride as does Mg.
(b) Both Li and Mg **readily** combine with carbon to form carbides; other IA metals do not.
(c) Both metals exhibit the +1 oxidation state in most of their compounds.
(d) The solubilities of Li compounds are closer to the corresponding Mg compounds than those of other IA metal compounds.
(e) Li and Mg form normal oxides when burned in air; other alkali metals form peroxides and superoxides.

d 11. Which of the following statements is **false**?

(a) When burned in air at 1 atm pressure, Li forms a normal oxide.
(b) When burned in air at 1 atm pressure, the IA metals (other than Li) form peroxides or superoxides.
(c) The IA metal oxides are basic.
(d) The IA metal oxides react with water to form weak bases.
(e) The IA metals react vigorously with water.

Uses of Group IA Metals and Their Compounds

c 12. Which one of the following compounds or elements is **not** correctly matched with a use?

	Substance	Use
(a)	$NaHCO_3$	baking soda
(b)	NaOH	production of soap
(c)	K_2CO_3	treatment for manic depression
(d)	KNO_3	a fertilizer
(e)	Li	heat transfer in experimental nuclear reactors

d 13. Which one of the following compounds is **not** correctly matched with its common name?

	Compound	Common Name
(a)	$NaHCO_3$	baking soda
(b)	$Na_2CO_3 \cdot 10H_2O$	washing soda
(c)	KNO_3	niter or saltpeter
(d)	$NaNO_3$	caustic soda
(e)	Na_2CO_3	soda or soda ash

a 14. What property of metallic lithium makes it a good choice as a heat transfer medium in experimental nuclear reactors?

(a) high heat capacity (b) low density (c) not corrosive
(d) high melting point (e) small atomic size for a IA metal

c 15. Which one of the following metals exhibits the photoelectric effect with the lowest energy radiation, thereby making it useful in photoelectric cells?

(a) Li (b) K (c) Cs (d) Be (e) Ba

Trace Elements and Life

b 16. Which of the following was the first trace element shown to be essential to human nutrition?

(a) Al (b) Fe (c) Zn (d) Co (e) Ni

b 17. Which one of the following has **not** been demonstrated to be essential to either human or animal nutrition?

(a) Cr (b) W (c) Mo (d) Se (e) As

c 18. Which one of the following has been shown to be present in at least 200 enzymes?

(a) Al (b) Cu (c) Zn (d) Co (e) Mo

d 19. Which metal is present in vitamin B_{12}?

(a) nickel (b) iron (c) zinc (d) cobalt (e) chromium

c 20. Which statement concerning essentiality of trace elements is **false**?

(a) Some elements are toxic, and it is difficult to convince health specialists that a toxic element could be a dietary essential at low levels.
(b) Activation analysis and electrothermal atomic absorption spectroscopy allow detection of trace elements in concentrations of a few parts per billion.
(c) Chromium is one trace element that can be easily measured in most laboratories.
(d) New laboratories are equipped to measure these elements with the necessary precision.
(e) Contamination is a serious problem both in analytical laboratories and for controlling conditions for animal studies.

e 21. Few laboratories in the world can accurately measure Cr in food and body tissue. Why?

(a) The samples containing Cr decompose before measurements can be made.
(b) The Cr reacts with air too quickly to allow accurate measurement.
(c) Neither activation analysis nor electrothermal atomic absorption spectroscopy can detect Cr.
(d) The sample cannot be prepared for analysis without destroying some of the Cr.
(e) The stainless steel commonly present in analytical laboratories contains Cr which easily contaminates samples.

Group IIA Metals: Properties and Occurrence

b 22. Which one of the following metals has the greatest tendency to form covalent compounds?

    (a) Ca   (b) Be   (c) Sr   (d) Ba   (e) Rb

a 23. The properties of beryllium compounds are likely to resemble those of ____ compounds.

    (a) Al   (b) B   (c) Li   (d) Sc   (e) Na

c 24. Which statement concerning alkaline earth (IIA) metals is **incorrect**?

    (a) The IIA metals are too reactive to occur free in nature.
    (b) The IIA metals are obtained by electrolysis of their molten chlorides.
    (c) Most oxides of IIA metals are amphoteric.
    (d) All known isotopes of radium are radioactive.
    (e) The IIA metals have a wider range of chemical properties than the IA metals.

Reactions of the Group IIA Metals

e 25. Which one of the following metals reacts **least** vigorously with water?

    (a) K   (b) Na   (c) Sr   (d) Ca   (e) Mg

a 26. Which of the Group IIA elements is **least** metallic?

    (a) Be   (b) Mg   (c) Ca   (d) Sr   (e) Ba

e 27. Which one of the following oxides has the most basic character?

    (a) Al_2O_3 (b) B_2O_3 (c) BeO   (d) MgO   (e) CaO

d 28. Which response contains all the amphoteric oxides listed below and no others?

    Li_2O, BaO, SrO, Al_2O_3, Ga_2O_3

    (a) BaO, Al_2O_3 (b) Li_2O, Ga_2O_3 (c) SrO
    (d) Al_2O_3, Ga_2O_3 (e) another combination

e 29. Which one of the Group IIA metals reacts with oxygen, O_2, almost exclusively according to the equation give below?

    $M + O_2 \rightarrow MO_2$

    (a) Be   (b) Mg   (c) Ca   (d) Sr   (e) Ba

b 30. Which one of the following is **most likely** to substitute for Ca^{2+} in compounds in animal (including human) bodies?

    (a) Ba^{2+} (b) Sr^{2+} (c) K^+ (d) Na^+ (e) Be^{2+}

Uses of Group IIA Metals and Their Compounds

a 31. Which one of the following compounds or elements and its uses do **not** match?

	Substance	Use
(a)	2MgSO$_4$·H$_2$O	plaster of Paris
(b)	strontium salts	red color in fireworks
(c)	BaSO$_4$	absorbs x-rays
(d)	Be	windows for x-ray tubes
(e)	Mg	photographic flash bulbs

a 32. Which Group IIA metal forms an impervious coating of oxide that protects it from further oxidation?

(a) Mg (b) Ca (c) Sr (d) Ba (e) Ra

c 33. Which one of the Group IIA elements is used as a reducing agent in the metallurgy of uranium, thorium and other metals?

(a) Be (b) Mg (c) Ca (d) Sr (e) Ba

The Post-Transition Metals

d 34. Which one of the following is a post-transition metal?

(a) strontium (b) cesium (c) chromium
(d) lead (e) mercury

d 35. The most abundant metal in the earth's crust is _____.

(a) sodium (b) calcium (c) potassium
(d) aluminum (e) iron

e 36. Which one of the following metals exhibits the **inert s-pair effect**?

(a) Na (b) K (c) Mg (d) Al (e) Sn

c 37. Which metal melts if held in the hand and has the largest liquid state temperature range of any element (29.8°C and 2403°C)?

(a) B (b) Al (c) Ga (d) In (e) Tl

c 38. Aluminum does **not** dissolve to a significant extent in concentrated HNO$_3$ because _____.

(a) It is an inactive metal.
(b) Its E^0 value for reduction of its cation is too negative.
(c) It forms a protective oxide coating that is resistant to further attack by HNO$_3$.
(d) Its E^0 value for reduction of its cation is too positive.
(e) Aluminum nitrate is insoluble in water.

e 39. Which of the following statements is **incorrect**?

  (a) Thallium is very toxic and has no important uses as a free metal.
  (b) The greatest use for indium is in electronics.
  (c) Gallium-67 was one of the first artificially produced isotopes used in medicine.
  (d) Aluminum is the most reactive of the post-transition metals.
  (e) Boron is a post-transition metal that crystallizes as a covalent solid.

d 40. Which one of the following statements about gallium (Ga) is **incorrect**?

  (a) Ga will melt if held in the hand.
  (b) Ga has the largest liquid state temperature range of any element (29.8° - 2403°C).
  (c) Ga is used in transistors and high-temperature thermometers.
  (d) Ga-67 was the first naturally occurring isotope to be used in medicine.
  (e) The atomic radius of Ga is less than that of Al which is the Group IIIA element just above it in the periodic table.

e 41. The radii of Ga, In and Tl are smaller than would be predicted from the radii of the preceding Group IIIA elements, B and Al, with the radius of Ga being even smaller than the radius of Al. What is the explanation for this?

  (a) Ga, In and Tl have an ns^2np^1 outer electron configuration.
  (b) Ga, In and Tl have no partially filled shells.
  (c) Ga, In and Tl have filled ns subshells.
  (d) Ga, In and Tl can experience the inert s-pair effect.
  (e) The increase in nuclear charge that accompanies the filling of the $(n-1)d$ subshell causes a contraction in the size of the atoms of Ga, In and Tl.

d*42. Which one of the following ions hydrolyzes the most extensively?

  (a) K^+ (b) Ca^{2+} (c) Mg^{2+} (d) Al^{3+} (e) Tl^{3+}

General Properties — the d-Transition Metals

c 43. Which one of the following is **not** a general property of transition elements?

  (a) All are metals.
  (b) Most of them exhibit multiple oxidation states.
  (c) Most are softer than nontransition metals.
  (d) Many of these metals and their compounds are effective as catalysts.
  (e) Many of their compounds are paramagnetic.

d 44. Which one of the following statements about the general properties of **most** d-transition metals is **false**?

  (a) They have higher melting points than most nontransition metals.
  (b) Compounds that contain their ions are often colored.
  (c) They form complex ions.
  (d) They each exhibit a single oxidation state.
  (e) Most are paramagnetic.

c*45. Which one of the following has the highest melting point?

(a) Sc (b) Ti (c) Cr (d) Cu (e) Zn

e 46. The nitrate of which one of the following cations would be expected to be **colorless** in aqueous solution?

(a) Cr^{3+} (b) Co^{2+} (c) Ni^{2+} (d) Cu^{2+} (e) Ca^{2+}

e*47. Substances that contain unpaired electrons are attracted weakly into magnetic fields and are said to be _____.

(a) diamagnetic (b) ferromagnetic (c) degenerate
(d) ferrimagnetic (e) paramagnetic

b 48. Given below are the electron configurations for several neutral atoms or ions. Which represent species that would **not** be paramagnetic?

I. $[Ar]3d^64s^2$ II. $[Ar]3d^{10}4s^2$ III. $[Ar]3d^64s^0$
IV. $[Ar]3d^74s^2$ V. $[Ar]3d^{10}4s^0$ VI. $[Ar]3d^54s^1$

(a) II and VI (b) II and V (c) I, II, III and V
(d) III and V (e) I, II and VI

d*49. Which **one** of the following species **must** be paramagnetic?

(a) Zn (b) Zn^{2+} (c) Sc^{3+} (d) Mn (e) Cu^+

Oxidation States — d-Transition Metals

b 50. Which one of the following statements is **false**?

(a) The maximum oxidation state of the Group IVB elements is +4.
(b) The most common oxidation states of the 3d-transition elements are +3 and +5.
(c) The elements in the middle of each d-transition series exhibit more oxidation states than those elements at either end.
(d) The maximum oxidation state of Group VIIB elements is +7.
(e) Elements of Group VB can exhibit the +3 oxidation state.

c 51. Which one of the following d-transition metals would be expected to exhibit the **largest** number of oxidation states in its compounds?

(a) Sc (b) Zn (c) Mn (d) Ni (e) Ti

a 52. Which one of the following would be expected to be the most basic?

(a) $Mn(OH)_2$ (b) $Mn(OH)_3$ (c) H_2MnO_3 (d) H_2MnO_4 (e) $HMnO_4$

e 53. Which one of the following is the strongest oxidizing agent?

(a) $Mn(OH)_2$ (b) $Mn(OH)_3$ (c) H_2MnO_3 (d) H_2MnO_4 (e) $HMnO_4$

352

b*54. What is the electron configuration for Co²⁺?

(a) [Ar]3d⁵4s² (b) [Ar]3d⁷4s⁰ (c) [Ar]3d⁶4s⁰
(d) [Ar]3d⁴4s² (e) [Ar]3d⁶4s¹

Chromium Oxides, Oxyanions, and Hydroxides

e 55. Which one of the following is most acidic?

(a) $Cr(OH)_2$ (b) $Cr(OH)_3$ (c) CrO (d) Cr_2O_3 (e) CrO_3

a 56. Which one of the following is most basic?

(a) $Cr(OH)_2$ (b) $Cr(OH)_3$ (c) CrO_3 (d) Cr_2O_3 (e) $H_2Cr_2O_7$

a 57. Consider a solution in which the chromate-dichromate equilibrium has been established. If the solution has a pH of 12, the _____ ion predominates and the solution is _____.

$$2CrO_4^{2-} + 2H^+ \rightleftharpoons Cr_2O_7^{2-} + H_2O \qquad K_c = 4.2 \times 10^{14}$$

(a) chromate, yellow   (b) dichromate, yellow   (c) chromate, orange
(d) dichromate, orange   (e) chromite, blue

d 58. Consider a solution in which the chromate-dichromate equilibrium has been established. If the solution has a pH of 2, the _____ ion predominates and the solution is _____.

$$2CrO_4^{2-} + 2H^+ \rightleftharpoons Cr_2O_7^{2-} + H_2O \qquad K_c = 4.2 \times 10^{14}$$

(a) chromate, yellow   (b) dichromate, yellow   (c) chromate, orange
(d) dichromate, orange   (e) chromite, blue

c 59. For the reversible reaction below, $K_c = 4.2 \times 10^{14}$

$$2CrO_4^{2-} + 2H^+ \rightleftharpoons Cr_2O_7^{2-} + H_2O$$

What is the concentration of CrO_4^{2-} ions in a solution of pH = 3.00, which is also 0.010 M in $Cr_2O_7^{2-}$ ions?

(a) 6.6×10^{-10} M   (b) 4.2×10^{-11} M   (c) 4.9×10^{-6} M
(d) 2.4×10^{-11} M   (e) 2.4×10^{-9} M

a 60. A sample of chromite ore is 28.0% $FeCr_2O_4$. What mass of sodium chromate can be obtained by heating 1.00 kg of the ore in air with excess sodium carbonate?

$$4FeCr_2O_4 + 8Na_2CO_3 + 7O_2 \rightarrow 2Fe_2O_3 + 8Na_2CrO_4 + 8CO_2$$

(a) 405 g   (b) 167 g   (c) 248 g   (d) 386 g   (e) 448 g

d 61. What volume of 0.100 M potassium dichromate solution is necessary to oxidize 1.00 g of $FeSO_4$ to $Fe_2(SO_4)_3$, in the presence of excess H_2SO_4? The dichromate ion is reduced to $Cr_2(SO_4)_3$.

(a) 22.0 mL   (b) 14.4 mL   (c) 18.7 mL   (d) 11.0 mL   (e) 43.8 mL

24 Some Nonmetals and Metalloids

Noble Gases — Occurrence, Uses, and Properties

d 1. Which noble gas is used in airport runway and approach lights?

(a) He   (b) Ne   (c) Ar   (d) Kr   (e) Xe

c 2. Which noble gas can be dangerous when it escapes from the soil and collects in the basements of dwellings?

(a) Kr   (b) Ar   (c) Rn   (d) Ne   (e) Xe

d 3. Which one of the following pairs of noble gases and uses is **not** correctly matched?

	Noble Gas	Use
(a)	xenon	mixed with krypton in high intensity flash tubes
(b)	argon	inert atmosphere for welding
(c)	helium	diluent for gaseous anesthetics
(d)	neon	radiotherapy for cancerous tissues
(e)	argon	filling incandescent light bulbs

d 4. Which of the following statements about noble gases is **incorrect**?

(a) Except for radon, they can be isolated by fractional distillation of air.
(b) Radon can be collected from the radioactive disintegration of radium salts.
(c) Helium is produced from some natural gas fields.
(d) The only forces of attraction among noble gas atoms are dipole-dipole attractions.
(e) The attractive forces of He are so small that it cannot be solidified at 1 atm.

Xenon Compounds

c 5. Which statement regarding the noble gases is **false**?

(a) All fluorine compounds of the noble gases involve participation by the d orbitals of the noble gas atoms.
(b) Xenon will react with fluorine to form the colorless solids XeF_2, XeF_4, and XeF_6.
(c) All of the xenon fluorides are formed in endothermic reactions.
(d) The oxidation state of xenon in Cs_2XeF_8 is +6.
(e) Oxygen compounds of xenon are known.

c 6. Which one of the following compounds (apparently) involves sp^3d^2 hybridization at the xenon atom?

(a) XeO_3 (b) XeF_2 (c) XeF_4 (d) XeF_6 (e) XeO_4

d 7. What is the molecular geometry of XeF_4?

(a) tetrahedral   (b) octahedral   (c) trigonal bipyramidal
(d) square planar   (e) square pyramidal

Halogens — Properties

d 8. Which of the following does **not** correctly describe the halogen at room temperature and atmospheric pressure?

(a) fluorine - a pale yellow gas
(b) chlorine - a yellow-green gas
(c) bromine - a red-brown liquid
(d) iodine - a dark violet liquid
(e) astatine - a solid

a 9. Which ion is **not** easily polarized by cations?

(a) F⁻ (b) Cl⁻ (c) Br⁻ (d) I⁻ (e) At⁻

e 10. Which halogen has the largest value for its electronegativity?

(a) astatine (b) iodine (c) bromine (d) chlorine (e) fluorine

e 11. Which property of halogens decreases with increasing atomic weight?

(a) ionic radius
(b) melting point
(c) boiling point
(d) polarizability
(e) first ionization energy

Halogens — Occurrence, Production, and Uses

b 12. Which one of the following does **not** describe the major natural sources of the corresponding halogen?

(a) fluorine - fluorspar, CaF_2; cryolite, Na_3AlF_6; and fluorapatite, $Ca_5(PO_4)_3F$
(b) chlorine - sodium hypochlorite, NaClO
(c) bromine - bromide salts in seawater, salt brines, or salt beds
(d) iodine - sodium iodate, $NaIO_3$, in deposits of sodium nitrate, $NaNO_3$
(e) astatine - no appreciable natural sources; produced artificially

b 13. Which halogen is the most important in terms of commercial use?

(a) fluorine (b) chlorine (c) bromine (d) iodine (e) astatine

c 14. Which of the following is the strongest oxidizing agent?

(a) Br_2 (b) Cl⁻ (c) F_2 (d) F⁻ (e) I_2

a 15. What mass of fluorine is contained in two tons of cryolite that is 42.0% pure Na_3AlF_6?

(a) 912 lb (b) 832 lb (c) 456 lb (d) 304 lb (e) 152 lb

a 16. The density of liquid Br_2 is 3.12 g/mL at 20.°C. What volume of Br_2 is produced at this temperature by the action of excess MnO_2 on 250. g of KBr in the presence of H_2SO_4?

$$2KBr + MnO_2 + 2H_2SO_4 \rightarrow Br_2 + K_2SO_4 + MnSO_4 + 2H_2O$$

(a) 53.8 mL (b) 80.7 mL (c) 168 mL (d) 424 mL (e) 524 mL

Reactions of the Free Halogens

d 17. Which halogen would react **most** rapidly with iron?

(a) Cl_2 (b) Br_2 (c) I_2 (d) F_2 (e) At_2

e*18. Which one of the following elements would react with chlorine **most** rapidly?

(a) Li   (b) Na   (c) K   (d) Rb   (e) Cs

e 19. Which of the following equations does **not** represent a correct reaction?

(a) $2Fe + 3F_2 \rightarrow 2FeF_3$ (b) $Cu + Cl_2 \rightarrow CuCl_2$

(c) $Fe + I_2 \rightarrow FeI_2$ (d) $2Fe^{3+} + 2I^- \rightarrow 2Fe^{2+} + I_2$

(e) $Cu + I_2 \rightarrow CuI_2$

b 20. Write and balance the **formula unit** equation for the reaction of iron with fluorine. The sum of the coefficients in the balanced equation (include one's) is _____.

(a) 6   (b) 7   (c) 3   (d) 4   (e) 5

c 21. Write and balance the **formula unit** equation for the reaction of iron with iodine. The sum of the coefficients in the balanced equation (include one's) is _____.

(a) 6   (b) 7   (c) 3   (d) 4   (e) 5

The Hydrogen Halides and Hydrohalic Acids

e 22. Which one of the following does **not** correctly describe one or all of the hydrogen halides, HX?

(a) They are all colorless gases.
(b) HF exhibits hydrogen bonding.
(c) The order of increasing boiling points is HCl<HBr<HI<HF.
(d) Their aqueous solutions are acidic.
(e) Aqueous HF is the strongest hydrohalic acid.

c 23. What mass of SiF_4 could be produced by the reaction of 15 grams of SiO_2 with an excess of HF? The equation for the reaction is

$$SiO_2 + 4HF \rightarrow SiF_4 + 2H_2O$$

(a) 1.04 g   (b) 12 g   (c) 26 g   (d) 104 g   (e) 8.7 g

The Oxoacids (Ternary Acids) of the Halogens

e 24. What is the effective bleaching and disinfecting agent in solutions of Cl_2 or hypochlorite?

(a) Cl_2 (b) Cl^- (c) HOCl   (d) H^+ (e) O radicals

356

d 25. Write and balance the **formula unit** equation for the reaction of chlorine with **cold** water. The sum of the coefficients in the balanced equation (include one's) is _____.

(a) 6   (b) 7   (c) 8   (d) 4   (e) 5

b*26. The anhydride of perchloric acid is _____.

(a) ClO_2 (b) Cl_2O_7 (c) Cl_2O_3 (d) ClO_3 (e) Cl_2O_5

d 27. Calculate the percentage of bromine in $Ca(BrO_3)_2$.

(a) 24.6%   (b) 27.0%   (c) 31.2%   (d) 54.0%   (e) 62.4%

Occurrence, Properties, and Uses — S, Se, and Te

a 28. Which one of the Group VIA elements is the most different chemically from the other members of the group?

(a) O   (b) S   (c) Se   (d) Te   (e) Po

d 29. What is the largest and most important use of sulfur?

(a) production of sulfurous acid
(b) vulcanization of rubber
(c) prevention of pollution
(d) production of sulfuric acid
(e) synthesis of sulfur-containing organic compounds

e 30. Which sulfur-containing ore is **incorrectly** named?

	Formula	Name of Ore
(a)	PbS	galena
(b)	FeS_2	iron pyrite
(c)	HgS	cinnabar
(d)	S_8	sulfur
(e)	$CaSO_4 \cdot 2H_2O$	barite

d 31. Which one of the following elements most commonly crystallizes in a brass-colored hexagonal crystal with a metallic luster?

(a) O   (b) S   (c) Se   (d) Te   (e) all VIA elements do

e 32. Which formula is **incorrect**?

(a) SF_4 (b) SF_6 (c) SO_2 (d) SO_3 (e) SCl_3

e 33. Calculate the mass of $KClO_3$ that would be required to produce 29.52 liters of oxygen measured at 127°C and 760. torr.

$$2KClO_3 \xrightarrow{heat} 2KCl + 3O_2$$

(a) 7.82 g   (b) 12.2 g   (c) 14.6 g   (d) 24.4 g   (e) 73.5 g

Reactions of Group VIA Elements

d 34. Write and balance the **formula unit** equation for the reaction of sulfur with potassium. Represent sulfur as S rather than S₈. The sum of the coefficients in the balanced equation (include one's) is _____.

(a) six   (b) seven   (c) three   (d) four   (e) five

e 35. Which of the equations below does **not** represent a correct reaction of a VIA element?

(a) $H_2 + S \rightarrow H_2S$
(b) $Se + 3F_2 \rightarrow SeF_6$
(c) $2S + Br_2 \rightarrow S_2Br_2$
(d) $Se_2Cl_2 + Cl_2 \rightarrow 2SeCl_2$
(e) $S + 2O_2 \rightarrow SO_4$

d*36. Write and balance the **formula unit** equation for the reaction of sulfur with nitric acid. What is the sum of the coefficients? (Two of the products are SO_2 and NO. Represent sulfur as S.)

(a) 10   (b) 12   (c) 14   (d) 16   (e) 18

Hydrides of Group VIA Elements

d 37. Which statement about the Group VIA hydrides is **false**?

(a) H_2S, H_2Se, and H_2Te are all gases at 25°C and atmospheric pressure.
(b) All are colorless.
(c) All except H_2O are toxic.
(d) The order of boiling points is $H_2O > H_2S > H_2Se > H_2Te > H_2Po$.
(e) All are covalent compounds.

Group VIA Oxides

b 38. The hybridization at sulfur in SO_2 is _____.

(a) sp (b) sp^2 (c) sp^3 (d) sp^3d (e) sp^3d^2

c 39. A substance produced during the combustion of many fossil fuels is _____.

(a) S_8 (b) H_2S (c) SO_2 (d) S_2Cl_2 (e) CuS

b 40. Which acid listed on the right **cannot** be obtained by adding water to the substance on the left?

(a) SO_3 sulfuric acid
(b) TeO_2 telluric acid
(c) $H_2S_2O_7$ sulfuric acid
(d) SO_2 sulfurous acid
(e) SeO_2 selenous acid

c*41. Which formula and name do **not** match?

	Formula	Name
(a)	H₂SeO₃	selenous acid
(b)	TeF₆	tellurium hexafluoride
(c)	H₂S₂O₃	"fuming sulfuric acid"
(d)	NaHSO₃	sodium hydrogen sulfite
(e)	H₆TeO₆	telluric acid

Oxoacids of Sulfur

b 42. Which statement regarding sulfuric acid is **false**?

(a) It is often used as a dehydrating agent because of its strong affinity for H₂O.
(b) Concentrated sulfuric acid is about 78% H₂SO₄ by mass and 10 molar.
(c) It is a strong acid with respect to its first step ionization.
(d) Its second step ionization is not complete in concentrated solutions.
(e) It is often present in "acid rain".

b 43. What maximum mass of sulfuric acid can be produced from the sulfur contained in 100. kilograms of iron pyrite that is 75.0% FeS₂?

(a) 84.4 kg (b) 123 kg (c) 136 kg (d) 144 kg (e) 168 kg

c*44. For selenic acid, H₂SeO₄, K_{a1} is very large and $K_{a2} = 1.15 \times 10^{-2}$. What is the pH of a 0.20 M H₂SeO₄ solution?

(a) 0.48 (b) 0.55 (c) 0.68 (d) 0.75 (e) 0.84

Group VA Elements

e 45. Which one of the following is the most basic oxide?

(a) N₂O₃ (b) P₄O₆ (c) As₄O₆ (d) Sb₄O₆ (e) Bi₂O₃

c 46. In which one of the following is the oxidation state of nitrogen given **incorrectly**?

(a) H₂N₂O₂, +1 (b) HNO₃, +5 (c) N₂H₄, +2
(d) N₂O₃, +3 (e) NaNO₂, +3

e 47. How many different oxidation states does nitrogen exhibit?

(a) 2 (b) 3 (c) 5 (d) 6 (e) more than 6

e*48. The reaction of calcium nitride, Ca₃N₂, with water produces two basic compounds, one of which is a gas. Write the **formula unit** equation for the reaction. What is the sum of all coefficients?

(a) 8 (b) 9 (c) 10 (d) 11 (e) 12

Chapter 24   359

Occurrence of Nitrogen

b 49. Which statement concerning nitrogen is **false**?

(a) Nitrogen is a colorless, odorless, tasteless gas.
(b) The high abundance of N_2 in the atmosphere and the low relative abundance of nitrogen compounds elsewhere is due to the high chemical reactivity of N_2.
(c) The bacteria in the root nodules of legumes convert N_2 into NH_3.
(d) Ammonia is the source of nitrogen in many fertilizers.
(e) The primary natural inorganic deposits of nitrogen are as KNO_3 and $NaNO_3$.

Hydrogen Compounds of Nitrogen

c 50. Ammonia can act as a(an) _____ .

I.   Brønsted-Lowry base   II. Arrhenius base
III. Lewis base   IV. solvent

(a) I and II   (b) I, II, and III   (c) I, III, and IV
(d) II and III   (e) I, II, and IV

b 51. The following reaction can be considered a(an) _____ reaction and a _____ reaction.

$[Ni(OH_2)_6]^{2+} + 6NH_3 \rightarrow [Ni(NH_3)_6]^{2+} + 6H_2O$

(a) redox; Brønsted-Lowry acid-base
(b) Lewis acid-base; displacement
(c) Arrhenius acid-base; displacement
(d) Lewis acid-base; Brønsted-Lowry acid-base
(e) redox; displacement

d 52. Which statement about amines is **false**?

(a) Amines are organic compounds that are structurally related to ammonia.
(b) All amines are weak bases.
(c) All amines have a lone pair (unshared pair) of electrons on the N atom.
(d) All amines involve a sp^2-hybridized N.
(e) Amines can be considered as derived from NH_3 by replacing one or more hydrogen atoms with organic groups.

Nitrogen Oxides

d 53. Write the molecular equation for the thermal decomposition of ammonium nitrate at 170° to 260°C to produce dinitrogen oxide and steam. What is the sum of the coefficients?

(a) 4   (b) 5   (c) 6   (d) 8   (e) 9

d 54. Write the molecular equation for the thermal decomposition of ammonium nitrate to nitrogen, oxygen, and steam above 260°C. What is the sum of the coefficients?

(a) 6   (b) 7   (c) 8   (d) 9   (e) 5

b 55. Which one of the following statements about NO is **false**?

    (a)   It is paramagnetic.
    (b)   It is quite unreactive.
    (c)   It can be produced by internal combustion engines.
    (d)   It can be converted to NO_2 in the air.
    (e)   It can be produced by combination of N_2 and O_2 in the presence of lightning.

c 56. Which compound gives photochemical smog a brownish color?

(a) NO   (b) HNO_2 (c) NO_2 (d) N_2O_4 (e) N_2O_3

b 57. According to the following equation, the volume of nitrogen (STP) required to produce 0.400 mol of NO is _____.

$$N_2 + O_2 \rightarrow 2NO$$

(a) 2.24 L   (b) 4.48 L   (c) 6.72 L   (d) 11.2 L   (e) 8.96 L

Some Oxoacids of Nitrogen and Their Salts

c 58. Which statement about the oxoacids of nitrogen is **false**?

    (a)   Nitric acid is a strong acid and a strong oxidizing agent.
    (b)   Nitrous acid is a weak acid.
    (c)   Pure nitric acid is a liquid with a yellow or light brown color.
    (d)   Nitrous acid acts as an oxidizing agent toward strong reducing agents.
    (e)   Nitrous acid acts as a reducing agent toward very strong oxidizing agents.

c 59. Write the equation for the reaction of gaseous ammonia with oxygen in the presence of red-hot platinum. (This is the first step of the Ostwald process.) What is the sum of the coefficients?

(a) 17   (b) 18   (c) 19   (d) 20   (e) 21

b 60. Which one of the following is used as a color preservative in meats?

(a) NH_2Cl (b) $NaNO_2$ (c) NH_4Cl (d) N_2O_4 (e) $Fe(NO_3)_3$

b 61. Potassium nitrate can be used as a fertilizer. Calculate the percentage of nitrogen in potassium nitrate.

(a) 8.46%   (b) 13.8%   (c) 16.5%   (d) 22.3%   (e) 26.2%

a 62. Given: $3Cu + 8HNO_3 \rightarrow 3Cu(NO_3)_2 + 2NO + 4H_2O$
What volume of NO (gas) at STP could be prepared by the reaction of 6.35 grams of Cu with an excess of HNO_3?

(a) 1.49 L   (b) 2.24 L   (c) 4.48 L   (d) 6.72 L   (e) 0.746 L

Phosphorus

a 63. What is the major mineral present in phosphate rock?

(a) $Ca_3(PO_4)_2$ (b) NaH_2PO_4 (c) Na_2HPO_4
(d) $Ca_5(PO_4)_3(OH)$ (e) $Ca_5(PO_4)_3F$

c 64. Superphosphate of lime is produced by the reaction of phosphate rock with _____.

(a) Na_2CO_3 (b) HCl (c) H_2SO_4 (d) $CaCO_3$ (e) $Ca(HCO_2)_3$

d 65. The tips of "strike anywhere" matches contain two substances. One is red phosphorus; what is the other?

(a) white phosphorus, P_4 (b) calcium phosphate, $Ca_3(PO_4)_2$
(c) silica (sand), SiO_2 (d) tetraphosphorus trisulfide, P_4S_3
(e) tetraphosphorus decoxide, P_4O_{10}

c 66. What are the most stable forms of nitrogen and phosphorus at room temperature and atmospheric pressure?

(a) N_2, P_2 (b) N, P_4 (c) N_2, P_4 (d) N_4, P_2 (e) N_4, P_4

c 67. The compound P_4O_{10} dissolves in water to form phosphoric acid as shown by the equation:

$$P_4O_{10} + 6H_2O \rightarrow 4H_3PO_4$$

If 28.4 grams of P_4O_{10} is dissolved in enough water to give 500. mL of solution, the resulting solution will be _____ molar in H_3PO_4.

(a) 0.100 (b) 0.400 (c) 0.800 (d) 1.60 (e) 0.200

c 68. What mass of Na_3PO_4 could be prepared by the reaction of 49.0 g of H_3PO_4 with 80.0 g of NaOH?

$$H_3PO_4 + 3NaOH \rightarrow Na_3PO_4 + 3H_2O$$

(a) 8.2 g (b) 16.4 g (c) 82.0 g (d) 109 g (e) 164 g

Silicon and Silicates

e 69. Which of the following does **not** accurately describe elemental silicon?

(a) semiconductor (b) shiny, blue-gray solid
(c) a brittle metalloid (d) high melting point
(e) crystallizes in a graphite-like lattice

c 70. Amethyst and agate are among the gems that consist of _____ with colored impurities.

(a) SiH_4 (b) SiF_4 (c) SiO_2 (d) Na_2SiF_6 (e) $Na_6Si_2O_7$

e 71. Which of the following statements concerning silicon and its bonding is **false**?

    (a) Silicon does not form stable double bonds.
    (b) Silicon does not form very stable Si–Si bonds unless the silicon atoms are bonded to very electronegative elements.
    (c) Silicon has vacant $3d$ orbitals in its valence shell that can accept electrons from donor atoms.
    (d) Silicon forms its strongest single bond with oxygen.
    (e) Silicon dioxide consists of linear molecules.

a 72. Clay minerals generally have _____ structures with _____ surface areas.

    (a) sheet-like, large   (b) sheet-like, small   (c) globular, large
    (d) globular, small   (e) irregular, small

e 73. Fused silicate glasses consist primarily of _____.

    (a) K_2SiO_3 and Ag_2SiO_3 (b) K_2SiO_3 and $MgSiO_3$
    (c) Na_2SiO_3 and K_2SiO_3 (d) $CaSiO_3$ and $MgSiO_3$
    (e) Na_2SiO_3 and $CaSiO_3$

a*74. Write the **formula unit** equation for the hydrolysis of silicon tetrachloride. One of the products is H_4SiO_4. What is the sum of the coefficients?

    (a) 10   (b) 11   (c) 12   (d) 13   (e) 14

c*75. What volume of SiF_4 (STP) can be obtained by the treatment of 100. grams of SiO_2 with excess HF?

$$SiO_2 + 4HF \rightarrow SiF_4 + 2H_2O$$

    (a) 26.6 L   (b) 34.6 L   (c) 37.3 L   (d) 39.2 L   (e) 42.4 L

25 Coordination Compounds

Coordination Compounds

e 1. Alfred Werner isolated the platinum(IV) compounds, PtCl$_4$•nNH$_3$ (n = 2 to 6). He added excess AgNO$_3$ to solutions of each and determined the moles of AgCl precipitated. He also measured and compared the conductances of solutions of each of these compounds with solutions of simple electrolytes. Which response below gives **all** true statements concerning his conclusions from these experiments?

    I.   From the moles of AgCl, he determined the **total** number of Cl$^-$ ions in each compound.
    II.   From the moles of AgCl, he determined the number of **uncoordinated** Cl$^-$ ions in each compound.
    III.   From the conductance of each solution, he determined the **total** number of ions present upon dissolution of each compound.

    (a) I   (b) II   (c) III   (d) I and II   (e) II and III

c 2. A Lewis base makes available an electron pair to be shared to form a(n) _____ bond.

    (a) ionic   (b) radical   (c) coordinate covalent
    (d) hydrogen   (e) basic

e 3. A Lewis acid accepts a share in an electron pair to form a(n) _____ bond.

    (a) hydrogen   (b) cationic   (c) anionic   (d) pi   (e) coordinate covalent

b 4. What is the charge on a complex consisting of 1 cobalt(III), 4 ammonia ligands and 2 chloride ion ligands?

    (a) –2   (b) +1   (c) –1   (d) +3   (e) 0

b 5. In which one of the following is the oxidation state of the **underlined** metal incorrect?

	Complex Ion	Oxidation State of Underlined Metal
(a)	[Mn(en)$_2$(F)(I)]NO$_3$	+3
(b)	[Pt(NH$_3$)$_2$(Cl)$_2$]Cl$_2$	+2
(c)	K$_3$[Co(NO$_2$)$_6$]	+3
(d)	K$_3$[Fe(CN)$_6$]	+3
(e)	[Cr(en)$_3$]$_2$(SO$_4$)$_3$	+3

c 6. In which one of the following is the oxidation state of the **underlined** metal incorrect?

	Complex Species	Oxidation State of Underlined Metal
(a)	[Cr(OH$_2$)$_3$Br$_3$]	+3
(b)	[Co(OH$_2$)$_6$](NO$_3$)$_2$	+2
(c)	[Co(NH$_3$)$_2$(en)(Cl)$_2$]$_2$SO$_4$	+1
(d)	[Fe(CO)$_5$]	0
(e)	[Cr(NH$_3$)$_6$]$_2$[NiCl$_4$]$_3$	+2

c 7. In which one of the following is the oxidation number of the **transition** metal incorrect?

	Complex Species	Oxidation Number of Transition Metal
(a)	[Co(en)(NH$_3$)$_2$(OH)$_2$]Cl	+3
(b)	K$_2$[CuCl$_4$]	+2
(c)	K$_4$[Fe(CN)$_6$]	+3
(d)	[Co$_2$(CO)$_8$]	0
(e)	[Pt(NH$_3$)$_2$(OH$_2$)Cl]$_3$PO$_4$	+2

The Ammine Complexes

b 8. Which one of the following ions forms a soluble complex in the presence of excess of aqueous ammonia?

(a) Al^{3+} (b) Cu^{2+} (c) Pb^{2+} (d) Be^{2+} (e) Fe^{2+}

c 9. Which one of the following ions does **not** form a soluble complex in the presence of excess aqueous ammonia?

(a) Zn^{2+} (b) Ni^{2+} (c) Mn^{2+} (d) Co^{2+} (e) Ag$^+$

d 10. Which is **not** one of the families (Groups) of metals that form soluble ammine complexes with excess aqueous NH$_3$?

(a) zinc   (b) copper   (c) nickel   (d) iron   (e) cobalt

Important Terms

c 11. The number of donor atoms to which an acceptor metal ion is bonded in a complex species is called the _____ number of the metal ion.

(a) oxidation   (b) ligand   (c) coordination
(d) donor   (e) chelating

d 12. The _____ sphere is enclosed in brackets in formulas for complex species, and it includes the central metal ion plus the coordinated groups.

(a) ligand   (b) donor   (c) oxidation   (d) coordination   (e) chelating

c 13. Which ligand formula is **incorrectly** matched with its **name as a ligand**?

(a) H$_2$O – aqua   (b) NH$_3$ – ammine   (c) CN$^-$ – cyanide
(d) Cl$^-$ – chloro   (e) OH$^-$ – hydroxo

d 14. Ligands may bond to a metal through one or more donor atoms. Which term below **incorrectly** designates the number of donor atoms given?

(a) 1 – monodentate   (b) 2 – bidentate   (c) 3 – tridentate
(d) 4 – tetradentate   (e) 6 – hexadentate

a 15. Which of the following ligands is a **polydentate** ligand?

(a) ethylenediamine (b) ammonia (c) cyanide
(d) water (e) nitrite

d*16. Which of the following ligands is a **monodentate** ligand?

(a) ethylenediamine (b) oxalate (c) diethylenetriamine
(d) ammonia (e) ethylenediaminetetraacetate

c 17. Which answer below lists the **incorrect** number of donor atoms for the given ligand?

(a) ethylenediamine – 2 (b) carbon monoxide – 1 (c) cyanide – 1
(d) nitrite – 1 (e) ethylenediaminetetraacetate – 4

Nomenclature

c 18. Which ligand formula is **incorrectly** matched with its **name as a ligand**?

	Ligand	Name
(a)	SO_4^{2-}	sulfato
(b)	CO_3^{2-}	carbonato
(c)	$S_2O_3^{2-}$	sulfido
(d)	F^-	fluoro
(e)	O^{2-}	oxo

c 19. Which of the following names for metals in complex anions is derived from the Latin stem?

(a) zincate (b) platinate (c) ferrate
(d) chromate (e) cobaltate

d 20. Below is a list of formulas for complex compounds; each is matched with its name. One formula – name combination contains an error. Which one?

	Formula	Name
(a)	[Co(en)$_2$Br$_2$]Br	dibromobis(ethylenediamine)cobalt(III) bromide
(b)	Fe(NH$_3$)$_5$(OH$_2$)]Cl$_3$	pentaammineaquairon(III) chloride
(c)	K$_3$[Co(NO$_2$)$_6$]	potassium hexanitrocobaltate(III)
(d)	[Cr(NH$_3$)$_3$(OH)$_3$]	triamminetrihydroxochromate(III)
(e)	[Ni(en)$_3$](NO$_3$)$_2$	tris(ethylenediamine)nickel(II) nitrate

c 21. Below is a list of formulas for complex compounds and ions; each is matched with its name. One formula – name combination contains an error. Which one?

	Formula	Name
(a)	[Co(NH$_3$)$_4$(OH$_2$)(I)]SO$_4$	tetraammineaquaiodocobalt(III) sulfate
(b)	K[Cr(NH$_3$)$_2$Cl$_4$]	potassium diamminetetrachlorochromate(III)
(c)	[Mn(CN)$_5$]$^{2-}$	pentacyanomanganate(II) ion
(d)	[Ni(CO)$_4$]	tetracarbonylnickel(0)
(e)	Ca[PtCl$_4$]	calcium tetrachloroplatinate(II)

d 22. Below is a list of formulas for complex compounds; each is matched with its name. One formula – name combination contains an error. Which one?

	Formula	Name
(a)	$K_3[Fe(CN)_6]$	potassium hexacyanoferrate(III)
(b)	$Na[CuCl_3]$	sodium trichlorocuprate(II)
(c)	$Na_3[Co(CO_3)_3]$	sodium tricarbonatocobaltate(III)
(d)	$[CoClBr(NH_3)_4]I$	tetraamminebromochlorocobalt(II) iodide
(e)	$Rb_3[FeF_6]$	rubidium hexafluoroferrate(III)

c 23. Below is a list of formulas for complex compounds; each is matched with its name. One formula – name combination contains an error. Which one?

	Formula	Name
(a)	$[Cr(NH_3)_3(OH)_3]$	triamminetrihydroxochromium(III)
(b)	$K_3[FeF_6]$	potassium hexafluoroferrate(III)
(c)	$Na[Cu(CN)_2]$	sodium dicyanocuprate(II)
(d)	$[Co(en)_3](NO_3)_3$	tris(ethylenediamine)cobalt(III) nitrate
(e)	$Na_2[SnCl_6]$	sodium hexachlorostannate(IV)

c 24. Below is a list of formulas for complex compounds and ions; each is matched with its name. One formula – name combination contains an error. Which one?

	Formula	Name
(a)	$[Co(NH_3)_6][Co(CN)_6]$	hexaamminecobalt(III) hexacyanocobaltate(III)
(b)	$Ca_3[Fe(C_2O_4)_3]_2$	calcium tris(oxalato)ferrate(III)
(c)	$[Co(en)_3]_3(SO_4)_2$	tris(ethylenediamine)cobalt(III) sulfate
(d)	$[Pt(NH_3)_4Cl_2]^{2+}$	tetraamminedichloroplatinum(IV) ion
(e)	$K[Cr(NH_3)_2Cl_4]$	potassium diamminetetrachlorochromate(III)

c 25. Below is a list of formulas for complex compounds and ions; each is matched with its name. One formula – name combination contains an error. Which one?

	Formula	Name
(a)	$[AlF_6]^{3-}$	hexafluoroaluminate ion
(b)	$[PtCl_4]^{2-}$	tetrachloroplatinate(II) ion
(c)	$K_3[SnCl_6]$	potassium hexachlorostannate(IV)
(d)	$Mg[PdCl_6]$	magnesium hexachloropalladate(IV)
(e)	$[Co(NH_3)_4Cl_2]_3PO_4$	tetraamminedichlorocobalt(III) phosphate

c 26. Below is a list of formulas for complex compounds and ions; each is matched with its name. One formula – name combination contains an error. Which one?

	Formula	Name
(a)	$[Ni(NH_3)_6]^{2+}$	hexaamminenickel(II) ion
(b)	$[Co(NH_3)_3(H_2O)_2Cl]_2SO_4$	triamminediaquachlorocobalt(II) sulfate
(c)	$K[Pt(NH_3)Cl_6]$	potassium amminepentachloroplatinate(IV)
(d)	$[Sn(C_2O_4)_3]^{2-}$	tris(oxalato)stannate(IV) ion
(e)	$[Co(NH_3)_4Cl_2]_3AsO_4$	tetraamminedichlorocobalt(III) arsenate

b 27. Below is a list of formulas for complex compounds and ions; each is matched with its name. One formula – name combination contains an error. Which one?

	Formula	Name
(a)	[Ag(NH$_3$)$_2$]NO$_3$	diamminesilver(I) nitrate
(b)	[Fe(CN)$_6$]$^{3-}$	hexacyanoiron(III) ion
(c)	[Cr(OH$_2$)$_6$]$_2$(SO$_4$)$_3$	hexaaquachromium(III) sulfate
(d)	Na$_2$[Ni(CN)$_4$]	sodium tetracyanonickelate(II)
(e)	Li[AgI$_2$]	lithium diiodoargentate(I)

a 28. Below is a list of formulas for complex compounds; each is matched with its name. One formula – name combination contains an error. Which one?

	Formula	Name
(a)	[Ni(en)$_2$Br$_2$]	dibromobis(ethylenediamine)nickelate(II)
(b)	Ca[Al(OH)$_4$]$_2$	calcium tetrahydroxoaluminate
(c)	[Fe(CN)]Cl$_2$	cyanoiron(III) chloride
(d)	[Mn(CO)$_5$]	pentacarbonylmanganese(0)
(e)	[Cr(OH$_2$)$_4$SO$_4$]Cl	tetraaquasulfatochromium(III) chloride

e 29. Below is a list of formulas for complex compounds and ions; each is matched with its name. One formula – name combination contains an error. Which one?

	Formula	Name
(a)	[Fe(CO)$_5$]	pentacarbonyliron(0)
(b)	[Cu(NH$_3$)$_4$]SO$_4$	tetraamminecopper(II) sulfate
(c)	Na[Al(OH)$_4$]	sodium tetrahydroxoaluminate
(d)	[Cu(CN)$_2$]$^-$	dicyanocuprate(I) ion
(e)	[Cr(NH$_3$)$_6$]Cl$_3$	trichlorohexaamminechromium(III)

Structures

d 30. Which idealized geometry is **not** matched correctly with the given coordination number?

	Coordination number	Idealized Geometry
(a)	2	linear
(b)	4	tetrahedral
(c)	5	square pyramidal
(d)	6	trigonal bipyramidal
(e)	4	square planar

d 31. Below is a list of complex ions and their structures. One pair is **incorrectly** matched. Which pair?

	Complex Ion	Structure
(a)	[Zn(NH$_3$)$_4$]$^{2+}$	tetrahedral
(b)	[CoF$_6$]$^{4-}$	octahedral
(c)	[Cu(NH$_3$)$_4$]$^{2+}$	square planar
(d)	[Cu(CN)$_2$]$^-$	angular
(e)	[Ni(OH$_2$)$_6$]$^{2+}$	octahedral

e 32. Below is a list of complex ions and their structures. One pair is **incorrectly** matched. Which pair?

	Complex Ion	Structure
(a)	$[Fe(CN)_6]^{3-}$	octahedral
(b)	$[PtCl_6]^{2-}$	octahedral
(c)	$[PtCl_4]^{2-}$	square planar
(d)	$[Zn(OH)_4]^{2-}$	tetrahedral
(e)	$[Co(en)_2(Cl)_2]^+$	distorted tetrahedral

c 33. Below is a list of complex ions and their structures. One pair is **incorrectly** matched. Which pair?

	Complex Ion	Structure
(a)	$[Co(en)_3]^{3+}$	octahedral
(b)	$[Cu(en)_2]^{2+}$	square planar
(c)	$[Co(EDTA)]^-$	tetrahedral
(d)	$[Ni(CN)_4]^{2-}$	square planar
(e)	$[Cd(CN)_4]^{2-}$	tetrahedral

d 34. Below is a list of complex species and their structures. One pair is **incorrectly** matched. Which pair?

	Complex Species	Structure
(a)	$[Fe(CO)_5]$	trigonal bipyramidal
(b)	$[Cr(OH_2)_6]^{3+}$	octahedral
(c)	$[Ni(CN)_4]^{2-}$	square planar
(d)	$[CdBr_4]^{2-}$	square pyramidal
(e)	$[Cu(NH_3)_4]^{2+}$	square planar

a 35. Below is a list of complex species and their structures. One pair is **incorrectly** matched. Which pair?

	Complex Species	Structure
(a)	$[Ni(CO)_4]$	square pyramidal
(b)	$[Co(OH_2)_6]^{2+}$	octahedral
(c)	$[Ag(NH_3)_2]^+$	linear
(d)	$[Hg(CN)_4]^{2-}$	tetrahedral
(e)	$[Zn(NH_3)_4]^{2+}$	tetrahedral

b 36. Below is a list of complex ions and their structures. One pair is **incorrectly** matched. Which pair?

	Complex Ion	Structure
(a)	$[AgCl_2]^-$	linear
(b)	$[CuCl_2]^-$	angular
(c)	$[ZnCl_4]^{2-}$	tetrahedral
(d)	$[Cu(NH_3)_4]^{2+}$	square planar
(e)	$[Zn(OH_2)_6]^{2+}$	octahedral

a 37. Below is a list of complex ions and their structures. One pair is **incorrectly** matched. Which pair?

	Complex Ion	Structure
(a)	$[CoCl_6]^{3-}$	trigonal bipyramidal
(b)	$[Cd(NH_3)_4]^{2+}$	tetrahedral
(c)	$[NiCl_4]^{2-}$	tetrahedral
(d)	$[Au(CN)_2]^-$	linear
(e)	$[Co(CN)_6]^{3-}$	octahedral

Structural (Constitutional) Isomers

d 38. Which of the following is **not** a type of structural (constitutional) isomer?

(a) ionization (b) linkage (c) hydrate
(d) geometric (e) coordination

a 39. Which of the following pairs of complex compounds are **not** ionization isomers?

(a) $[Pt(NH_3)_4Cl_2]Br_2$ and $[Pt(NH_3)_4Br_2]I_2$
(b) $[Co(NH_3)_5NO_2]SO_4$ and $[Co(NH_3)_5SO_4]NO_2$
(c) $[Cr(NH_3)_5SO_4]Br$ and $[Cr(NH_3)_5Br]SO_4$
(d) $[Co(NH_3)_5Br]SO_4$ and $[Co(NH_3)_5SO_4]Br$
(e) $[Pt(NH_3)_4Br_2]Cl_2$ and $[Pt(NH_3)_4Cl_2]Br_2$

e 40. The following are examples of what kind of isomers?

$[Co(NH_3)_4(OH_2)Cl]Cl_2$ and $[Co(NH_3)_4Cl_2]Cl \cdot H_2O$

(a) linkage (b) optical (c) coordination
(d) ionization (e) hydrate

e 41. The following are examples of what kind of isomers?

$[Co(NH_3)_4Cl_2][Cr(NH_3)_2(CN)_4]$ and $[Cr(NH_3)_4(CN)_2][Co(NH_3)_2(CN)_2Cl_2]$

(a) linkage (b) optical (c) hydrate (d) ionization (e) coordination

c 42. Two complex ions consisting of a Co^{3+} ion, 5 NH_3 ligands, and a CN^- ligand can exist. For one of these complex ions, the C of the CN^- ligand acts as the donor atom; and for the other complex ion, the N of the CN^- ligand acts as the donor atom. What kind of isomers are these two complex ions?

(a) geometric (b) optical (c) linkage (d) ionization (e) coordination

d 43. Which of the following is not the name (as a ligand) for a ligand that can bind to metal ions in more than one way; that is, using two different donor atoms?

(a) cyano (b) nitro (c) isocyano (d) aqua (e) nitrito

Stereoisomers

e 44. The following square-planar complexes are what kind of isomers?

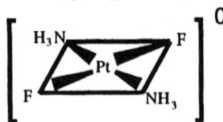

    (a) linkage   (b) coordination   (c) ionization
    (d) optical   (e) geometric

b 45. Which response includes all of the following statements about the neutral square-planar complexes shown below that are **true** and no false statements?

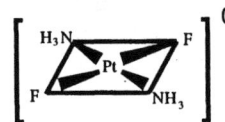

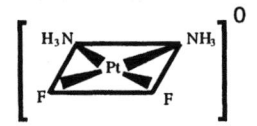

    I.   The compounds are optical isomers.
    II.   The isomer on the left is the *trans* isomer.
    III.   The isomer on the left is the *cis* isomer.
    IV.   The isomers are nonsuperimposable mirror images.
    V.   The isomers have different properties.

    (a) I, IV, and V   (b) II and V   (c) I, III, and IV
    (d) I, II, and IV   (e) III and V

c 46.

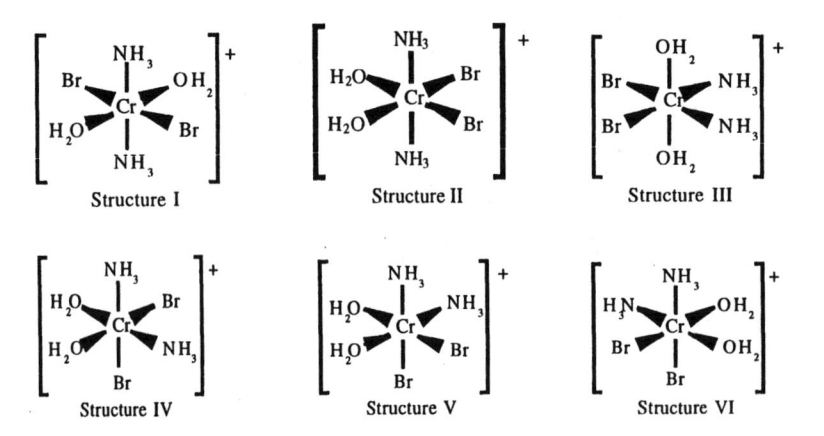

Which of the above structures are identical?

    (a) IV and V   (b) V and VI   (c) IV and VI
    (d) II and IV   (e) none are identical

a 47.

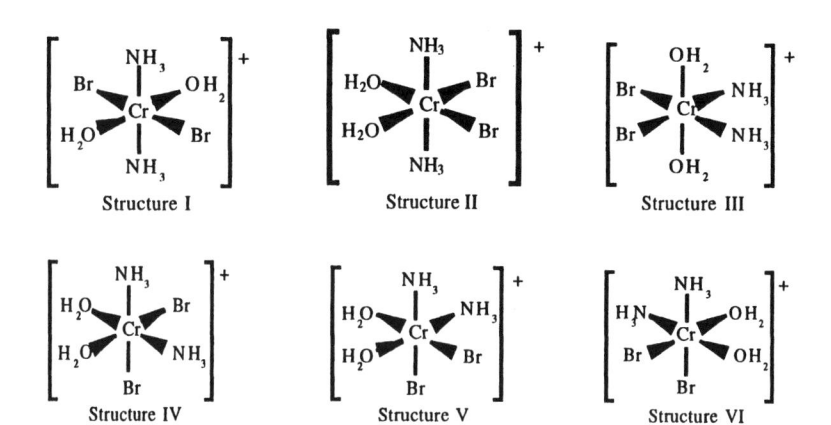

Which of the above structures has each of the three pairs of like ligands in the *trans* position to each other?

(a) I   (b) II   (c) III   (d) IV   (e) V

e 48.

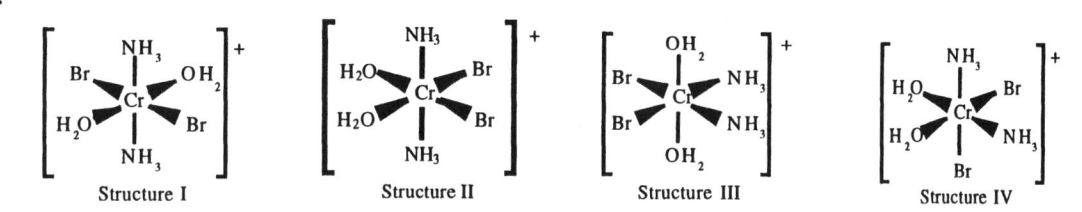

Which of the above structures has each of the three pairs of like ligands in the *cis* position to each other?

(a) I   (b) II   (c) III   (d) II and III   (e) IV

a 49.

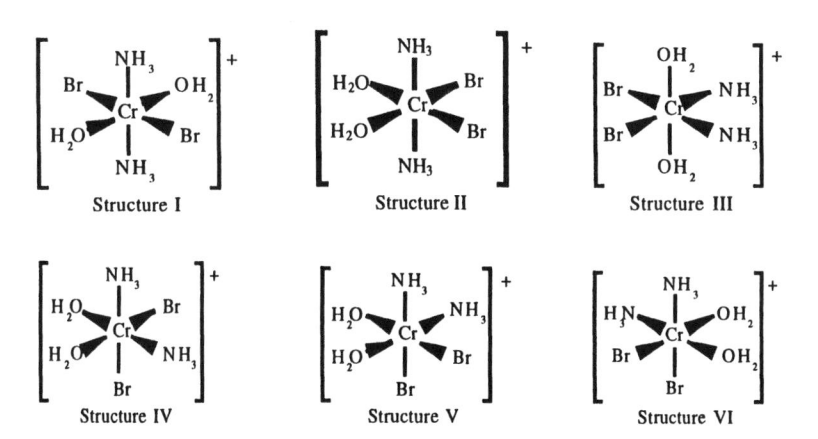

Which of the above structures are optical isomers?

(a) V and VI   (b) II and III   (c) I and IV
(d) I and III   (e) none are optical isomers

b 50. For which of the following are *cis-trans* isomers the **only kind** of isomers that can exist?

(a) [Co(NH$_3$)$_5$Br]$^{2+}$　　(b) [Co(NH$_3$)$_4$Cl$_2$]$^+$　　(c) [Co(NH$_3$)$_2$(OH$_2$)Cl$_2$]$^+$
(d) [Co(en)$_3$]$^{3+}$　　(e) [Co(NH$_3$)$_6$]$^{3+}$

c 51. What is a general formula for octahedral complexes that exhibit *mer - fac* isomerism?
(M = metal; A,B and C = ligands)

(a) MA$_5$B　　(b) MA$_4$B$_2$　　(c) MA$_3$B$_3$　　(d) MA$_2$B$_2$　　(e) MA$_2$B$_2$C$_2$

e 52. Which of the following answers contains **all** of the true statements concerning the tris(ethylenediamine)cobalt(III) ion, [Co(en)$_3$]$^{3+}$, and no false statements?

I.　Each bidentate ethylenediamine ligand bonds in *cis* positions.
II.　The [Co(en)$_3$]$^{3+}$ exists in the form of a pair of nonsuperimposable mirror images.
III.　The coordination number of the Co metal ion is 3.
IV.　The [Co(en)$_3$]$^{3+}$ exists as a pair of optical isomers.
V.　The two optical isomers each rotate a plane of polarized light by equal amounts but in opposite directions.

(a) I, II and III　　(b) I, III and V　　(c) IV and V
(d) I, II, III, IV and V　　(e) I, II, IV and V

Crystal Field Theory

c 53. If cobalt(II) nitrate is obtained by crystallization from aqueous solution, how many water molecules should be associated with each formula unit of cobalt(II) nitrate in the solid state?

(a) 7　　(b) 2　　(c) 6　　(d) 4　　(e) 9

a 54. The total number of electrons in the 3d orbitals of Co^{3+} is _____.

(a) 6　　(b) 7　　(c) 3　　(d) 4　　(e) 5

c 55. In which one of the following does the **transition metal** have a 3d^6 electronic configuration?

(a) [Cr(OH$_2$)$_6$]$^{3+}$　　(b) [Fe(CN)$_6$]$^{3-}$　　(c) [Fe(CN)$_6$]$^{4-}$
(d) [Ni(NH$_3$)$_6$]$^{2+}$　　(e) [Zn(NH$_3$)$_4$]$^{2+}$

e 56. In which one of the following does the **transition metal** have a 3d^8 electronic configuration?

(a) [Fe(NCS)(OH$_2$)$_5$]$^{2+}$　　(b) [FeF$_6$]$^{4-}$　　(c) [CuCl$_4$]$^{2-}$
(d) [Co(NH$_3$)$_6$]$^{3+}$　　(e) [Ni(NH$_3$)$_6$]$^{2+}$

e 57. In which one of the following does the **transition metal** have a 3d^9 electronic configuration?

(a) [Zn(NH$_3$)$_4$]$^{2+}$　　(b) [Co(OH$_2$)$_6$]$^{2+}$　　(c) [Co$_2$(CO)$_8$]
(d) [Fe(CO)$_5$]　　(e) [Cu(NH)$_4$]$^{2+}$

d 58. Which one of the following octahedral configurations has a low spin configuration?

(a) d^3 (b) d^9 (c) d^8 (d) d^5 (e) d^2

c 59. Which one of the following octahedral configurations has **no** low spin configuration?

(a) d^4 (b) d^7 (c) d^8 (d) d^6 (e) d^5

b 60. How many unpaired electrons would you expect for each formula unit of $[V(CN)_6]^{3-}$?

(a) 1 (b) 2 (c) 3 (d) 4 (e) 5

d 61. How many unpaired electrons does one formula unit of $[MnF_6]^{3-}$ contain?

(a) 1 (b) 2 (c) 3 (d) 4 (e) 5

a 62. Which one of the following statements is **false**?

(a) Paramagnetic metal ions must have an odd number of electrons.
(b) Diamagnetic metal ions cannot have an odd number of electrons.
(c) Low spin complexes can be paramagnetic.
(d) In high spin octahedral complexes, Δ_{oct} is less than the electron pairing energy.
(e) Low spin complexes contain strong field ligands.

d 63. Consider the complex ion $[Co(NH_3)_6]^{3+}$. Which response contains **all** of the following statements that are **true**, and no false statements?

I. It is paramagnetic. II. It is a high spin complex.
III. It is a low spin complex. IV. It has octahedral geometry.
V. It does not exist as geometric isomers.

(a) I and II (b) III and IV (c) II, IV, and V
(d) III, IV, and V (e) I and IV

e 64. Consider the complex ion $[CoCl_6]^{4-}$. Which response contains **all** of the following statements that are **true**, and no false statements?

I. The complex is a d^7 complex. II. The Co has two unpaired electrons.
III. It is paramagnetic. IV. It is a low spin complex.
V. It has tetrahedral geometry. VI. It is a high spin complex.

(a) II, IV, and V (b) I, III, and V (c) II, V, and VI
(d) III, V, and VI (e) I, III, and VI

a 65. Consider the complex ion $[FeF_6]^{3-}$. Which response includes **all** of the following statements that are **true**, and no false statements?

I. It is paramagnetic. II. It is a low spin complex.
III. It is a high spin complex. IV. The oxidation number of iron is +3.

(a) I, III, and IV (b) II and III (c) IV
(d) I and II (e) II and IV

c 66. Consider the complex ion [Cr(CN)$_6$]$^{3-}$. Which response includes **all** of the following statements that are **true**, and no others?

    I.   It is diamagnetic.
    II.   The chromium has three unpaired electrons.
    III.   It is a high spin complex.
    IV.   The oxidation number of the chromium is –3.

(a) III and IV   (b) I and IV   (c) II and III
(d) I, II, and III   (e) I, II, and IV

b*67. Consider the complex ion [Mn(OH$_2$)$_6$]$^{2+}$. Which response includes **all** of the following statements that are **true**, and no false statements?

    I.   It is diamagnetic.
    II.   It is a low spin complex.
    III.   It is a high spin complex.
    IV.   The ligands are weak field ligands.
    V.   The manganese has 5 unpaired electrons.

(a) I, II and V   (b) III, IV, and V   (c) I, II and IV
(d) II and V   (e) III and IV

e*68. Consider the complex ion [Fe(CN)$_6$]$^{4-}$. Which of the following responses includes **all** of the **true** statements with respect to this complex ion and the ions from which it was formed, and no false statements?

    I.   The complex ion is octahedral.
    II.   Fe^{2+} is a d^5 ion.
    III.   CN$^-$ is a strong field ligand.
    IV.   CN$^-$ is a weak field ligand.
    V.   The complex ion is a low spin complex.
    VI.   The complex ion is a high spin complex.
    VII.   The complex ion contains no unpaired electrons.
    VIII.   The complex ion contains four unpaired electrons.

(a) I, II, III, V, and VII   (b) I, II, III, V, and VII   (c) II, IV, VI, and VIII
(d) I, II, IV, VI, and VIII   (e) I, III, V, and VII

Color and the Spectrochemical Series

d 69. The nitrate of which one of the following cations would be colored in aqueous solutions?

(a) Al^{3+} (b) Ca^{2+} (c) Na$^+$ (d) Cr^{3+} (e) Pb^{2+}

d 70. Which one of the following is always a strong field ligand?

(a) Br$^-$ (b) H$_2$O   (c) F$^-$ (d) CN$^-$ (e) Cl$^-$

c 71. Which of the following statements concerning octahedral complexes is incorrect?

(a) Strong field ligands produce large crystal field splittings.
(b) Weak field ligands produce high spin complexes.
(c) Halide ions are strong field ligands.
(d) Weak field ligands result in relatively small values for Δ_{oct}.
(e) A relatively large value for Δ_{oct} causes a complex ion to absorb relatively high energy (shorter wavelength) light.

26 Nuclear Chemistry

Nuclear Reactions

b 1. Which one of the following statements about nuclear reactions does **not** correctly distinguish nuclear reactions from ordinary chemical reactions?

(a) Particles within the nucleus are involved.
(b) No new elements can be produced.
(c) Rate of reaction is independent of the presence of a catalyst.
(d) Rate of reaction is independent of temperature.
(e) They are often accompanied by the release of enormous amounts of energy.

a 2. Which statement is **false**?

(a) Dalton's theory explained how one element could be converted into another.
(b) Marie Curie is the only person to receive a Nobel Prize in both physics and chemistry.
(c) Henri Becquerel discovered natural radioactivity.
(d) Ernest Rutherford discovered that atoms of one element may be converted into atoms of other elements by spontaneous nuclear disintegrations.
(e) Marie Curie's daughter Irene and her husband also received a Nobel Prize in chemistry.

Neutron-Proton Ratio and Nuclear Stability

a 3. A term that is used to describe (only) different nuclear forms of the same element is ____.

(a) isotopes   (b) shells   (c) nucleons
(d) nuclides   (e) nuclei

c 4. The **most** stable nuclides, in general, have ____ numbers of neutrons and ____ numbers of protons.

(a) odd, even   (b) odd, odd   (c) even, even
(d) even, odd   (e) odd, equal

d 5. Which statement concerning stable nuclides and/or the "magic numbers" (such as 2, 8, 20, 28, 50, 82 or 128) is false?

(a) Nuclides with their number of protons equal to a "magic number" are especially stable.
(b) Nuclides with their number of neutrons equal to a "magic number" are especially stable.
(c) Nuclides with the sum of the numbers of their protons and neutrons equal to a "magic number" are especially stable.
(d) Above atomic number 20, the most stable nuclides have more protons than neutrons.
(e) The existence of "magic numbers" suggests an energy level (shell) model for the nucleus.

Nuclear Stability and Binding Energy

e 6. The difference between the sum of the masses of the electrons, protons and neutrons of an atom (calculated mass) and the actual measured mass of the atom is called the _____.

(a) nuclear mass
(b) isotopic mass
(c) decay mass
(d) critical mass
(e) mass deficiency

****

Masses of subatomic particles that may be useful for questions 7 - 16 and 27 - 29.

electron   0.00055 amu   proton   1.0073 amu   neutron   1.0087 amu

****

d 7. The actual mass of a $^{37}_{17}$Cl atom is 36.9659 amu. Calculate the mass deficiency (amu/atom) for a $^{37}_{17}$Cl atom.

(a) 0.623 amu
(b) 0.388 amu
(c) 0.263 amu
(d) 0.341 amu
(e) 0.302 amu

e 8. Calculate the mass deficiency of $^{17}_{8}$O. The actual mass of an $^{17}_{8}$O atom is 16.9991 amu.

(a) 0.153 amu
(b) 0.152 amu
(c) 0.148 amu
(d) 0.147 amu
(e) 0.142 amu

d 9. The actual mass of an atom of $^{23}_{11}$Na is 22.9898 amu. What is its mass deficiency?

(a) 0.202 amu
(b) 0.808 amu
(c) 0.196 amu
(d) 0.201 amu
(e) 0.402 amu

c 10. The mass of one atom of $^{27}_{13}$Al is 26.9815 amu. What is its mass deficiency?

(a) 0.766 amu
(b) 0.414 amu
(c) 0.242 amu
(d) 0.143 amu
(e) 0.176 amu

c 11. Calculate the mass deficiency for an atom of $^{103}_{45}$Rh whose actual mass is 102.9055 amu.

(a) 0.951 amu
(b) 0.934 amu
(c) 0.952 amu
(d) 0.941 amu
(e) 0.937 amu

e 12. The isotopic mass of $^{59}_{27}$Co is 58.9332 amu. What is its mass deficiency in amu per atom?

(a) 0.285 amu
(b) 0.396 amu
(c) 0.429 amu
(d) 0.498 amu
(e) 0.557 amu

d 13. The atomic mass of $^{75}_{33}$As is 74.9216 amu. What is its mass deficiency?

(a) 0.665 amu
(b) 0.718 amu
(c) 0.690 amu
(d) 0.703 amu
(e) 0.687 amu

Harcourt, Inc

c 14. The actual mass of a $^{55}_{25}$Mn atom is 54.9381 amu. Calculate the mass deficiency for $^{55}_{25}$Mn.

(a) 0.436 amu/atom   (b) 0.488 amu/atom   (c) 0.519 amu/atom
(d) 0.533 amu/atom   (e) 0.537 amu/atom

a 15. The actual mass of an atom of $^{64}_{28}$Ni is 63.9280 amu. Calculate its mass deficiency.

(a) 0.605 amu/atom   (b) 0.596 amu/atom   (c) 0.614 amu/atom
(d) 0.621 amu/atom   (e) 0.590 amu/atom

a 16. The actual mass of an atom of $^{20}_{10}$Ne is 19.9924 amu. Calculate the mass deficiency in g/mol of atoms.

(a) 0.173 g/mol   (b) 0.265 g/mol   (c) 0.099 g/mol
(d) 0.143 g/mol   (e) 0.191 g/mol

****
Values that may be useful for questions 17 - 29.

speed of light = 3.00 x 10^8 m/s   1 J = 1 kg•m^2/s^2 1 cal = 4.18 J
****

a 17. The mass deficiency for an isotope was found to be 0.410 amu/atom. Calculate the binding energy in J/mol of atoms.

(a) 3.69 x 10^{13} J/mol   (b) 1.23 x 10^{20} J/mol   (c) 3.69 x 10^3 J/mol
(d) 1.23 x 10^3 J/mol   (e) 1.23 x 10^{23} J/mol

a 18. The mass deficiency of $^{30}_{14}$Si is 0.2755 amu/atom. Calculate the binding energy in J/mol of atoms.

(a) 2.48 x 10^{13} J/mol   (b) 1.79 x 10^{27} J/mol   (c) 1.06 x 10^{14} J/mol
(d) 6.19 x 10^{10} J/mol   (e) 5.93 x 10^9 J/mol

c 19. If the mass deficiency for a particular isotope is 0.250 amu/atom, what is the binding energy of the isotope expressed in joules per mole?

(a) 4.68 x 10^{12} J/mol   (b) 4.81 x 10^{12} J/mol   (c) 2.25 x 10^{13} J/mol
(d) 2.72 x 10^{13} J/mol   (e) 5.55 x 10^{12} J/mol

c 20. The mass deficiency for $^{75}_{33}$As is 0.7029 amu/atom. What is the binding energy of in J/mol?

(a) 9.96 x 10^{12} J/mol   (b) 5.44 x 10^{12} J/mol   (c) 6.33 x 10^{13} J/mol
(d) 2.95 x 10^{14} J/mol   (e) 1.20 x 10^{15} J/mol

c 21. The mass deficiency of $^{20}_{10}$Ne is 0.1731 g/mol. Calculate the binding energy for $^{20}_{10}$Ne in kJ/mol of atoms.

(a) 2.73 x 10^{11} kJ/mol   (b) 1.16 x 10^{10} kJ/mol   (c) 1.56 x 10^{10} kJ/mol
(d) 1.69 x 10^{10} kJ/mol   (e) 2.77 x 10^{11} kJ/mol

b 22. What is the binding energy of an atom having a mass deficiency of 0.4721 amu per atom? Express your answer in kJ/mol of atoms.

(a) 5.26 kJ/mol (b) 4.25 x 10^{10} kJ/mol (c) 1.42 kJ/mol
(d) 4.25 x 10^{20} kJ/mol (e) 2.77 x 10^{11} kJ/mol

a 23. What is the binding energy in kJ/mol of atoms with a mass deficiency of 0.1064 amu/atom?

(a) 9.58 x 10^9 kJ/mol (b) 3.20 x 10^{-1} kJ/mol (c) 5.48 x 10^{33} kJ/mol
(d) 2.58 x 10^{17} kJ/mol (e) 2.62 x 10^5 kJ/mol

a 24. The mass deficiency for $^{64}_{28}$Ni is 0.6050 amu/atom. Calculate its binding energy in kJ/mol.

(a) 5.44 x 10^{10} kJ/mol (b) 2.28 x 10^{21} kJ/mol (c) 1.82 x 10^3 kJ/mol
(d) 2.59 x 10^{16} kJ/mol (e) 6.99 x 10^8 kJ/mol

b 25. The mass deficiency of $^{79}_{35}$Br is 0.7393 amu/atom. Calculate its binding energy in kJ/mol of atoms.

(a) 1.55 x 10^{11} kJ/mol (b) 6.65 x 10^{10} kJ/mol (c) 2.72 x 10^{10} kJ/mol
(d) 1.98 x 10^{10} kJ/mol (e) 2.60 x 10^{10} kJ/mol

d 26. The mass deficiency of an isotope is 0.2482 amu per atom. Determine its binding energy in kcal/mol.

(a) 7.45 x 10^9 kcal/mol (b) 2.67 x 10^{10} kcal/mol (c) 4.62 x 10^{10} kcal/mol
(d) 5.34 x 10^9 kcal/mol (e) 6.18 x 10^{10} kcal/mol

c 27. Calculate the binding energy (in J/mol) for $^{103}_{45}$Rh whose isotopic mass is 102.9055 amu.

(a) 2.86 x 10^{13} J/mol (b) 1.22 x 10^{14} J/mol (c) 8.57 x 10^{13} J/mol
(d) 2.86 x 10^5 J/mol (e) 8.75 x 10^{16} J/mol

b 28. Calculate the binding energy (in J/mol) for $^{64}_{28}$Ni whose isotopic mass is 63.9280 amu.

(a) 4.45 x 10^{16} J/mol (b) 5.45 x 10^{13} J/mol (c) 1.82 x 10^8 J/mol
(d) 5.54 x 10^{19} J/mol (e) 5.75 x 10^{15} J/mol

c 29. Calculate the binding energy (in kJ/mol) for $^{22}_{10}$Ne whose isotopic mass is 21.9914 amu.

(a) 1.56 x 10^{10} J/mol (b) 1.12 x 10^{10} J/mol (c) 1.72 x 10^{10} J/mol
(d) 2.53 x 10^{11} J/mol (e) 2.75 x 10^{11} J/mol

d 30. Which isotope below has the highest nuclear binding energy per gram? (No calculations are necessary.)

(a) $^{4}_{2}$He (b) $^{16}_{8}$O (c) $^{32}_{16}$S (d) $^{55}_{25}$Mn (e) $^{238}_{92}$U

a 31. Which isotope below has the smallest nuclear binding energy per gram? (No calculations are necessary.)

(a) $^{4}_{2}He$ (b) $^{16}_{8}O$ (c) $^{32}_{16}S$ (d) $^{55}_{25}Mn$ (e) $^{238}_{92}U$

e 32. Which statement concerning average nuclear binding energy per gram of nuclei is **false**?

(a) For nuclei with small mass numbers, it increases rapidly with increasing mass number.
(b) It reaches a maximum around mass number 50.
(c) It decreases slowly with increasing mass number for nuclei with mass numbers greater than 50.
(d) The nuclei with the highest binding energies (mass numbers 40 to 150) are the most stable.
(e) Nuclei with mass numbers greater than 210 can easily be decomposed into protons and neutrons.

Radioactive Decay

c 33. A positron has a mass number of _____, a charge of _____, and a mass equal to that of a(an) _____.

(a) 0, 1+, proton   (b) 1, 2+, proton   (c) 0, 1+, electron
(d) 1, 2+, electron   (e) 0, 0, proton

c 34. A beta particle has a mass number of _____, a charge of _____, and a mass equal to that of a(n) _____.

(a) 1, 1+, proton   (b) 1, 0, neutron   (c) 0, 1–, electron
(d) 0, 1+, electron   (e) 4, 2+, helium nucleus

e 35. Which of the following particles or rays has the greatest penetrating ability?

(a) beta particles   (b) alpha particles   (c) protons
(d) positrons   (e) gamma rays

Neutron-Rich Nuclei

b 36. If a radioisotope lies above the "band of stability," one would predict that it would decay by _____.

(a) alpha emission   (b) beta emission   (c) positron emission
(d) electron capture   (e) K capture

b 37. Complete and balance the following equation. The missing term is _____.

$$^{85}_{36}Kr \rightarrow \underline{} + ^{0}_{-1}\beta$$

(a) $^{85}_{35}Kr$ (b) $^{85}_{37}Rb$ (c) $^{84}_{36}Kr$ (d) $^{85}_{35}Br$ (e) $^{85}_{35}Rb$

d 38. Complete and balance the following equation. The missing term is _____.
$$^{80}_{32}Ge \rightarrow \underline{} + ^{0}_{-1}\beta$$

(a) $^{80}_{33}Hg$ (b) $^{80}_{31}Ga$ (c) $^{79}_{33}As$ (d) $^{80}_{33}As$ (e) $^{80}_{32}Ge$

c 39. Complete and balance the following equation. The missing term is _____.
$$^{92}_{37}Rb \rightarrow \underline{} + ^{0}_{-1}\beta$$

(a) $^{92}_{38}U$ (b) $^{92}_{36}Kr$ (c) $^{92}_{38}Sr$ (d) $^{92}_{37}Rb$ (e) $^{91}_{36}K$

e 40. Complete and balance the following equation. The missing term is _____.
$$^{239}_{94}Pu + ^{4}_{2}He \rightarrow \underline{} + ^{1}_{0}n$$

(a) $2\,^{115}_{47}Ag$ (b) $2\,^{106}_{45}Rh$ (c) $^{235}_{92}U$ (d) $^{233}_{91}Pa$ (e) $^{242}_{96}Cm$

d 41. Complete and balance the following equation. The missing term is _____.
$$^{14}_{7}N + ^{1}_{1}H \rightarrow ^{15}_{8}O + \underline{}$$

(a) $^{0}_{-1}\beta$ (b) $^{0}_{+1}\beta$ (c) $^{4}_{2}He$ (d) $^{0}_{0}\gamma$ (e) $^{1}_{0}n$

a 42. Complete and balance the following equation. The missing term is _____.
$$^{44}_{20}Ca + ^{2}_{1}H \rightarrow \underline{} + ^{1}_{0}n$$

(a) $^{45}_{21}Sc$ (b) $^{43}_{22}Ti$ (c) $^{47}_{20}Ca$ (d) $^{46}_{21}Sc$ (e) $^{43}_{28}Ar$

b 43. Complete and balance the following equation. The missing term is _____.
$$^{246}_{96}Cm + ^{12}_{6}C \rightarrow \underline{} + 4\,^{1}_{0}n$$

(a) $^{254}_{96}Cm$ (b) $^{254}_{102}No$ (c) $^{257}_{102}No$ (d) $^{254}_{100}Fm$ (e) $^{258}_{102}No$

Neutron-Poor Nuclei

d 44. Complete and balance the following equation. The missing term is _____.
$$^{55}_{29}Cu \rightarrow \underline{} + ^{0}_{+1}\beta$$

(a) $^{55}_{30}Zn$ (b) $^{56}_{29}Cu$ (c) $^{55}_{28}Ni$ (d) $^{55}_{28}Cu$ (e) $^{55}_{28}Cs$

c 45. The alpha emission by lead-204 results in the product isotope _____.

(a) $^{200}_{82}Pb$ (b) $^{203}_{81}Tl$ (c) $^{200}_{80}Hg$ (d) $^{204}_{83}Bi$ (e) $^{204}_{80}Hg$

c 46. The conversion of $^{222}_{86}Rn$ to $^{218}_{84}Po$ occurs via _____.

(a) beta emission (b) K capture (c) alpha emission
(d) positron emission (e) fission

e 47. If the nucleus $^{106}_{47}$Ag decays by an electron capture, the resulting isotope would be _____.

(a) $^{107}_{47}$Ag (b) $^{106}_{48}$Cd (c) $^{110}_{49}$In (d) $^{105}_{46}$Pd (e) $^{106}_{46}$Pd

c 48. A radioisotope of argon, $^{37}_{18}$Ar, lies below the "band of stability." One would predict that it decays via _____.

(a) neutron emission (b) beta emission (c) electron capture
(d) alpha emission (e) fission

b 49. Which one of the following represents *K* capture?

(a) $^{211}_{84}$Po → $^{211}_{84}$Po + $^{0}_{0}\gamma$
(b) $^{22}_{11}$Na + $^{0}_{-1}$e → $^{22}_{10}$Ne
(c) $^{1}_{1}$p → $^{1}_{0}$n + $^{0}_{+1}\beta$
(d) $^{14}_{7}$N + $^{1}_{0}$n → $^{14}_{6}$C + $^{1}_{1}$H
(e) $^{239}_{94}$Pu + $^{4}_{2}$He → $^{242}_{96}$Cm + $^{1}_{0}$n

Detection of Radiations

b 50. A Geiger-Muller counter is a _____.

(a) photographic detector (b) gas ionization counter (c) cloud chamber
(d) fluorescence detector (e) spectrophotometer

b 51. A substance that absorbs high energy radiation such as gamma rays and subsequently emits visible light is the basis for _____.

(a) photographic detection (b) a scintillation counter (c) a cyclotron
(d) a Geiger-Muller counter (e) a linear accelerator

Rates of Decay and Half-life

d 52. The half-life of Kr-79 is 35.0 hours. What is the value of the specific rate constant, *k*?

(a) 0.639 h^{-1} (b) 0.0571 h^{-1} (c) 24.3 h^{-1} (d) 0.0198 h^{-1} (e) 0.0086 h^{-1}

a 53. The half-life of Tc-99 is 2.13 x 10^5 years. What is the value of the specific rate constant, *k*?

(a) 3.25 x 10^{-6} y^{-1} (b) 1.41 x 10^{-6} y^{-1} (c) 4.69 x 10^{-6} y^{-1}
(d) 0.693 y^{-1} (e) 1.48 x 10^5 y^{-1}

e 54. The specific rate constant for the decay of Tc-95 is 0.0346 h^{-1}. What is its half-life?

(a) 10.6 h (b) 8.70 h (c) 28.9 h (d) 0.0499 h (e) 20.0 h

c 55. The specific rate constant for the decay of Bi-210 is 0.138 d^{-1}. What is its half-life?

(a) 0.20 d (b) 2.18 d (c) 5.02 d (d) 0.096 d (e) 3.84 d

a 56. The half-life of $^{231}_{91}$Pa is 3.25 x 10⁴ y. How much of an initial 10.40 microgram sample remains after 3.25 x 10⁵ y?

(a) 0.0102 µg   (b) 0.240 µg   (c) 0.0240 µg
(d) 1.02 µg   (e) 1.04 µg

c 57. The half-life of Sr-83 is 32.4 hours. How much of a 20.0-mg sample of $^{83}_{38}$Sr will be left after 75.0 hours?

(a) 3.68 mg   (b) 0.249 mg   (c) 4.02 mg   (d) 0.497 mg   (e) 4.62 mg

b 58. Nitrogen-13 has a half-life of 9.97 minutes. How much of a 10.0-g sample remains after 60.0 minutes?

(a) 9.2 g   (b) 0.15 g   (c) 0.35 g   (d) 1.2 g   (e) 2.5 g

c 59. How long would it take for 2.8 micrograms of a 10.4 microgram sample of $^{231}_{91}$Pa to decay? The half-life of $^{231}_{91}$Pa is 3.25 x 10⁴ y.

(a) 1.2 x 10⁵ y   (b) 6.2 x 10⁴ y   (c) 1.5 x 10⁴ y
(d) 8.1 x 10⁶ y   (e) 4.0 x 10⁵ y

b 60. Cobalt-60 has a half-life of 5.27 years. How long will it require for 10.5 mg of a 22.8-mg sample of $^{60}_{27}$Co to decay?

(a) 3.98 y   (b) 4.71 y   (c) 2.05 y   (d) 2.57 y   (e) 5.92 y

a 61. The half-life of ^{33}P is 25.3 days. How long will it take for 64.0 g to decay to 1.0 g?

(a) 150 d   (b) 350 d   (c) 210 d   (d) 120 d   (e) 100 d

d 62. A 20.0-g sample of ^{277}Th decays to 17.0 g in 3.00 days. What is the rate constant for the decay of this isotope?

(a) 1.9 d⁻¹   (b) 0.49 d⁻¹   (c) 1.50 d⁻¹   (d) 0.054 d⁻¹   (e) 0.12 d⁻¹

e 63. If two-thirds of a radon-86 sample decays in 6.1 days, what is the half-life of this isotope?

(a) 1.18 d   (b) 0.41 d   (c) 0.10 d   (d) 4.2 d   (e) 3.8 d

Disintegration Series

e 64. The radionuclides ^{238}U, ^{235}U, and ^{232}Th cannot attain nuclear stability by only one nuclear reaction. Instead they decay in a series of disintegrations. All end with a stable isotope of the element _____.

(a) Ar   (b) Co   (c) Bi   (d) Rn   (e) Pb

Uses of Radionuclides

b 65. What is **not** a practical use of radionuclides?

    (a) Radioactive dating to estimate the age of an article.
    (b) Irradiation of some foods in order to transmutate poisonous elements into nonpoisonous elements.
    (c) To be power sources of heart pacemakers.
    (d) To act as radioactive tracers in medicine, in chemical research, and plant studies.
    (e) Serves as radioactive source in smoke detectors.

b 66. Charcoal found in an old tomb has a carbon-14 activity that is 0.366 times that of present dry wood. Estimate the age of the wood. The half-life of C-14 is 5730 years.

    (a) 1.64×10^3 y   (b) 8.31×10^3 y   (c) 3.77×10^3 y
    (d) 3.61×10^3 y   (e) 6.98×10^2 y

c 67. A rock contains 3.008 mg of ^{238}U and 0.624 mg of ^{206}Pb. The half-life of ^{238}U is 4.51×10^9 years. Estimate the age of the rock.

    (a) 1.02×10^{10} y   (b) 3.74×10^4 y   (c) 1.39×10^9 y
    (d) 6.12×10^8 y   (e) 1.12×10^9 y

Artificial Transmutations of Elements

d 68. A cyclotron cannot be used to accelerate _____.

    (a) protons   (b) alpha particles   (c) electrons
    (d) neutrons   (e) deuterons

Nuclear Fission

c 69. Which of the following would you expect to have the highest binding energy per nucleon?

    (a) ^2_1H (b) ^4_2He (c) $^{56}_{26}Fe$ (d) $^{141}_{56}Ba$ (e) $^{235}_{92}U$

d 70. Which one of the following reactions represents fission?

    (a) $^{238}_{92}U + ^{12}_6C \rightarrow ^{246}_{98}Cf + 4\,^1_0n$
    (b) $^{11}_5B + ^4_2He \rightarrow ^{12}_6C + ^3_1H$
    (c) $^{233}_{90}Th \rightarrow ^{233}_{91}Pa + ^{\ \ 0}_{-1}e$
    (d) $^{235}_{92}U + ^1_0n \rightarrow ^{146}_{57}La + ^{87}_{35}Ba + 3\,^1_0n$
    (e) $^3_1H + ^2_1H \rightarrow ^4_2He + ^1_0n$

Nuclear Fission Reactors

e 71. Which one of the components of a light water reactor listed below is described by an incorrect function?

	Component	Function
(a)	moderator	slows neutrons
(b)	fuel	supplies neutrons plus heat
(c)	cooling system	coolant for the reactor
(d)	control rods	absorb some neutrons to control rate of fission
(e)	shielding	prevents the absorption of beta gamma rays by the fuel and cooling system

d 72. Which of the following is **not** a hazard or problem resulting from the use of nuclear reactors to produce power?

(a) Proper shielding precautions must be taken to protect people.
(b) The possibility of a "meltdown" if the cooling system fails.
(c) The safe handling and storage of spent fuel for hundreds of thousands of years because it contains long-lived radionuclides.
(d) The possibility of a nuclear explosion occurring.
(e) The possibility of theft for use in constructing atomic weapons.

Nuclear Fusion

e 73. Which one of the following represents fusion?

(a) $^{238}_{92}U + ^{12}_{6}C \rightarrow ^{246}_{98}Cf + 4\,^{1}_{0}n$
(b) $^{11}_{5}B + ^{4}_{2}He \rightarrow ^{12}_{6}C + ^{3}_{1}H$
(c) $^{233}_{90}Th \rightarrow ^{233}_{91}Pa + ^{0}_{-1}e$
(d) $^{235}_{92}U + ^{1}_{0}n \rightarrow ^{146}_{57}La + ^{87}_{35}Ba + 3\,^{1}_{0}n$
(e) $^{3}_{1}H + ^{2}_{1}H \rightarrow ^{4}_{2}He + ^{1}_{0}n$

a 74. Which one of the following would be most likely to undergo fusion under the proper conditions?

(a) $^{2}_{1}H$ (b) $^{4}_{2}He$ (c) $^{56}_{26}Fe$ (d) $^{141}_{56}Ba$ (e) $^{235}_{92}U$

c 75. The major problem associated with the development of fusion for controlled energy generation is _____.

(a) containment of radioactive products of fusion
(b) relatively low energy yield per gram of fuel
(c) containment of extremely high temperature plasmas
(d) depletion of water reserves which serve as a source of fuel
(e) the resulting air pollution

27 Organic Chemistry I: Formulas, Names, and Properties

Alkanes and Cycloalkanes

d 1. What is the molecular formula for heptane?

(a) C_7H_{14} (b) C_7H_{12} (c) C_9H_{18} (d) C_7H_{16} (e) C_9H_{20}

e 2. In the homologous series of alkanes, each member differs from the next member by a _____ group.

(a) CH   (b) CH_3 (c) C_nH_{2n+2} (d) C_2H_5 (e) CH_2

a 3. Which one of the following **cannot** exist in isomeric forms?

(a) C_3H_8 (b) C_4H_{10} (c) C_5H_{12} (d) C_6H_{14} (e) C_7H_{16}

c 4. How many possible constitutional isomers of C_6H_{14} exist?

(a) 6   (b) 3   (c) 5   (d) 4   (e) 9

d 5. Which one of the following formulas could represent a cycloalkane?

(a) C_2H_6 (b) C_3H_8 (c) C_4H_{10} (d) C_6H_{12} (e) C_7H_{16}

c 6. Which one of the following statements is **false**?

(a) Unsubstituted saturated hydrocarbons may contain even or odd numbers of carbon atoms.
(b) The boiling points of normal alkanes increase with increasing molecular weight.
(c) Most hydrocarbons are polar.
(d) The carbon atoms in saturated hydrocarbons are best described as sp^3 hybridized.
(e) An alkyl **group** may be represented, in general, as C_nH_{2n+1}.

b*7. Which of the following structural formulas contains an error?

(a) 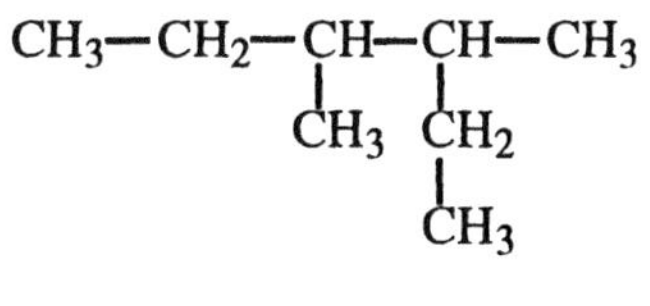 $CH_3-CH_2-CH-CH-CH_3$
           | |
          CH_3 CH_2
                |
               CH_3

(b) 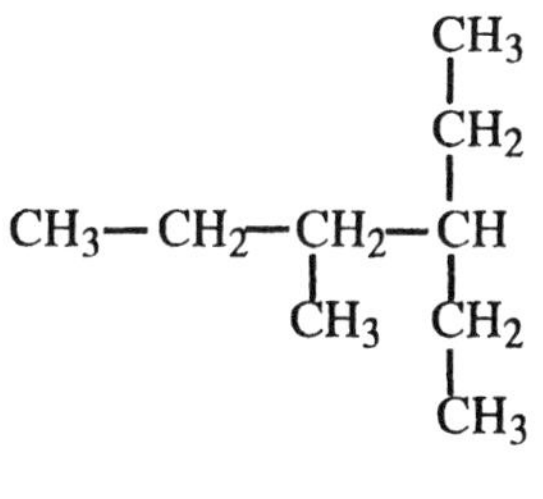 $CH_3-CH_2-CH_2-CH$
              |
             CH_3 CH_2
                     |
                    CH_3
with CH_3-CH_2 branch going up: CH_3 / CH_2 above CH

(c) 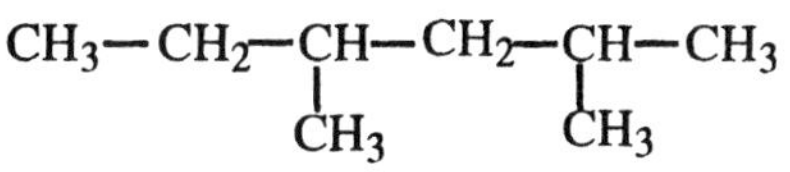 $CH_3-CH_2-CH-CH_2-CH-CH_3$
              | |
             CH_3 CH_3

(d) 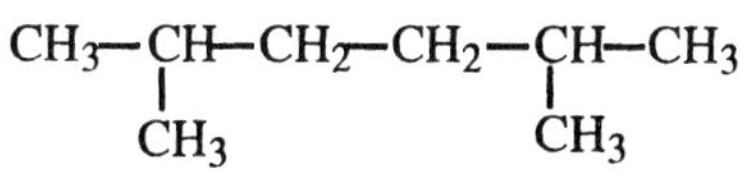 $CH_3-CH-CH_2-CH_2-CH-CH_3$
           | |
          CH_3 CH_3

(e) 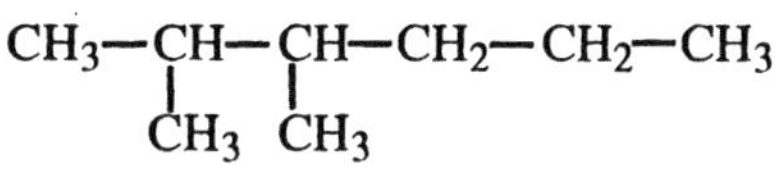 $CH_3-CH-CH-CH_2-CH_2-CH_3$
           | |
          CH_3 CH_3

385
Harcourt, Inc

Naming Saturated Hydrocarbons

c 8. The correct IUPAC name for the compound shown below is _____.

$$\begin{array}{c} \text{CH}_3 \\ | \\ \text{CH}_3 \quad\quad \text{CH}_2 \\ | \quad\quad\quad | \\ \text{CH}_3-\text{CH}-\text{CH}_2-\text{C}-\text{CH}_3 \\ | \\ \text{CH}_3 \end{array}$$

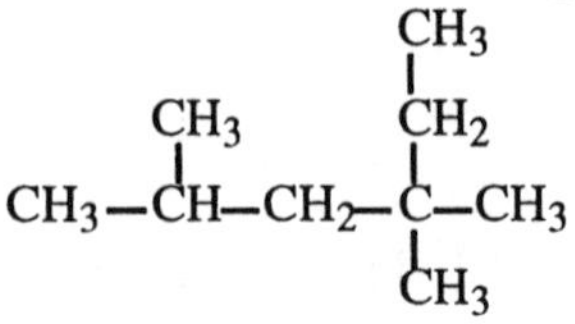

(a) 3,5-dimethylhexane
(b) 2-ethyl-4-methylpentane
(c) 2,4,4-trimethylhexane
(d) 2-methyl-4-ethylpentane
(e) 2-ethyl-2,4-dimethylpentane

b 9. The correct IUPAC name for the compound shown below is _____.

$$\begin{array}{c} \text{CH}_3 \\ | \\ \text{CH}_2 \\ | \\ \text{CH}_2 \quad\quad\quad\quad\quad\quad \text{CH}_3 \\ | \quad\quad\quad\quad\quad\quad\quad | \\ \text{CH}_3-\text{CH}_2-\text{CH}-\text{CH}_2-\text{CH}_2-\text{CH}-\text{CH}_2-\text{C}-\text{CH}_3 \\ \quad\quad\quad\quad\quad\quad\quad\quad | \quad\quad\quad\quad | \\ \quad\quad\quad\quad\quad\quad\quad\quad \text{CH}_3 \quad\quad\quad \text{CH}_3 \end{array}$$

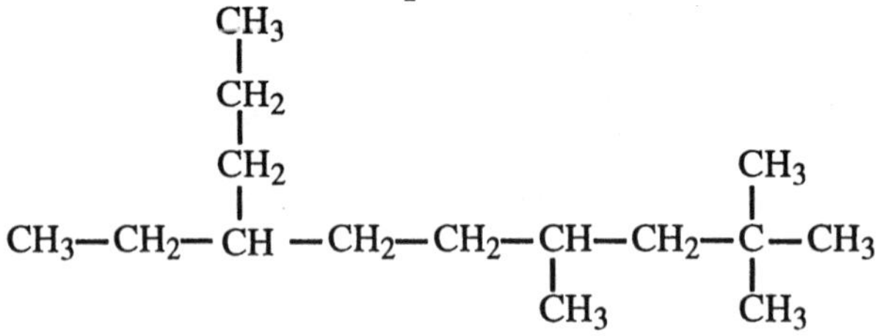

(a) 2,2,4-trimethyl-7-propylnonane
(b) 7-ethyl-2,2,4-trimethyldecane
(c) 3-propyl-6,8,8-trimethylnonane
(d) 4-ethyl-7,9,9-trimethylnonane
(e) 4-ethyl-7,9,9-trimethyldecane

a 10. What is the structure for 3,4-dimethylhexane?

(a) 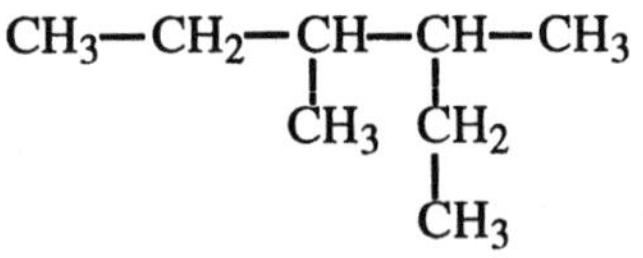 $\text{CH}_3-\text{CH}_2-\text{CH}-\text{CH}-\text{CH}_3$
       $\quad\quad\quad\quad\quad | \quad\quad |$
       $\quad\quad\quad\quad\quad \text{CH}_3 \quad \text{CH}_2$
       $\quad\quad\quad\quad\quad\quad\quad\quad |$
       $\quad\quad\quad\quad\quad\quad\quad\quad \text{CH}_3$

(b) 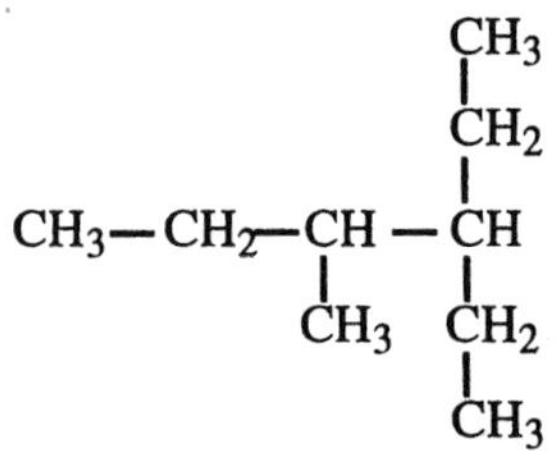 $\text{CH}_3-\text{CH}_2-\text{CH}-\text{CH}$
       $\quad\quad\quad\quad\quad | \quad\quad |$
       $\quad\quad\quad\quad\quad \text{CH}_3 \quad \text{CH}_2$
       $\quad\quad\quad\quad\quad\quad\quad\quad |$
       $\quad\quad\quad\quad\quad\quad\quad\quad \text{CH}_3$
       with additional CH3 at top

(c) 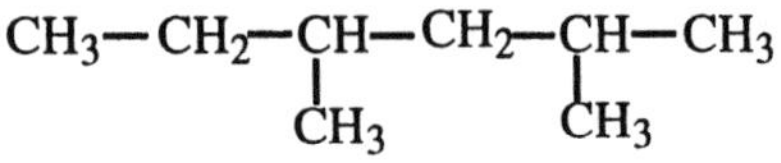 $\text{CH}_3-\text{CH}_2-\text{CH}-\text{CH}_2-\text{CH}-\text{CH}_3$
       $\quad\quad\quad\quad\quad | \quad\quad\quad\quad |$
       $\quad\quad\quad\quad\quad \text{CH}_3 \quad\quad\quad \text{CH}_3$

(d) 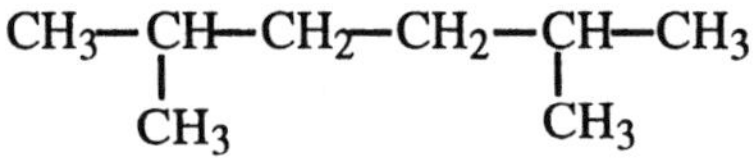 $\text{CH}_3-\text{CH}-\text{CH}_2-\text{CH}_2-\text{CH}-\text{CH}_3$
       $\quad\quad\quad | \quad\quad\quad\quad\quad\quad |$
       $\quad\quad\quad \text{CH}_3 \quad\quad\quad\quad\quad \text{CH}_3$

(e) 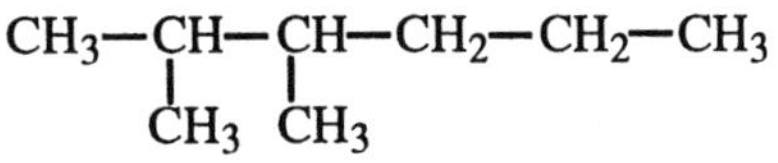 $\text{CH}_3-\text{CH}-\text{CH}-\text{CH}_2-\text{CH}_2-\text{CH}_3$
       $\quad\quad\quad | \quad\quad |$
       $\quad\quad\quad \text{CH}_3 \quad \text{CH}_3$

c 11. What is the correct IUPAC name for $(\text{CH}_3)_2\text{CH}(\text{CH}_2)_3\text{CH}_3$?

(a) heptane
(b) 2,2-dimethylpentane
(c) 2-methylhexane
(d) 1,1-dimethylpentane
(e) hexane

e 12. What is the correct condensed formula for 2,2,4-trimethylpentane?

(a) $(\text{CH}_3)_3\text{CCH}_2\text{C}(\text{CH}_3)_3$
(b) $(\text{CH}_3)_3\text{C}(\text{CH}_2)_2\text{CH}_3$
(c) $(\text{CH}_3)_3\text{CHCH}_2\text{CH}(\text{CH}_3)_2$
(d) $(\text{CH}_3)_2\text{CHCH}(\text{CH}_3)_2$
(e) $(\text{CH}_3)_3\text{CCH}_2\text{CH}(\text{CH}_3)_2$

d 13. Give the correct IUPAC name for the compound with the following structural formula.

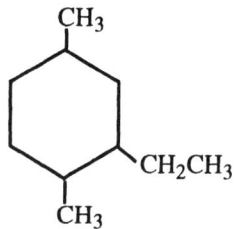

(a) 2-ethyl-1,4-dimethylcyclopentane
(b) 3-ethyl-1,4-dimethylcyclohexane
(c) 1-ethyl-2,5-dimethylcyclohexane
(d) 2-ethyl-1,4-dimethylcyclohexane
(e) cyclic-2-ethyl-1,4-dimethylhexane

Alkenes

a 14. A hydrocarbon containing one double bond per molecule is called a(an) _____.

(a) alkene (b) alkyne (c) alkane
(d) methylene (e) saturated hydrocarbon

d 15. Which one of the following formulas could represent an unsaturated hydrocarbon?

(a) C_2H_6 (b) C_3H_8 (c) C_4H_{10} (d) C_6H_{12} (e) C_7H_{16}

b 16. Which of the following hydrocarbons can exist as *cis-trans* isomers?

(a) 1-butene (b) 2-butene (c) 2-methyl-2-butene
(d) methylpropene (e) 2-methyl-1-butene

d 17. Give the correct IUPAC name for the following compound.

(a) 3,4-dimethyl-3-pentene
(b) 3,4-dimethyl-*cis*-3-pentene
(c) 3,4-dimethyl-*trans*-3-pentene
(d) 2,3-dimethyl-2-pentene
(e) 2-ethyl-3-methyl-2-butene

a 18. Which of the following is the correct formula for 2-methyl-1-butene?

(a) $CH_3-CH_2-\underset{\underset{CH_3}{|}}{C}=CH_2$ (b) $CH_2=CH-\underset{\underset{CH_3}{|}}{CH}-CH_3$ (c) $CH_3-\underset{\underset{CH_3}{|}}{C}=CH_2$

(d) $CH_3-CH=\underset{\underset{CH_3}{|}}{C}-CH_3$ (e) none of these

c 19. The formula for cyclohexene is _____.

(a) C_6H_{14} (b) C_6H_{12} (c) C_6H_{10} (d) C_6H_9 (e) C_6H_6

Harcourt, Inc

a 20. Give the correct IUPAC name for the following compound.

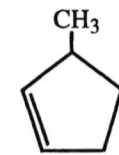

(a) 3-methylcyclopentene
(b) 5-methyl-1-cyclopentene
(c) 1-methyl-2-cyclopentene
(d) 5-methylcyclohexene
(e) 3-methyl-*cis*-cyclopentene

Petroleum

d 21. Which one of the following statements regarding petroleum products is false?

(a) Each oil field produces petroleum with its own set of characteristics.
(b) Paraffin and asphalt are among the highest boiling (melting) fractions of petroleum.
(c) The lubricating oil fraction of petroleum consists of higher molecular weight components than the gasoline fraction.
(d) Straight chain hydrocarbons like octane generally burn more smoothly than branched hydrocarbons.
(e) Cracking involves heating higher molecular weight hydrocarbons in the absence of air and in the presence of a catalyst.

Alkynes

c 22. Give the correct IUPAC name for the following alkyne.

$$CH_3CHCH_2CHC\equiv CCH_2CH_3$$
$$\quad\;\;\;|\qquad\;\;\;|$$
$$\quad\;\;CH_3\;\;\;CH_3$$

(a) 2,4-dimethyl-5-octyne
(b) 4-methyl-5-isooctyne
(c) 5,7-dimethyl-3-octyne
(d) 1,3-dimethylbutylethylacetylene
(e) 5,7-dimethyloctyne

d 23. What is the correct structure for 2,7-dimethyl-4-nonyne?

(a) $CH_3CHCH_2CH\equiv CHCH_2CHCH_2CH_3$
      | |
      CH_3 CH_3

(b) $CH_3CH_2CHC\equiv CCH_2CH_2CHCH_3$
          | |
          CH_3 CH_3

(c) $CH_3CHCH_2C\equiv CCH_2CH_2$
      | |
      CH_3 CH_3

(d) $CH_3CHCH_2C\equiv CCH_2CHCH_2CH_3$
      | |
      CH_3 CH_3

(e) None of the preceding structures is correct.

e*24. Which of the following alkynes can exist as *cis-trans* isomers?

(a) propyne
(b) 2-butyne
(c) 3-methyl-1-butyne
(d) 1-butyne
(e) Alkynes cannot have geometric isomerism.

Aromatic Hydrocarbons

d 25. What are **two** possible names for the following compound?

H₃C—⌬—CH₃

(a) 1,3-dimethylbenzene; *p*-xylene
(b) 1,3-dimethylbenzene; *m*-xylene
(c) 1,4-dimethylbenzene; toluene
(d) 1,4-dimethylbenzene; *p*-xylene
(e) 1,4-dimethylbenzene; *o*-xylene

b 26. What is the name for the following compound?

(a) 4-ethyl-1,3,5-trimethylbenzene   (b) 2-ethyl-1,3,5-trimethylbenzene
(c) 1,3,5-trimethyl-4-ethylbenzene   (d) 6-ethyl-1,3,5-trimethylbenzene
(e) 1-ethyl-2,4,6-trimethylbenzene

Hydrocarbons: A Summary

d 27. Which of the hydrocarbons below is classified **incorrectly**?

(a) methane   alkane   saturated   aliphatic
(b) propene   alkene   unsaturated   aliphatic
(c) naphthalene ———   ———   aromatic
(d) benzene   ———   unsaturated   aliphatic
(e) acetylene   alkyne   unsaturated   aliphatic

Organic Halides

a 28. Which one of the following is **not** an alkyl chloride?

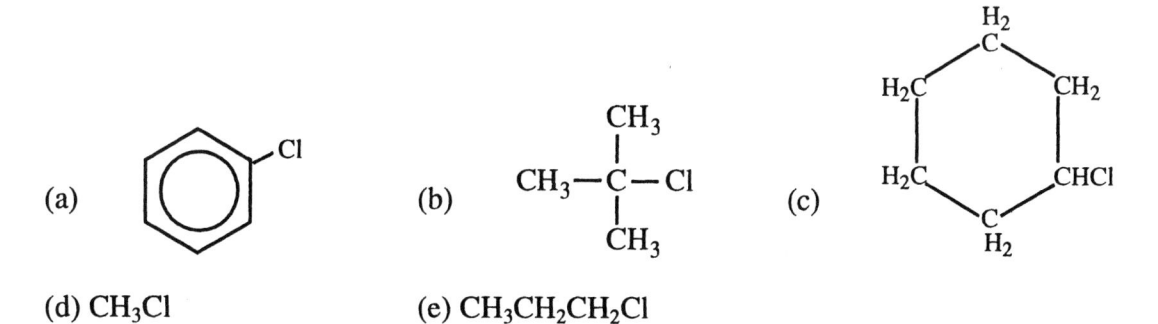

(a) C₆H₅Cl (chlorobenzene)

(b) (CH₃)₃C—Cl

(c) cyclohexyl chloride (C₆H₁₁Cl)

(d) CH₃Cl

(e) CH₃CH₂CH₂Cl

d 29. What is the correct name for the compound below?

$$Br-\underset{\underset{H}{|}}{\overset{\overset{Br}{|}}{C}}-\underset{\underset{H}{|}}{\overset{\overset{H}{|}}{C}}-\underset{\underset{H}{|}}{\overset{\overset{H}{|}}{C}}-H$$

(a) 1,1-dibromoethane (b) 1,2-dibromopentane (c) 1,2-dibromopropane
(d) 1,1-dibromopropane (e) None of these answers is correct.

e 30. Which combination of formula, IUPAC name, and common name below is **incorrect**?

	Formula	IUPAC Name	Common Name
(a)	$CHCl_3$	trichloromethane	chloroform
(b)	CCl_4	tetrachloromethane	carbon tetrachloride
(c)	C_6H_5I	iodobenzene	phenyl iodide
(d)	CH_3Cl	chloromethane	methyl chloride
(e)	CH_2Cl_2	dichloromethane	methene chloride

d 31. What is the correct name for the following compound?

$$CH_3-\underset{\underset{Cl}{|}}{\overset{\overset{Cl}{|}}{CH}}-CH-CH_2-\underset{\underset{Br}{|}}{CH}-CH_3$$

(a) 2-bromo-4,5-dichlorohexane (b) 5-bromo-*trans*-2,3-dichlorohexane
(c) 4-bromo-1,2-dichloropentane (d) 5-bromo-2,3-dichlorohexane
(e) 2,3,5-trihalohexane

d 32. Which of the following statements concerning freons is **incorrect**?

(a) Freons are chlorofluorocarbon compounds.
(b) Freon™ is a DuPont trademark for certain chlorofluorocarbons
(c) The use of freons as propellants in aerosol cans was banned in the U.S. in 1978.
(d) Freons are very poisonous to animals and plants.
(e) Freons damages the earth's ozone layer.

Alcohols and Phenols

b 33. How many structural isomers of the alcohol with the formula $C_5H_{12}O$ exist?

(a) 6 (b) 8 (c) 5 (d) 12 (e) 3

e 34. Which one of the following is a **primary** alcohol?

(a) 2-propanol (b) 2-butanol (c) 2-methyl-2-propanol
(d) cyclohexanol (e) 2-methyl-1-propanol

c 35. Which one of the following is a **secondary** alcohol?

(a) CH_3CH_2OH (b) CH_3OH (c) $CH_3CH(OH)CH_3$
(d) $(CH_3)_3COH$ (e) None of these answers is a secondary alcohol.

e 36. Which one of the following is a **secondary** alcohol?

(a) 2-methyl-1-pentanol
(b) 2,2-dimethyl-1-pentanol
(c) 2-methyl-2-pentanol
(d) 2,3,4-trimethyl-2-pentanol
(e) 3,3,4-trimethyl-2-pentanol

d 37. Which one of the following is a **tertiary** alcohol?

(a) CH_3CH_2OH
(b) CH_3OH
(c) $CH_3CH(OH)CH_3$
(d) $(CH_3)_3COH$
(e) None of these answers is a tertiary alcohol.

c*38. There are 8 structural isomers of the alcohol with the formula $C_5H_{12}O$. How many are primary (1°); how many are secondary (2°); and how many are tertiary (3°) alcohols?

(a) 5(1°), 2(2°), 1(3°)
(b) 5(1°), 3(2°), 1(3°)
(c) 4(1°), 3(2°), 1(3°)
(d) 4(1°), 2(2°), 2(3°)
(e) 3(1°), 4(2°), 1(3°)

c 39. Which of the following statements about alcohols and phenols is **incorrect**?

(a) Phenols are weakly acidic.
(b) Polyhydric alcohols that contain 2 OH groups per molecule are called glycols.
(c) Ethanol is very toxic and causes permanent blindness if a small amount is taken internally.
(d) The properties of alcohols depend on the number of OH groups per molecule and the size of the nonpolar portion of the molecule.
(e) Ethylene glycol is miscible with H_2O and used in commercial permanent antifreeze.

d 40. What is the correct IUPAC name for the following compound?

$$CH_3-CH-CH_2-CH_2-CH-CH_3$$
$$\quad\quad |\quad\quad\quad\quad\quad\quad\quad |$$
$$\quad\quad CH_3\quad\quad\quad\quad\quad OH$$

(a) 2-methyl-5-hexanol
(b) 1,4-dimethyl-1-pentanol
(c) 5,5-dimethyl-2-pentanol
(d) 5-methyl-2-hexanol
(e) 2,5-dimethyl-5-pentanol

d 41. Which one of the following is *m*-cresol?

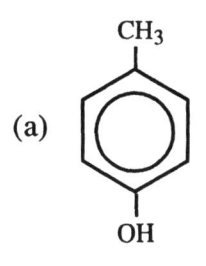

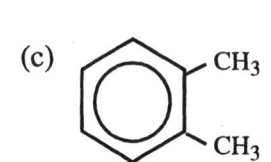

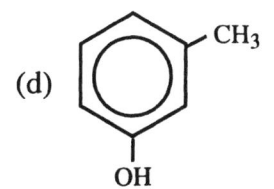

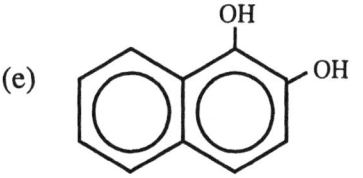

a 42. Which one of the following is *p*-cresol?

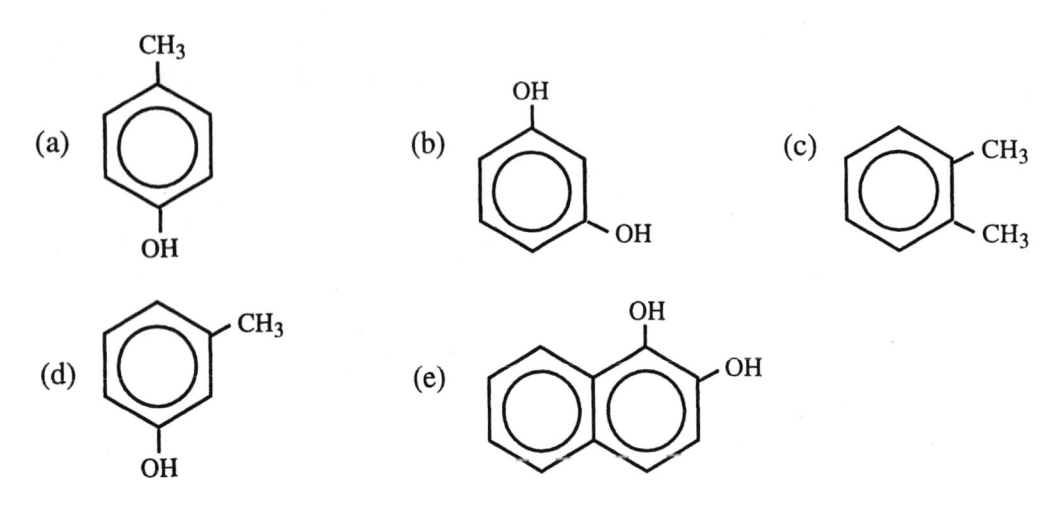

(a) [structure: benzene with CH₃ on top and OH on bottom]
(b) [structure: benzene with OH and OH meta]
(c) [structure: benzene with two adjacent CH₃ groups]
(d) [structure: benzene with CH₃ and OH para]
(e) [structure: naphthalene with two adjacent OH groups]

Ethers

e 43. Which of the following is a **mixed** ether?

(a) dimethyl ether (b) diethyl ether (c) methyl ethyl ether
(d) diphenyl ether (e) methyl phenyl ether

d 44. Which of the following is a **mixed** ether?

(a) methoxymethane (b) ethoxyethane (c) methoxyethane
(d) methoxybenzene (e) phenoxybenzene

d 45. Which of the following statements about ethers is **incorrect**?

(a) Ethers are not very polar and are chemically rather unreactive.
(b) The physical properties of ethers are similar to those of the corresponding alkanes.
(c) Diethyl ether is a very low boiling liquid.
(d) Ethers are very soluble in water because their structure is like water with organic groups substituted for the two hydrogens.
(e) Diethyl ether is a good solvent for organic compounds.

Aldehydes and Ketones

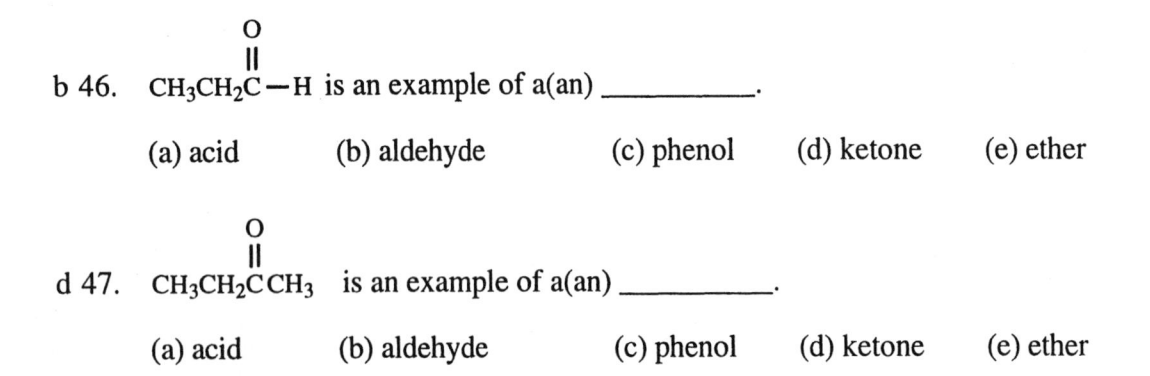

b 46. $CH_3CH_2\overset{O}{\overset{\|}{C}}-H$ is an example of a(an) _____.

(a) acid (b) aldehyde (c) phenol (d) ketone (e) ether

d 47. $CH_3CH_2\overset{O}{\overset{\|}{C}}CH_3$ is an example of a(an) _____.

(a) acid (b) aldehyde (c) phenol (d) ketone (e) ether

e 48. Which is the correct name for the compound given below?

$$CH_3CHCH_2CH_2CHCH_2\overset{\overset{O}{\|}}{C}-H$$
$$||$$
$$CH_2CH_3CH_2CH_3$$

(a) 3,6-diethylheptanal
(b) 6-aldoethyl-3-methyloctane
(c) 2,5-diethyl-7-heptanal
(d) 6-ethyl-3-methyl-8-octanal
(e) 3-ethyl-6-methyloctanal

c 49. The systematic name for $CH_3CH_2-\overset{\overset{O}{\|}}{C}-\underset{\underset{CH_3}{|}}{CH}-CH_3$ is _____.

(a) 2-methyl-3-propanol
(b) 2-methyl-3-pentanal
(c) 2-methyl-3-pentanone
(d) 2-methyl-3-propanone
(e) ethyl isopropyl ketone

Amines

b 50. Strychnine and nicotine are examples of naturally occurring _____.

(a) carboxylic acids
(b) amines
(c) amino alcohols
(d) esters
(e) steroids

d 51. Which one of the following compounds is **not** an amine?

(a) $(CH_3)_2NH$
(b) $CH_3CH_2NH_2$
(c) $C_6H_5NH_2$

(d) $CH_3C\overset{\nearrow O}{\underset{\searrow NH_2}{}}$
(e) $(CH_3)_3N$

e 52. Which one of the following is a **heterocyclic** amine?

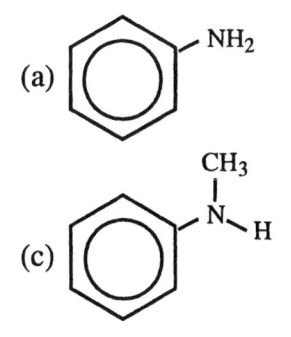

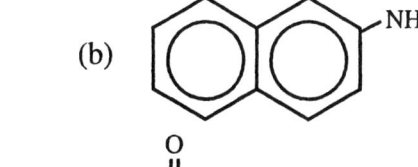

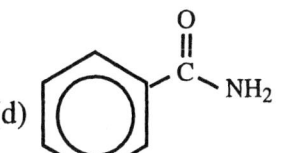

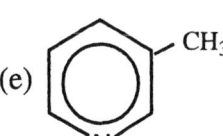

(a) C₆H₅–NH₂
(b) naphthyl–NH₂
(c) C₆H₅–N(CH₃)H
(d) C₆H₅–C(=O)–NH₂
(e) pyridine with CH₃

a 53. Which is the structural formula for a secondary amine?

(a) $(CH_3)_2NH$
(b) $CH_3CH_2NH_2$
(c) $C_6H_5NH_2$
(d) $HOCH_2CH_2NH_2$
(e) $(CH_3)_3N$

c 54. Which is the structural formula for a tertiary amine?

(a) CH₃CH₂CH₂—NH₂

(b) CH₃CH₂—N—CH₃
             |
             H

(c) CH₃—N—CH₂CH₃
         |
         CH₃

(d) CH₃—N—H
         |
         H

(e) C₆H₅—N—CH₃
          |
          H

d 55. Which of the responses is a correct name for the following compound?

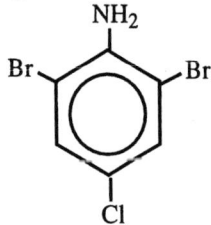

[structure: benzene ring with NH₂ on top, Br at positions 2 and 6, Cl at position 4]

(a) 4-chloro-2,6-dibromotoluene
(b) 2-amino-1,3-dibromo-5-chlorotoluene
(c) 1,3-dibromo-4-chloro-2-aniline
(d) 2,6-dibromo-4-chloroaniline
(e) 3,5-dibromo-1-chloroaniline

Carboxylic Acids

d 56. CH₃CH₂CH₂COOH is _____.

(a) isobutyl alcohol
(b) butyl alcohol
(c) propionic acid
(d) butanoic acid
(e) oxalic acid

b 57. Which of the responses is an accepted name for the following compound?

CH₃—CH₂—CH—CH—C(=O)OH
           | |
           Br   CH₃

(a) γ-bromo-β-methylpentanoic acid
(b) 3-bromo-2-methylpentanoic acid
(c) 3-bromohexanoic acid
(d) 2-bromo-1-methylpentanoic acid
(e) 3-bromo-3-ethyl-2-methylpropanoic acid

c 58. Which of the following pairs of names are both acceptable for

CH₃—CH—CH₂—C(=O)OH
     |
     CH₃

(a) 3-methylbutanoic acid — γ-methylbutyric acid
(b) 2-methylbutanoic acid — β-methylbutyric acid
(c) 3-methylbutanoic acid — β-methylbutyric acid
(d) 2-methylbutanoic acid — γ-methylbutyric acid
(e) 3-methylpropanoic acid — β-methylpropionic acid

e 59. Which one of the following is *m*-methylbenzoic acid (also called *m*-toluic acid?

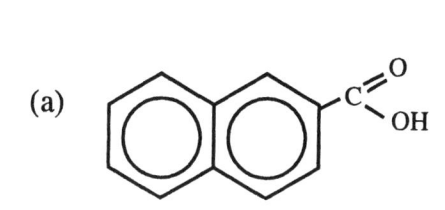

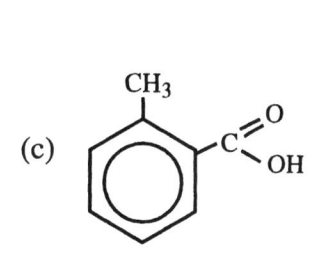

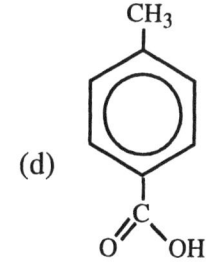

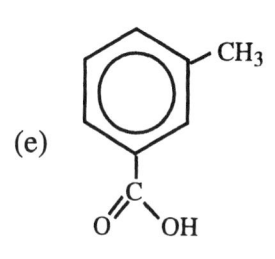

(a) 2-naphthoic acid structure
(b) phthalic acid structure (1,2-benzenedicarboxylic acid)
(c) *o*-methylbenzoic acid
(d) *p*-methylbenzoic acid
(e) *m*-methylbenzoic acid

d 60. Which of the following is **not** a carboxylic acid containing more than one –COOH group per molecule?

(a) oxalic acid (b) tartaric acid (c) citric acid
(d) benzoic acid (e) lactic acid

Some Derivatives of Carboxylic Acids

a 61. Which one of the following is an aliphatic acyl group?

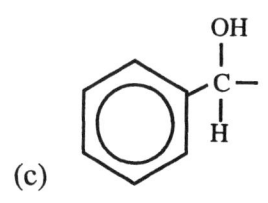

(a) CH$_3$–C(=O)–
(b) C$_6$H$_5$–C(=O)–
(c) C$_6$H$_5$–CH(OH)–
(d) CH$_3$–CH(OH)–
(e) CH$_3$CH$_2$COO$^-$

c 62. Why have acid halides **not** been observed in nature?

(a) All halogen compounds are very uncommon in nature.
(b) Carboxylic acids, and therefore their derivatives, are uncommon in nature.
(c) Acid halides are too reactive to exist in nature.
(d) Their –NH$_2$ group causes them to decompose rapidly.
(e) No one has looked for them.

d 63. Glyceryl tristearate is a(an) _____.

(a) acid (b) alcohol (c) amide (d) fat (e) oil

b 64. CH₃CH₂CH₂CH₂C(=O)OCH₂CH₃ is called _____.

(a) butyl acetate (b) ethyl pentanoate (c) propyl pentanoate
(d) ethyl butanoate (e) butyl ethanoate

d*65. The formulas for the compounds below are labelled as types of acid derivatives. Which is **incorrectly** labelled?

(a) CH₃C(=O)—O—C(=O)CH₃ acid anhydride
(b) CH₃CH₂C(=O)—Cl acid halide
(c) CH₃C(=O)—OCH₂CH₃ ester

(d) CH₂—OH / CH—OH / CH₂—OH glyceride (triester)
(e) CH₃C(=O)—NH₂ amide

Functional Groups

b*66. Which one of the following functional groups is the most polar?

(a) carbonyl (b) carboxyl (c) ester
(d) amide (e) ether

c 67. Which classification by functional group for the given compound is **incorrect**?

(a) ethyl acetate — ester (b) 2-butanol — alcohol (c) acetylene — alkene
(d) propyne — alkyne (e) chloroform — organic halide

d 68. Which classification by functional group for the given compound is **incorrect**?

(a) ethylene — alkene (b) m-cresol — phenol (c) 2-propanol — alcohol
(d) acetone — ester (e) aniline — amine

Substitution Reactions

e 69. The reaction of methane with chlorine in the presence of ultraviolet radiation can produce _____.

(a) chloromethane (b) dichloromethane (c) trichloromethane
(d) carbon tetrachloride (e) all of the products listed in the first four responses

c 70. Benzene and other aromatic compounds undergo _____ reactions readily.

(a) addition (b) hydrogenation (c) substitution
(d) isomerization (e) dehydration

c 71. Anhydrous iron(III) chloride can be used as a catalyst for the halogenation of an aromatic ring. What kind of compound is iron(III) chloride in this reaction?

(a) Lewis base (b) Brønsted-Lowry base (c) Lewis acid
(d) Brønsted-Lowry acid (e) reducing agent

d 72. The reaction of 1-butanol with nitric acid is an example of a(an) _____ reaction.

(a) aromatic (b) addition (c) alkylation
(d) substitution (e) olefinic

b 73. Sodium lauryl sulfate (or sodium dodecyl sulfate) is an effective _____.

(a) Lewis acid (b) detergent (c) vasodilator
(d) electron transfer agent (e) catalyst for hydrogenation

d 74. Alfred Nobel became rich by discovering how to make the very sensitive explosive nitroglycerine (or glycerol trinitrate) into the more safe-to-handle explosive, dynamite. What was his method?

(a) He carefully distilled out the impurities that made it sensitive.
(b) He reacted it with sulfuric acid and replaced the nitrate groups with sulfate groups.
(c) He froze the nitroglycerine into a solid.
(d) He absorbed the nitroglycerine into diatomaceous earth or wood meal.
(e) He added water to the nitroglycerine to keep it wet.

e 75. Why is it difficult to prepare only methyl chloride when reacting methane with chlorine?

(a) The reaction is very slow under all possible conditions.
(b) The methyl chloride is very unstable and quickly decomposes.
(c) The conditions required to start the reaction are so extreme that most of the reactants are destroyed before they can react.
(d) The methane is so reactive with other methane molecules that carbon chains tend to form.
(e) It is difficult to stop the reaction after only one chlorine has been substituted, therefore the result is a mixture of compounds with one, two, three, or four chlorines per molecule.

c 76. What fundamental class of organic reactions is the nitration of an aromatic ring?

(a) polymerization (b) addition (c) substitution
(d) elimination (e) cyclization

Addition Reactions

b 77. Which classes of hydrocarbons react rapidly at room temperature by addition reactions?

(a) aromatics and alkenes (b) alkenes and alkynes (c) alkanes and aromatics
(d) alkanes and alkenes (e) alkanes and alkynes

c 78. Vegetable oils can be converted to fats by _____ in the presence of a catalyst under high pressures and at high temperatures.

(a) substitution (b) alkylation (c) hydrogenation
(d) photochemical reaction (e) bromination

b 79. The reaction of propylene (propene) with cold sulfuric acid, followed by treatment of the product with steam, produces _____.

(a) ethyl alcohol (b) 2-propanol (c) propyl sulfate
(d) ethyl hydrogen sulfate (e) glycerol

d*80. Dilute aqueous hypochlorous acid is sometimes called chlorine water because the reaction of HOCl with water produces Cl_2. Treatment of allyl alcohol, $CH_2(OH)$-$CH=CH_2$, with chlorine water produces _____, an intermediate in the production of glycerol.

(a) $CH_2(OH)CH_2(OH)$ (b) CH_3OH
(c) CH_3CH_2Cl (d) $CH_2(OH)CH(OH)CH_2Cl$
(e) $CH_2(Cl)CH(OH)CH_2Cl$

c 81. The addition of Br_2 is used as the reaction to distinguish between alkanes and alkenes. What is the observation that accompanies this test?

(a) A bright red color is produced when Br_2 reacts with an alkene.
(b) Bromine dissolves in alkanes but not in alkenes.
(c) Bromine is a dark red liquid. When it adds to the double bond of an alkene to make the dibromide, it becomes colorless.
(d) The red color of bromine disappears when it dissolves in alkanes.
(e) Bromine reacts with alkanes to form a precipitate.

b 82. What is the product of the reaction given below?

$$CH_3-\underset{\underset{}{}}{\overset{\overset{O}{\|}}{C}}-CH_3 + HCN \xrightarrow{NaOH\,(aq)} \underline{\qquad}$$

(a) $CH_3-\underset{H}{\overset{OCN}{\underset{|}{\overset{|}{C}}}}-CH_3$
(b) $CH_3-\underset{CN}{\overset{OH}{\underset{|}{\overset{|}{C}}}}-CH_3$
(c) $CH_3-\underset{H}{\overset{CN}{\underset{|}{\overset{|}{C}}}}-CH_3$

(d) $CH_3-\overset{OH}{\underset{}{\overset{|}{C}}}=CH_2$
(e) $CH_3-\overset{CN}{\underset{}{\overset{|}{C}}}=CH_2$

c 83. What fundamental class of organic reactions is a hydrogenation reaction?

(a) substitution (b) polymerization (c) addition
(d) elimination (e) cyclization

Harcourt, Inc

b 84. What fundamental class of organic reactions is a hydration reaction?

    (a) substitution   (b) addition   (c) elimination
    (d) polymerization   (e) cyclization

Elimination Reactions

e 85. What is the missing product for the reaction below?

$$CH_3-CH_2Br + Na^+OH^- \longrightarrow \underline{\hspace{1cm}} + H_2O + NaBr$$

    (a) $HC\equiv CH$ (b) $CH_3=CH$ (c) CH_3CH_2
    (d) $CH_3CH_2CH_2CH_3$ (e) $CH_2=CH_2$

c 86. What fundamental class of organic reactions is a dehalogenation reaction?

    (a) substitution   (b) addition   (c) elimination
    (d) polymerization   (e) hydrolysis

d 87. What fundamental class of organic reactions is a dehydration reaction?

    (a) substitution   (b) addition   (c) polymerization
    (d) elimination   (e) cyclization

d 88. What fundamental class of organic reactions results in the increasing of the degree of unsaturation of the reacting compound?

    (a) substitution   (b) addition   (c) polymerization
    (d) elimination   (e) hydrolysis

Polymerization

e 89. Below is a list of polymers that are each matched with its class. Which match is **incorrect**?

	Polymer	Class
(a)	proteins	natural
(b)	wool	natural
(c)	teflon	synthetic
(d)	nylon	synthetic
(e)	silk	synthetic

e 90. The reaction by which tetrafluoroethylene is converted into teflon is called _____.

    (a) fluorination   (b) fluoridation   (c) vulcanization
    (d) elastation   (e) polymerization

d 91. Below is a list of substances each matched with its function in natural rubber. Which match is **incorrect**?

	Substance	Function
(a)	latex (sap of rubber tree)	natural source for rubber
(b)	sulfur	vulcanization
(c)	carbon black	reinforcing agent
(d)	antimony (V) sulfide	cross-linking
(e)	zinc oxide	filler

b 92. Which of the following is **not** an addition polymer?

(a) polyvinyl chloride (b) nylon (c) teflon
(d) styrofoam (e) polypropylene

a 93. Proteins are examples of

(a) natural condensation polymers.
(b) natural addition polymers.
(c) synthetic addition polymers.
(d) natural polyesters.
(e) synthetic condensation polymers.

e*94. Which of the following is not a condensation polymer correctly classified?

(a) protein — polypeptide
(b) nylon — polyamide
(c) Dacron — polyester
(d) Mylar — polyester
(e) Teflon — polyamide

28 Organic Chemistry II: Shapes, Selected Reactions, and Biopolymers

Constitutional Isomers

c 1. What are two types of constitutional (or structural) isomers?

(a) positional and geometrical
(b) functional group and geometrical
(c) functional group and positional
(d) geometrical and optical
(e) positional and optical

c 2. Below are five pairs of constitutional isomers. One pair has different functional groups. Which pair has the different functional groups?

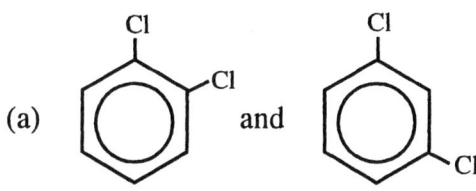

(a) [dichlorobenzene isomers]

(b) $CH_2=CHCH=CHCH_3$ and $CH_2=CHCH_2CH=CH_2$

(c) CH_3CH_2OH and CH_3OCH_3

(d) $CH_3CH_2CH_2CH_3$ and CH_3CHCH_3 with CH_3 branch

(e) $CH_3CH_2CH_2Br$ and CH_3CHCH_3 with Br branch

d 3. Which of the following is a constitutional isomer of 2-pentanone that has a different functional group?

(a) 3-pentanone
(b) 2-butanone
(c) 2-pentanal
(d) pentanal
(e) 2-pentanol

e 4. Which of the following compounds **does not** have a constitutional isomer with a different functional group?

(a) propanone (acetone)
(b) 3-pentanone
(c) ethanol
(d) dimethyl ether
(e) ethanal (acetaldehyde)

Stereoisomers

a 5. What are the two types of stereoisomers?

(a) geometrical and optical
(b) optical and positional
(c) geometrical and positional
(d) geometrical and functional group
(e) chiral and optical

c 6. Which of the following is a **chiral** molecule?

(a) CH$_3$CBr$_2$Cl
(b) CBrCl$_2$I
(c) CH$_3$CBrClI
(d) (CH$_3$)$_2$CHCl
(e) CH$_2$BrI

b 7. Does the following compound have any asymmetric carbons? If so, which one (or ones)?

$$Br-\overset{1}{C}=\overset{2}{C}-\overset{3}{C}-\overset{4}{C}-H$$

with H on C1 (bottom), H on C2 (bottom), H and H on C3 (top), I on C3 (bottom), H and H on C4 (top and bottom).

(a) 1 (b) 3 (c) 2 (d) 1 and 2 (e) 1, 2, and 3

d 8. Which of the following compounds can have geometrical isomers?

(a) 1,1-dichloroethene
(b) 2,3-dichloropropene
(c) 1,2-dichloropropane
(d) 1,2-dichlorocyclobutane
(e) 3-ethyl-3-hexene

e 9. Which statement concerning optical isomers is **false**?

(a) Optical isomers have identical physical properties, except for their interaction with polarized light.
(b) A racemic mixture is a single sample containing equal amounts of the two optical isomers of a compound.
(c) Each optical isomer of a compound rotates polarized light by equal amounts but in opposite directions.
(d) A racemic mixture does not rotate the plane of polarized light because the effects of the two isomers exactly cancel.
(e) Optical isomers react exactly the same with any particular compound.

Conformations

e 10. Which of the following statements concerning isomers and conformations is **false**?

(a) Conformations of a compound differ in the extent of rotation about one or more single bonds.
(b) A staggered conformation is slightly more stable than the eclipsed conformation.
(c) At least one chemical bond must be broken and reformed to convert one isomer into another.
(d) In the staggered conformation, there is less repulsive interaction between H atoms on adjacent C atoms.
(e) The chair and twist boat forms are two isomers of cyclohexane.

d 11. Conversion of one form of a molecule into another form by rotation about a C—C single bond occurs very rapidly at room temperature. These different forms of the molecule are known as _____.

(a) constitutional isomers
(b) optical isomers
(c) geometric isomers
(d) conformations
(e) enantiomers

d 12. Cyclohexane can assume two different geometries, but to do this the ring becomes nonplanar. What are these two arrangements of cyclohexane called?

(a) *cis* and *trans*
(b) *vic* and *gem*
(c) *strained* and *unstrained*
(d) twist boat and chair
(e) *meta* and *para*

Reactions of Brønsted-Lowry Acids and Bases

d 13. The strengths of monocarboxylic acids greatly increase when electronegative substituents replace hydrogen on the α-carbon. What are the 2 main reasons for this increase?

I. The electronegative substituents pull electron density from the carboxylic acid group.
II. The replacement of the hydrogen greatly increases the molecular weight.
III. The replacement of the hydrogen decreases the solubility of the molecule.
IV. The more electronegative substituents help to stabilize the resulting carboxylate anion by spreading the negative charge over more atoms.

(a) I and II   (b) I and III   (c) II and III   (d) I and IV   (e) III and IV

c 14. What are the products of the reaction of ethanol and metallic sodium?

(a) $CH_3CH_2CH_2CH_2$ and NaOH
(b) $CH_3CH_2OCH_2CH_3$, NaH, and NaOH
(c) $CH_3CH_2O^-$, Na^+, and H_2
(d) CH_3CH_3 and NaOH
(e) CH_3O^-, Na^+, and H_2

c 15. Which of the following classes of organic compounds has the greatest acid strength?

(a) phenols
(b) alcohols
(c) carboxylic acids
(d) amines
(e) substituted ammonium ions

c 16. The reaction of methylamine with hydrochloric acid produces _____.

(a) ammonia
(b) ammonium methylate
(c) methylammonium chloride
(d) chloromethane
(e) chloromethylamine

c 17. The pK_b for CH_3NH_2 is 3.30. What is the value for the K_a of $CH_3NH_3^+$?

(a) 2.0×10^3
(b) 5.0×10^{-4}
(c) 2.0×10^{-11}
(d) 5.0×10^{10}
(e) 5.0×10^{-18}

b 18. The K_a for formic acid (HCOOH) is 1.8×10^{-4}. What is the pK_b for the formate ion (HCOO$^-$)?

(a) 3.74   (b) 10.26   (c) 1.87   (d) 12.13   (e) 5.6×10^{-11}

c 19. Which of the following classes of organic compounds has the greatest strength as a base?

(a) aromatic amines
(b) aliphatic amines
(c) alkoxides
(d) carboxylates
(e) phenoxides

Oxidation-Reduction Reactions

d 20. When primary alcohols are oxidized to aldehydes, the reaction mixture is heated to a temperature slightly above the boiling point of the aldehyde so that the aldehyde distills out as soon as it is formed. Why must aldehydes be removed from the reaction mixture as soon as they are formed?

    (a)   The aldehydes might polymerize.
    (b)   The aldehydes might be reduced back to the alcohol.
    (c)   The aldehydes might oxidize to CO_2 and H_2O.
    (d)   Aldehydes are easily oxidized to carboxylic acids.
    (e)   Aldehydes readily react with unreacted alcohol.

c 21. The reaction of 1-propanol with $K_2Cr_2O_7$ in acidic solution produces _____.

(a) propyne   (b) propene   (c) propanal
(d) propanone   (e) cyclopropanone

e 22. The reaction of a secondary alcohol with potassium dichromate in an acidic solution produces a(n) _____.

(a) carboxylic acid   (b) aldehyde   (c) amine
(d) primary alcohol   (e) ketone

d 23. Which one of the following cannot be easily oxidized to a carboxylic acid by an acidic solution of $K_2Cr_2O_7$?

(a) methanol   (b) ethanol   (c) acetaldehyde
(d) 2-butanol   (e) propanal

a 24. The reduction of an aldehyde produces a(n) _____.

(a) primary alcohol   (b) secondary alcohol   (c) tertiary alcohol
(d) ketone   (e) carboxylic acid

d 25. Catalytic reduction of acetaldehyde produces _____.

(a) acetic acid   (b) methanol   (c) formic acid
(d) ethanol   (e) ethene

b 26. The reduction of a ketone produces a(n) _____.

(a) primary alcohol   (b) secondary alcohol   (c) tertiary alcohol
(d) aldehyde   (e) carboxylic acid

b 27. The reduction of acetone produces _____.

(a) ethanol   (b) 2-propanol   (c) 1-propanol
(d) propanal   (e) propene

a 28. The oxidation of toluene by a basic solution of KMnO₄ can be used in the production of _____.

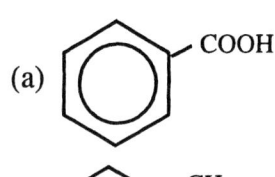

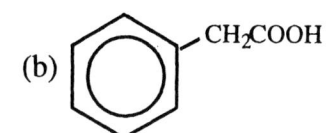

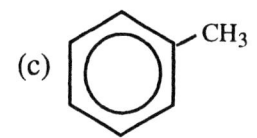

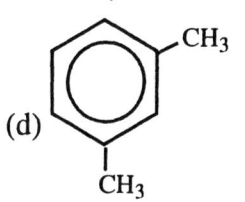

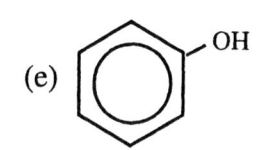

(a) C₆H₅-COOH
(b) C₆H₅-CH₂COOH
(c) C₆H₅-CH₃
(d) 2,6-dimethylbenzene (CH₃ groups on ring)
(e) C₆H₅-OH

d 29. Which statement concerning the combustion of organic compounds is **false**?

(a) Combustion reactions of organic compounds are highly exothermic.
(b) The heat of combustion is the amount of energy liberated per mole of hydrocarbon burned.
(c) Aromatic hydrocarbons burn incompletely **in air**.
(d) The large amounts of heat produced by the combustion of hydrocarbons in an internal combustion engine expands the gases present in the engine, thereby driving the pistons and producing power.
(e) Incomplete combustion of hydrocarbons produces the poisonous gas CO.

a 30. The reaction of calcium carbide, CaC_2, with water produces calcium hydroxide and _____.

(a) acetylene (b) ethanol (c) carbon monoxide
(d) ethylene (e) carbon dioxide and carbonic acid

Formation of Carboxylic Acid Derivatives

e 31. The reaction of propionic acid with _____ produces propionyl chloride.

(a) sodium chloride (b) hydrochloric acid (c) hypochlorous acid
(d) chlorine (e) phosphorus pentachloride

a 32. The reaction of propanoic acid with ethyl alcohol in the presence of sulfuric acid produces _____.

(a) CH₃CH₂C(=O)OCH₂CH₃
(b) CH₃CH₂C(=O)OCH₂CH₂CH₃
(c) CH₃C(=O)OCH₂CH₃
(d) CH₃C(=O)OCH₂CH₂CH₃
(e) CH₃CH₂CH₂C(=O)OCH₂CH₃

e 33. Which of the following is **not** an advantage for the preparation of esters by the reaction of an acyl halide and alcohol *versus* the preparation by the reaction of an organic acid and alcohol?

(a) Reactions between acids and alcohols are slow and require prolonged boiling.
(b) Reactions between acyl halides and alcohols are rapid and require no catalyst.
(c) Reactions between acids and alcohols require a strong inorganic acid catalyst.
(d) Reactions between acids and alcohols establish an equilibrium with both reactants and products present.
(e) Acyl halides are much cheaper than organic acids because they are readily available from natural fats and oils.

b 34. What is the missing product for the following reaction?

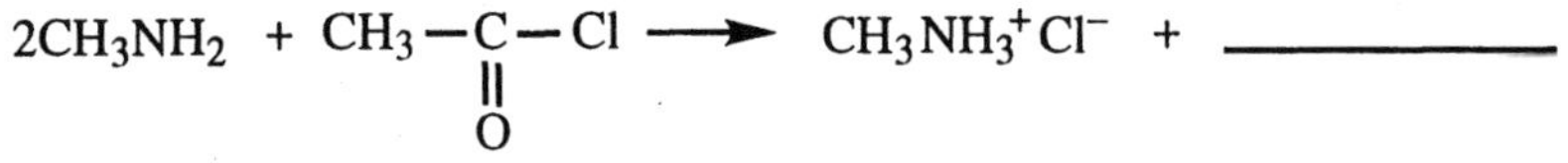

$$2CH_3NH_2 + CH_3-\underset{O}{\underset{\|}{C}}-Cl \longrightarrow CH_3NH_3^+ Cl^- + \underline{\qquad}$$

(a) 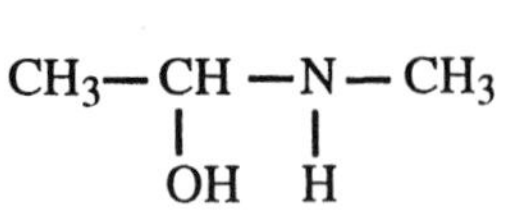 CH$_3$—CH—N—CH$_3$
         | |
         OH H

(b) 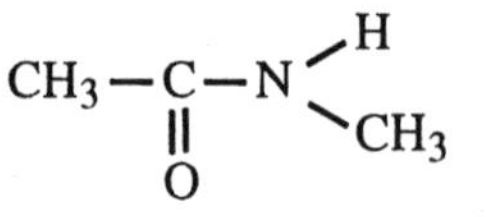 CH$_3$—C(=O)—N(H)(CH$_3$)

(c) CH$_3$—CH$_2$—O—NH$_2$

(d) CH$_3$—CH$_2$—O—N(CH$_3$)$_2$

(e) CH$_3$—NH—NH—CH$_3$

a 35. Which of the following can usually be used as the reactants (especially in the laboratory) to prepare amides?

I. a primary amine and an acyl halide
II. a tertiary amine and an organic acid
III. a secondary amine and an acyl halide
IV. a primary amine and an aldehyde
V. a tertiary amine and an acyl halide

(a) I and III   (b) II and IV   (c) I, III and V
(d) I and II   (e) I, II, III and V

e 36. The reaction of an acyl chloride with an amine to produce an amide requires 2 moles of amine per each mole of acyl chloride. Why?

(a) One mole of amine is required to buffer the solution.
(b) One mole of amine is required to make the solution basic.
(c) Each amide molecule contains 2 nitrogen atoms and therefore 2 amine molecules are required to supply the nitrogen.
(d) An excess of amine is required to shift the equilibrium toward the products.
(e) Half of the amine is used to convert to the amide and the other half reacts with the displaced hydrogen and chlorine to form an ammonium salt.

b*37. A reaction in which a small molecule, such as H$_2$O or HCl, is eliminated and two larger molecules are bonded together is called a(n) _____ reaction.

(a) elimination   (b) condensation   (c) addition
(d) hydration   (e) oxidation

Harcourt, Inc

Hydrolysis of Esters

e 38.   Which statement below concerning the given reaction is **false**?

$$R-\overset{O}{\underset{\|}{C}}-O-R' + Na^+OH^- \xrightarrow{heat} R-\overset{O}{\underset{\|}{C}}-O^-Na^+ + \underline{\qquad}$$
ester   salt of acid

(a)   The sodium salts of long-chain fatty acids are soaps.
(b)   Strong reagents are required for the reactions of esters because esters are not very reactive.
(c)   Esters are hydrolyzed by refluxing with solutions of strong bases.
(d)   The hydrolysis of esters by strong bases is called saponification.
(e)   The missing product in the equation above is $R'-H$.

Carbohydrates

d 39.   What is the general formula for carbohydrates?

(a) $(CH_2O)_n$ (b) $C_n(H_2O)_{n+2}$ (c) $C(H_2O)_n$
(d) $C_n(H_2O)_m$ (e) $C_n(H_2O)_{n+1}$

c 40.   Glucose is a monosaccharide that consists of a 6 carbon chain. The second, third, fourth and fifth carbons are each bonded to four different groups. Therefore glucose is _____.

(a) a ketose   (b) a glycogen   (c) chiral
(d) a polymer   (e) an oligosaccharide

e 41.   A disaccharide is formed when two monosaccharides are linked by the elimination of a molecule of water and the formation of an ether bond. The newly formed C—O bond linking the monosaccharides is called the _____ bond.

(a) aldose   (b) ketose   (c) glycogen   (d) cyclic   (e) glycosidic

Polypeptides and Proteins

d 42.   Which statement concerning the twenty different α-amino acids widely found in nature is true?

(a)   Each α-amino acid has a different number of carbon atoms.
(b)   Each α-amino acid has one amino (–NH₂) group.
(c)   Each α-amino acid has one carboxylic acid (–COOH) group.
(d)   These α-amino acids differ in the R groups attached to the α-carbon.
(e)   None of these α-amino acids are aromatic compounds.

a 43.   The amino acids are usually classified by their R groups according to what two criteria?

(a) polarity and acidity - basicity   (b) polarity and ring-containing
(c) polarity and chirality   (d) acidity - basicity and chirality
(e) ring-containing and attachment to α- or β-carbon

d 44. The functions of proteins do **not** include _____.

(a) enzymes (b) hormones (c) toxins (d) sugars (e) structural

b 45. The order of the amino acids of a protein is termed the _____ structure.

(a) isomeric (b) primary (c) secondary (d) tertiary (e) quaternary

Nucleic Acids

c 46. Nucleic acids are composed of what three different monomers?

(a) phosphate group – protein – selected organic base
(b) enzyme – protein – carbohydrate unit
(c) phosphate group – carbohydrate unit – selected organic base
(d) enzyme – hormone – selected organic acid
(e) protein – carbohydrate unit – selected organic acid

e 47. Which statement about RNA and DNA is **incorrect**?

(a) The carbohydrates in nucleic acids are ribose in RNA and 2-deoxyribose in DNA.
(b) The bases adenine, guanine and cytosine are found in both RNA and DNA.
(c) The fourth base is uracil in RNA and thymine in DNA.
(d) In DNA adenine and thymine always pair by forming two hydrogen bonds.
(e) In both RNA and DNA, guanine and cytosine always pair by forming one hydrogen bond.